D1156435

# BATS

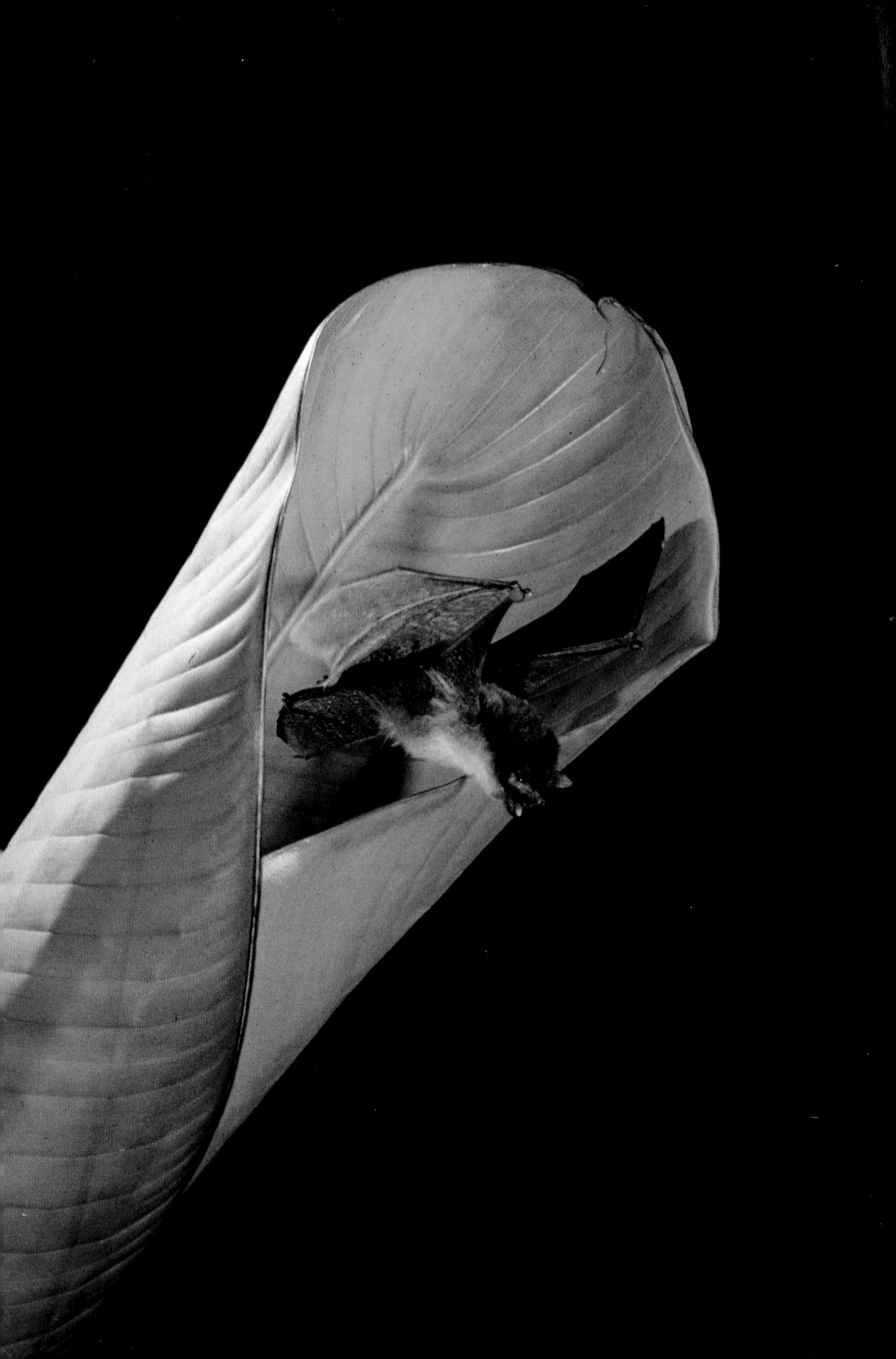

# BATS

*an illustrated guide to all species*

MARIANNE TAYLOR

Merlin D. Tuttle

SCIENCE EDITOR & PHOTOGRAPHER

Smithsonian Books

WASHINGTON, DC

This book may be purchased for educational, business, or sales promotional use. For information, please write: Special Markets Department, Smithsonian Books, P. O. Box 37012, MRC 513, Washington, DC 20013

ISBN 978-1-58834-647-6

This book was conceived, designed and produced by
The Bright Press, an imprint of The Quarto Group.
The Old Brewery, 6 Blundell Street,
London N7 9BH, United Kingdom.
T (0)20 7700 6700
www.quarto.com

Publisher Susan Kelly
Creative Director Michael Whitehead
Editorial Director Tom Kitch
Art Director James Lawrence
Commissioning Editor Kate Shanahan
Project Editor Elizabeth Clinton
Science Editor Merlin D. Tuttle
Designer Heather Bowen
Illustrators Richard Peters, Ginny Zeal
Editorial Assistant Niamh Jones

Published in North America by Smithsonian Books
Director Carolyn Gleason
Senior Editor Christina Wiginton
Managing Editor Jaime Schwender
Creative Director Jody Billert

Library of Congress Cataloging-in-Publication Data
Names: Taylor, Marianne, 1972- author. | Tuttle, Merlin D., photographer.
Title: Bats : an illustrated guide to all species / Marianne Taylor ; photographs by Merlin D. Tuttle.
Description: Washington, DC : Smithsonian Books, 2019 | Includes indexes.
Identifiers: LCCN 2018043061 | ISBN 9781588346476 (hardcover)
Subjects: LCSH: Bats. | Bats—Ecology.
Classification: LCC QL737.C5 T386 2018 | DDC 599.4—dc23 LC record available at https://lccn.loc.gov/2018043061

Manufactured in Singapore, not at government expense
23 22 5

# CONTENTS

# FOREWORD

I've been passionate about bats ever since discovering thousands of Gray Myotis in a cave near my Tennessee home in 1959. Almost nothing was yet known about them. However, stained cave ceilings, guano deposits, and stories told by old-timers confirmed they had once been among the most abundant mammals of eastern North America. Health officials were erroneously blaming them for a rabies outbreak in foxes, and cave owners were pouring kerosene into roosts and incinerating thousands of bats at a time. They were in such precipitous decline that leading experts were predicting extinction.

Beginning as a high school student, I went on to band more than 40,000 of them, trace their migratory movements, and document their value and urgent need for protection as a part of my eventual doctoral thesis. As I explained the value and needs of bats, fear began to be replaced with understanding and appreciation. Cave owners and explorers began to protect key roosting caves. Today, as a direct result, there are millions more Gray Myotis than when their extinction was predicted. For many other species, we are dangerously late in protecting them.

Worldwide, countless bat species remain in alarming decline, victims of needless fear, persecution, and neglect. In recent decades we have finally begun to document their essential roles in maintaining the health of whole ecosystems, keeping insect populations in check, dispersing seeds, and pollinating flowers. Our real fears should be about losing them. Yet, far too often, sensational headlines warn of potential disease pandemics instead. Bats will leave us alone if we leave them alone, but toxic pesticides won't.

Most of the world's bats have barely been studied beyond having been described as species. When it comes to amazing research discoveries, they are a long-neglected goldmine. However, it is incumbent upon us scientists to go beyond exploitation for knowledge alone. Bats urgently need help, and we who study them bear responsibility for their survival. We must strive to leave more than we find.

Some bats have gone extinct, unnoticed until years later. As an ecologist, I am most concerned about the decline of species that traditionally have formed large aggregations, especially those with narrow cave requirements. These are vulnerable to so-called "Passenger Pigeon effects." For example, when numbers fall below critical thresholds, they may be unable to heat roosts sufficiently to rear young. As you will learn in this book, their loss can threaten the health of whole ecosystems and economies.

Too often, we focus mostly on saving the rarest species, often ones whose distribution is limited to a single small area. Though we may love and want to save all bats, such species have the least impact in ensuring the long-term health of our planet. Conservation status listings can be dangerously misleading. As traditionally and here applied, they often refer only to a species' perceived risk of extinction. However, long before extinction, a species can be reduced to biological irrelevance. Keeping traditionally common species sufficiently abundant to maintain ecosystem health should be a top priority.

It has often been said that a picture is worth a thousand words, and never has this been truer than in the case of bats. Hopefully, the pages that follow will help more people to recognize bats as the intelligent, curious, comical, even essential animals I have personally known for the past sixty years.

*Merlin D. Tuttle*

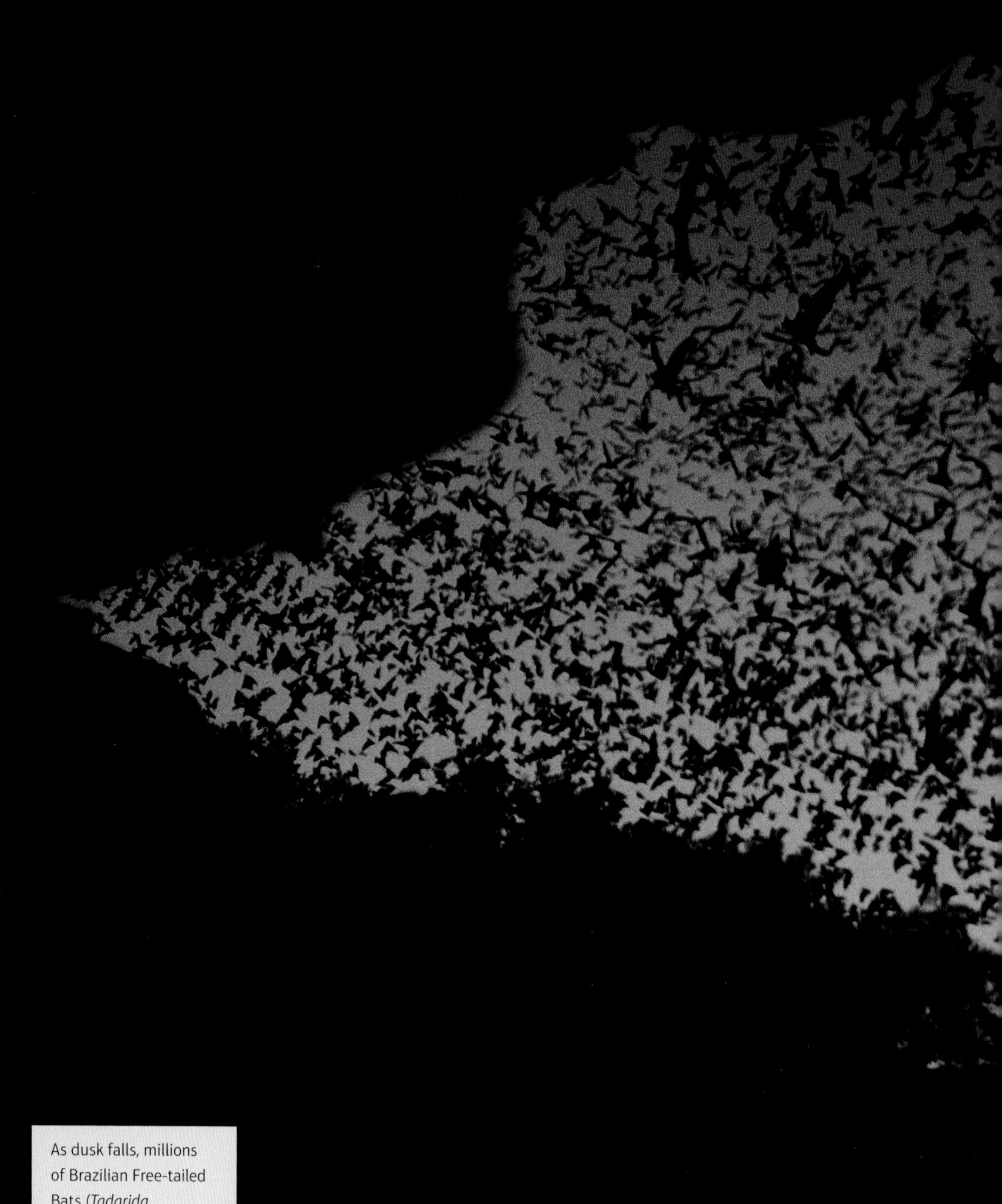

As dusk falls, millions of Brazilian Free-tailed Bats (*Tadarida brasiliensis*) leave their maternity roost at Bracken Cave, Texas, in one of the world's most unforgettable spectacles.

# INTRODUCTION

Flying almost blindly but unerringly through the night in their three-dimensional maze of sound, bats appear to be supernatural creatures to us. Stepping into their world means entering darkness—by night into the air, and by day into the earth itself, the deep and foreboding caves where they sleep away the daylight hours. Bats have long featured as sinister figures and icons of horror in folklore, literature, and cinema, and there is no denying that they are, to most of us, deeply mysterious. Yet above all, they are deeply misunderstood. This book seeks to illuminate the dark world of bats, and reveal their true nature as exceptionally diverse, fascinating, intelligent, and endlessly enchanting animals.

# BAT RESEARCH

A summer evening, warm and balmy. The sky is shading from turquoise to indigo, and nocturnal insects are out—from moths, buzzing like bees as they speed past your face, to clouds of mosquitoes, discouraging you from lingering too long in one spot. Against the darkening sky, a black shape flickers. Not a bird: you can make out the smooth convexities between the fingertips of its wings, plus it twists and turns in a crazy path across the sky, in movements more extreme than even the most agile bird could manage. The bat dips across the silhouette of a tree and you lose it in the darkness.

Encounters with bats in the wild are often over almost as soon as they have begun. The moment is magical, but frustrating, leaving you longing to know more. Bats are difficult for the amateur to study. Even just working out which species you are looking at can be impossible, when all you have to go on is a fast-moving silhouette, its calls beyond the register of what your ears can detect. However, modern technology is becoming ever more helpful here. You can use an electronic bat detector, which not only hears the calls, but plays them back at a slower speed so that you can hear them too. It can also produce a visual representation of the calls, and even tell you which species is making them.

Those seriously involved in bat research have a fast-growing range of tools at their disposal. Infrared-sensing night vision goggles help in observing bat behavior, while examination of feces, assisted by genetic barcoding, can identify the insects that they have consumed. Bats can be captured in mist nets, spun from ultra-fine thread, and have tiny radio transmitters attached, which will track them to their roosts. Another technique, satellite tagging, helps researchers to follow bats' foraging forays or migratory journeys. Motion sensors can trigger video cameras, capturing footage of bats pollinating flowers. Slow-motion video cameras record their flight maneuvers in frame-by-frame detail, allowing every move to be analyzed. This technology has been used in southeast Asia to prove that *Pteropus hypomelanus* does not damage durian flowers, as was formerly believed. All around the world, researchers are studying bat biology and behavior in ever more imaginative and sophisticated ways, some of which are discussed here. Their insights are helping to fill in the gaps in our knowledge—a task both fascinating and urgent, given the threats faced by many bat populations today. Armed with that knowledge, every bat moment you experience will be that little bit richer.

Researchers Merlin Tuttle and Teresa Nichta take footage of Pygmy Round-eared Bats (*Lophostoma brasiliense*) roosting in a termite nest in Trinidad.

The Andersen's Leaf-nosed Bat (*Hipposideros pomona*) has acute enough hearing to accurately pinpoint insects moving on the ground.

# WHAT IS A BAT?

Bats, like us, are mammals—warm-blooded, hairy, and equipped with mammary glands that produce milk for their young. On the most superficial examination, they resemble mice or shrews, but have wings rather than forelimbs—the "fingers" greatly elongated and connected by membranes to support flight. Look a little more closely, though, and you will realize that bats are nothing like these small earthbound mammals. Look at a few different kinds of bats and you will see that bats as a group show a huge variety of forms.

The defining characteristic of all bats that sets them apart from all other mammals, and most other living things, is, of course, that they can fly. They engage in real, self-powered flight, an ability shared only with birds and insects. The freedoms this affords have allowed bats to adapt and spread across almost the entire world, and to pursue many different ways of life. Another special trait shared by nearly all bats is their ability to echolocate—they generate sounds, using the resultant echoes to choose their flight path around obstacles, and to hone in on prey. This biological sonar system enables bats to be active at night, and thus avoid competing with—and being hunted by—day-flying birds.

## A WORD ABOUT CLASSIFICATION

Bats form the taxonomic order Chiroptera, meaning "hand wings." Every bat is instantly recognizable as a bat, thanks to the modification of its forelimbs into wings (see Anatomy, page 26). This trait makes Chiroptera one of the most clearly defined of the mammal orders. With more than 1,300 species worldwide, it is also the second largest, holding about 20 percent of the world's extant mammals. In diversity, bats are outnumbered only by rodents (of which there are well over 2,000 species). By comparison, another familiar mammalian order, Carnivora, holds just 260 or so species—that is all of the world's cats, dogs, seals, weasels, bears, and other flesh-eaters.

Traditionally, the order Chiroptera has been divided into two suborders: Microchiroptera (the small, mainly insect-eating "microbats," comprising about seventeen families) and Megachiroptera (the mainly larger fruit bats and flying foxes—the "megabats," with just one family, Pteropodidae). This division reflects the obvious physical differences between microbats and megabats: their size, the different shape of their heads, ears, and snouts, and microbats' unclawed second fingers. However, recent research into bat molecular genetics suggests that this division may not properly express how different bats are related to each other. Certain microbats appear to be more closely related to megabats than they are to the rest of the microbats. Accordingly, some experts now classify bats into two different suborders: Pteropodiformes (or Yinpterochiroptera) and Vespertilioniformes (or Yangochiroptera). The former would hold the megabats (Pteropodidae), plus the horseshoe bats (Rhinolophidae), leaf-nosed bats (Hipposideridae), false vampire bats (Megadermatidae), Kitti's Hog-nosed Bat (Craseonycteridae), and mouse-tailed bats (Rhinopomatidae), and the latter all the remaining microbats. This new division seems to be supported by other evidence, but research is still ongoing.

A mother and pup Ethiopian Epauletted Fruit Bat (*Epomophorus labiatus*).

remarkably long those lives can be. Anyone who has kept pet hamsters, gerbils, or other small rodents will know that these little animals are not around for very long—three years is a good age. In that brief time, though, a female rodent can have several litters of young, replacing herself many times over. Over her whole life, a female mouse could have anything from thirty to 140 babies, each able to breed at about one month old. For bats, things are very different. Even small species can live thirty to forty years in the wild. The oldest record for a small *Myotis* bat is at least forty-one years. However, their reproductive rate is much slower, with females in most species having just one pup a year. This means that the parental investment in each pup is much higher, and maternal care is more protracted and involved than with other small mammals.

In common with most other animals that enjoy long lives and have slow reproductive rates, bats are highly intelligent, and socially sophisticated. Researchers working with wild bats have found that they learn very quickly and have long memories. Frog-eating bats learn new frog calls in just a few trials, and can remember them when recaptured by scientists years later. The ability to be social has allowed bats to share information and thrive in a wide range of circumstances. Bonds of friendship

## FLYING PRIMATES?

As early as the mid-eighteenth century, some biologists had explored the idea that the megabats were completely separate from the microbats, and were instead related to the primates (based on general appearance, megabats' lack of laryngeal echolocation, and various other anatomical details). This would mean that microbats and megabats evolved their flying "hands" quite independently of one another.

It cannot be denied that a megabat's face, with its big, bright eyes and foxy muzzle, looks very different to the small-eyed, complex faces of most microbats, and the similarity between a fruit-bat's face and a lemur's is obvious and striking. However, it is not unusual in nature for unrelated animals to evolve a similar body-plan, when those animals pursue similar ways of life: consider dolphins and fish, hummingbird hawkmoths and hummingbirds, or indeed bats and birds. These examples demonstrate what is called convergent evolution, which could explain the lemur-like traits of megabats. Of course, convergent evolution could also explain wings evolving in two unrelated mammal groups, but the similarity between megabat and microbat wing anatomy is so great that convergent evolution seemed unlikely. Modern DNA research now appears to indicate a common origin for all bats, mega and micro alike, and the "flying primate" hypothesis now seems less likely.

between bats can persist over many years, even when they travel long distances. Some bats live as faithful, monogamous pairs. They are also known to show altruism, as discussed in Behavior (page 28).

We tend to think of most bats as small animals, but even so the extreme tininess of Kitti's Hog-nosed Bat or Bumblebee Bat (*Craseonycteris thonglongyai*) still comes as a surprise. Weighing under ⅛oz (2g), with a head and body measuring just 1⅛–1⁵⁄₁₆in (2.9–3.3cm), this bat could comfortably rest on a human finger. It is arguably the world's smallest mammal by length (the Etruscan Shrew is longer-bodied, but weighs fractionally less). The world's biggest bat (based on wingspan) is the Large Flying-fox (*Pteropus vampyrus*); it weighs up to 2lb 3oz (1kg) or more and is about 11¾in (30cm) long, with a wingspan of up to 5½ft (1.6m). However, some megabats are much smaller. The most petite is the Spotted-winged Fruit Bat (*Balionycteris maculata*), just 2³⁄₁₆in (5.5cm) in length and weighs about ½oz (14g); the largest microbat, the Spectral Bat (*Vampyrum spectrum*) is more than ten times as heavy.

Bats are, very obviously, quite different from all other mammals. Although some may look superficially like rodents or shrews (or lemurs, in the case of the fruit bats), molecular genetic studies reveal that they are not closely related to any of these. Instead, their closest relatives are the carnivores, the hoofed mammals, the whales, and the pangolins (a small group of scaly-skinned, ant-eating creatures).

The Kitti's Hog-nosed Bat (*Craseonycteris thonglongyai*) is probably the smallest bat and the smallest mammal on earth.

# EVOLUTION

Bats have existed on earth for more than fifty million years—and likely far longer. The earliest known fossils of true bats are dated at around 52 to 55.5 million years old, and have been found in North America and Eurasia. Bat fossils, however, are in short supply, especially complete ones. Complete fossils can form only when an organism is buried in some kind of sediment, gently enough that its body parts are not crushed and dispersed beyond recognition. This is unlikely to happen to small, fragile, land-dwelling animals such as bats. It is no coincidence that the vast majority of animal fossils represent aquatic and/or very large-bodied creatures—even then, fossilization is still an extremely rare process.

These early bat fossils are superficially very similar to modern bats. However, they do show less advanced adaptations for efficient flight than the bats seen today. They had shorter forelimbs and longer legs, and claws on each "fingertip" (rather than on just one or two, as modern bats have). They would have been weaker flyers than today's bats, but more capable of climbing and scrambling through the trees. It is likely that earlier "protobats" descended from small arboreal (tree-dwelling) mammals that were skilled climbers, gradually evolving the ability to glide from branch tip to branch tip. Many other kinds of tree-dwelling animals, from frogs and snakes to other mammal groups, have independently evolved ways to glide through the air; their rate of descent is slowed to a safe level because of air resistance against panels of extended, flattened skin or webbing. In most gliding mammals, this takes the form of a patagium—a skin membrane stretching from front paw to hind paw. In bats, the membrane stretches between elongated forelimb digits, meaning that, unlike the gliders, bats can flap and thus achieve true powered flight.

A well-preserved, 52-million-year-old fossil of *Onychonycteris finneyi*—a North American microbat discovered in the west of the United States in 2003—had anatomy capable of flight. However, the structure of its inner ears suggested that it did not possess the ability to echolocate, indicating that of the two great bat talents, flight came first. Echolocation followed not long afterward, in evolutionary terms. Fossils of *Paleochiropteryx tupaidon* and its relative *P. spiegeli*, both from Germany, are dated at about forty-eight million years old. They show evidence that these species were laryngeal echolocators, just as modern microbats are.

Of modern bats, only megabats do not use laryngeal echolocation, the remarkable navigational system unique to Chiroptera. However, their closest relatives among the microbats possess highly sophisticated laryngeal echolocation systems. This suggests that megabats' ancestors were the same, but the ability and associated anatomy was lost somewhere along the line. Interestingly, one genus of megabats (*Rousettus*) does echolocate, albeit in a much more rudimentary way than the microbats (using tongue-clicks rather than larynx-generated ultrasound). It is believed that this trait has evolved independently in *Rousettus*, rather than being a vestigial one from an ancestor capable of full laryngeal echolocation.

## What drives evolutionary change?

At its heart, evolution is a very simple concept. All members of a species have a slightly different genetic make-up, which gives them different traits. Some traits are better for survival and successful breeding in a given environment than others. The survivors pass on to their offspring those genes that give them advantages, while the more poorly

The Spotted Bat (*Euderma maculatum*) has evolved a striking coat pattern that may provide camouflage in dappled light.

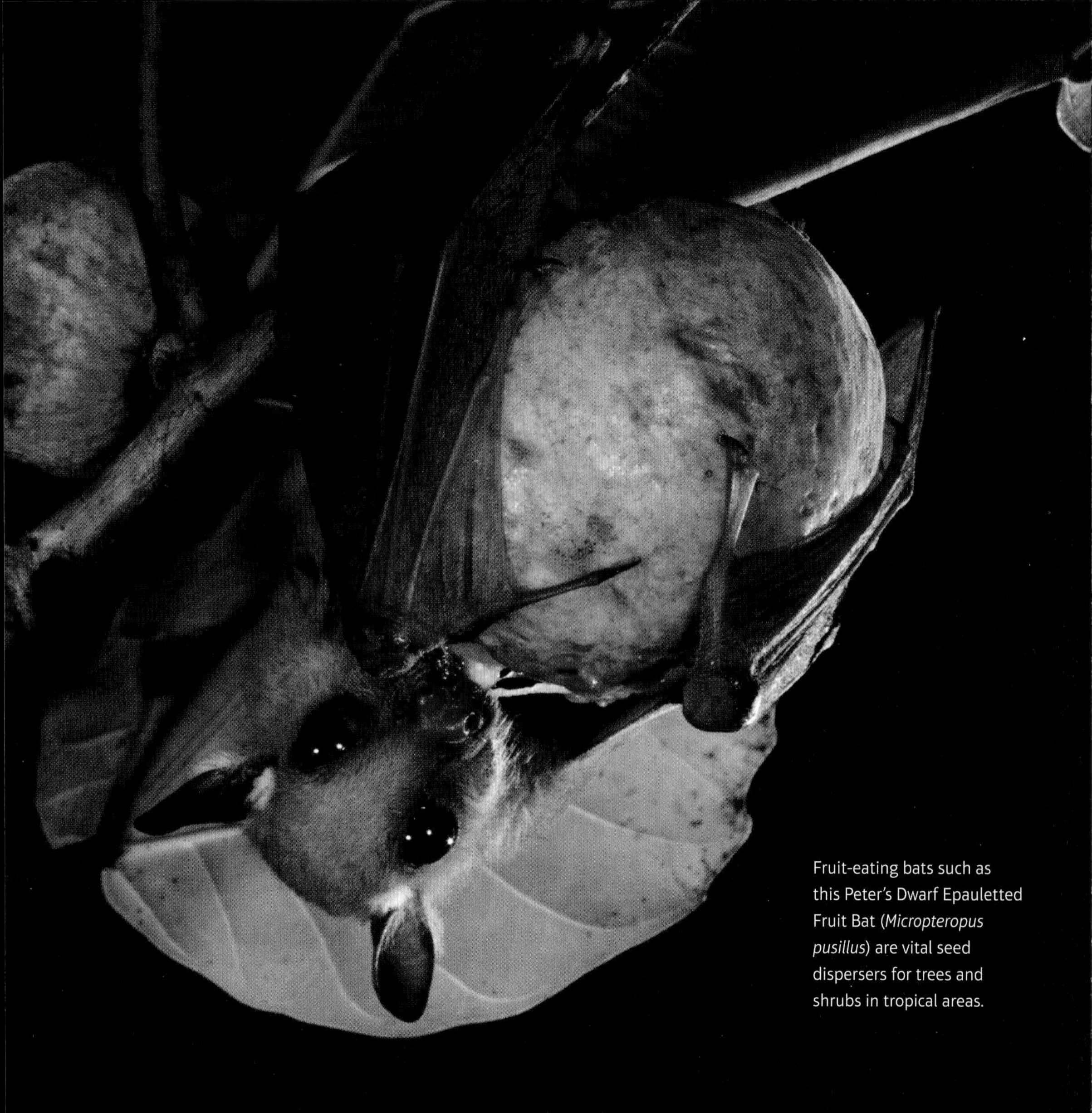

Fruit-eating bats such as this Peter's Dwarf Epauletted Fruit Bat (*Micropteropus pusillus*) are vital seed dispersers for trees and shrubs in tropical areas.

breeding. In this way, successive generations become better adapted to their environment, and over time species gradually change and new forms evolve.

Scientists are increasingly confident that the fruit-eating, non-echolocating megabats evolved from fully echolocating, insect-catching microbats. Once the ancestors of megabats had begun to feed on immobile fruit rather than flying insects, many new selective pressures were imposed. For fruit-eaters, it helps to have good eyesight and good color vision, so the eyes of the early megabats grew larger. However, a sophisticated larygeal echolocation system for hunting insects was apparently no longer necessary, and gradually this ability was lost (though different, simpler forms of echolocation for navigation have subsequently evolved in at least some of the group).

# DIVERSITY

Over the last fifty-five or so million years, many species of bats have existed and died. It is known that at least fifty million years ago, at least three distinct bat families existed. It is likely that the bulk of early diversification took place in Africa, following its colonization of primitive species from Eurasia. Among the early African bats that showed marked specialization is the genus *Aegyptonycteris*, consisting of unusually large and omnivorous bats. Another genus of giants called *Witwatia* possessed large shearing teeth and strong jaws, revealing them to be fearsome carnivores.

Bat fossils from elsewhere include *Icarops*, which lived in Australia. This genus is related to the modern New Zealand Short-tailed Bats (*Mystacina*). The anatomy of *Icarops* suggests that it, like *Mystacina*, foraged on the ground. This discovery undermined the generally accepted idea that ancestors of *Mystacina*, which colonized New Zealand one to two million years ago, only evolved as ground-foragers because New Zealand had no other mammals, and so no obvious competitors. *Icarops* lived in Australia, where ground-dwelling marsupial mammals are plentiful, and was happily foraging on the ground twenty-three million years ago—long before the ancestors of *Mystacina* colonized New Zealand. This is an example of how the fossils of long-extinct animals can cast new light on our understanding of modern animals and their ecology.

The bat species on Earth today are amazingly diverse, in shape, size, and habits, and there is much disagreement on how best to classify them into a taxonomic order. This book recognizes 1,384 species, in 219 genera and 21 families. Vespertilionidae, the "vesper" or evening bats, is the largest family, with 485 species—125 of which are in the largest bat genus *Myotis*, sometimes known as the "mouse-eared bats." There are also a few monotypic families (containing just one genus), and of those the family Craseonycteridae contains just one species. This book recognizes 200 species in the megabat family (Pteropodidae) in 45 genera. The most diverse of these genera is *Pteropus*, with 65 species.

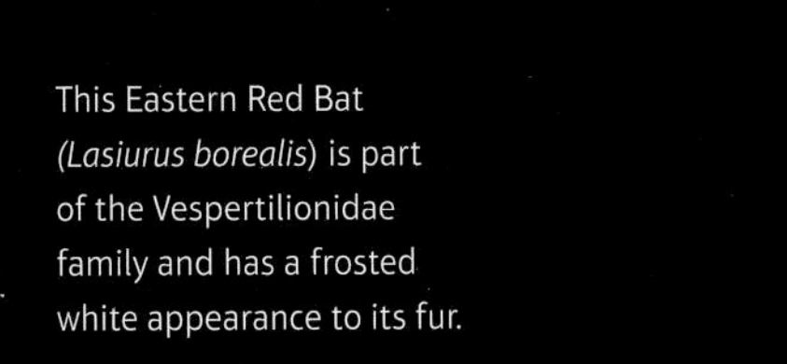

This Eastern Red Bat (*Lasiurus borealis*) is part of the Vespertilionidae family and has a frosted white appearance to its fur.

The number of species in a grouping is a measure of how successful that group is, but it is not the only one. Distribution and abundance are also part of the story. Vespertilionidae are the most common and best known, occurring on all continents except Antarctica. Within Vespertilionidae, *Myotis* is the most diverse and widespread genus, occurring across North and South America, Eurasia and Africa. Curiously, it did not flourish in Australia, where just one of its species, the Large-footed Myotis (*Myotis macropus)* is known to occur (there may be another Australian Myotis, *M. australis*, but this is known only from its type specimen, which is of uncertain origin). The family Pteropodidae and its largest genus, *Pteropus*, are only found in Madagascar, south and southeast Asia, parts of Australia, and on islands between these land masses.

Some individual bat species have populations many millions strong and extensive world ranges. Far more species are highly range-restricted, with small populations. While more than half of the bat species that have been assessed by the International Union for Conservation of Nature (IUCN) are not considered to be in any immediate danger of extinction, twenty-five are considered to be Critically Endangered (the highest level of threat category as defined by the IUCN). About fifty are classed as Endangered, 100 Vulnerable and almost eighty Near Threatened. Some 200 are classed as Data Deficient. This means that not enough is known about these species to assign them to a category, but it is certain that at least some of them are threatened, to a greater or lesser extent. Unfortunately for species that play key ecosystem roles, the strong emphasis on endangered status is far from sufficient. Such animals can become ecologically irrelevant long before becoming sufficiently rare to be listed as endangered. The challenges involved in studying and protecting some of these species are immense; see Research and Conservation, page 53.

## Mega & micro

To start with the obvious, megabats have large, wide eyes, often with amber or brown-colored irises. In contrast, microbats tend to have smaller, black, sometimes "pinprick" eyes. With ears the size difference can be reversed;

Megabats, such as Geoffroy's Rousette (*Rousettus amplexicaudatus*), generally have large eyes, small ears, and long but simple snouts.

Some microbats have very large outer ears of complex structure, including a pronounced inner projection called a tragus. The ears of megabats are of modest size and simple design, and have no tragus. Megabats also have simple but well-projecting muzzles with a pointed, foxy shape, designed to give a well-developed sense of smell. Microbats tend to have short muzzles and sometimes sport elaborate fleshy appendages on their noses—adaptations to help refine and direct the sounds they make.

There are other differences in appearance. Microbats have a claw on the first digit only, the thumb; megabats have a claw on the second digit as well. And while most microbats have a tail (either contained within the section of the patagium that links the feet, or sometimes partly free), most megabats have no tail, or only a rudimentary one, and no continuous patagium between the feet. Both groups have pointed teeth, but those of microbats are pin-sharp; megabats possess larger but blunter canines, and a strong bite for penetrating or crushing tough fruit skins.

Most bats have brown fur, its length and thickness varying according to the prevailing climate where the species reside. All bats are born naked, and a couple of species remain bare-skinned throughout adulthood. Usually bats are a shade paler on the underside than the upperside—a common feature in the animal world known as counter-shading. It functions as a form of camouflage, working against the natural condition of being lit from above and shadowed beneath, to make the body seem less obviously a solid object. Some bats have contrasting golden, pale, or even white markings. These are also adaptations to help with camouflage in their particular habitat. In some cases they are not apparent when the bat is at rest, but are "flashed" as part of a display to attract a mate (see page 37).

Unlike most microbats, which roost in caves or other cavities, many megabats roost out in the open. Others, such as the genus *Rousettus*, however, do roost in caves. This genus also uses a form of echolocation based on tongue clicks. Being fruit-eaters, they do not need to be able to target moving prey and their form of sonar is less sophisticated than the laryngeal echolocation of the microbats—but sufficient to help them navigate through their dark roosting spaces.

Microbats, such as Lesser Woolly Bat (*Kerivoula lanosa*), usually have very small eyes, big ears, and short snouts that often bear folds.

# BIOLOGY

Based on size differences, it is not surprising that biologists traditionally split the order Chiroptera into the suborders Megachiroptera and Microchiroptera. Now that genetic research appears to link some microbats more closely to megabats than to other microbats, many authors prefer a new division: megabats plus some microbats (suborder Pteropodiformes) on one side, and the remaining microbats (suborder Vespertilioniformes) on the other. However, this does not negate the clear differences between megabats and microbats in both appearance and habits.

## Senses

Prominent among the facial features of a typical microbat are its big or huge ears, often with a complex structure. It may also display leaf-like projections around its nose, or distinct folds and wrinkles on its face.

Hearing is the most important sense for the laryngeal echolocating species. The fleshy structures adorning the nose in some bat groups are positioned not to enhance the sense of smell, but to direct and modify the sounds they produce through their mouths or noses (see Echolocation page 24).

However small-eyed they may be, the microbats are not blind: In fact many appear to see quite well. Being nocturnal creatures, the retinas of their eyes are dominated by rod cells, which detect contrast differences, whereas diurnal animals have a higher proportion of color-detecting cone cells. However, scientists have established that nectar-feeding bats are able to detect ultraviolet light as well as light in (the human) visible spectrum. Many flowers reflect ultraviolet light in distinctive patterns to attract their pollinators. It has been known for many years that insect pollinators such as bees respond to these patterns, and the same appears to be true of pollinating bats.

Unlike microbats, megabats have large eyes and sight is their dominant sense, although their sense of smell is also well developed. Their adaptations to see well in poor light include what is called a tapetum lucidum—

The extraordinary nose of Tomes's Sword-nosed Bat (*Lonchorhina aurita*) enables it to direct its echolocation calls accurately.

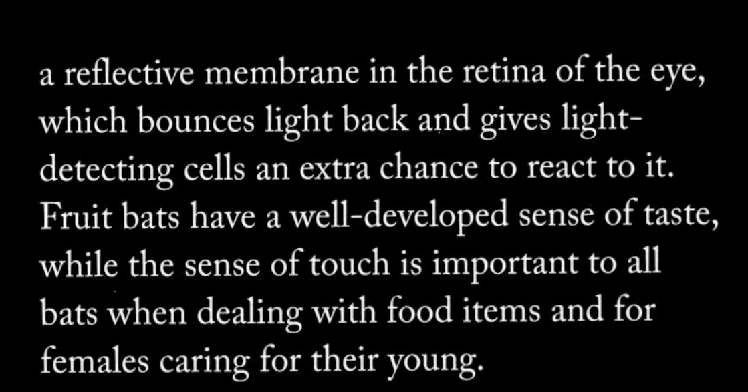

a reflective membrane in the retina of the eye, which bounces light back and gives light-detecting cells an extra chance to react to it. Fruit bats have a well-developed sense of taste, while the sense of touch is important to all bats when dealing with food items and for females caring for their young.

## Flight

Bats are amazingly accomplished and agile fliers, easily exceeding even the most aerobatic birds. The principles of bat flight are the same as in other flying animals—the wingbeats create thrust and the wing shape creates lift, which overcome the opposing forces of weight and drag. The wingbeats describe a circular, as well as up-and-down, motion. They are drawn in on the upstroke to reduce air resistance, and open out and curve over on the downstroke to "collect" a pocket of air to push against. The downstroke is powered primarily by chest muscles, the upstroke by back muscles. Miniscule muscles in the patagium also make small adjustments to the wing shape, allowing for quick turns.

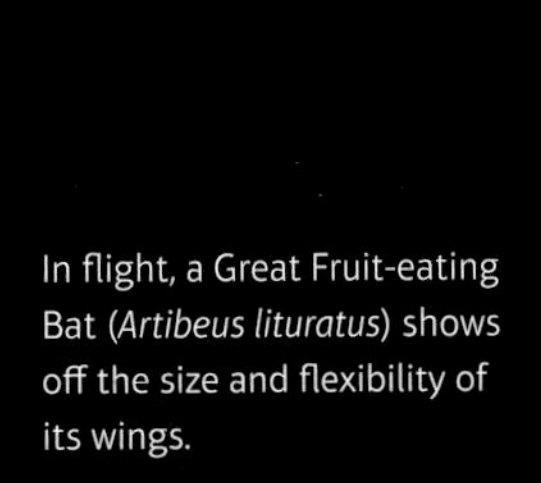

In flight, a Great Fruit-eating Bat (*Artibeus lituratus*) shows off the size and flexibility of its wings.

The two layers of skin that form the patagium envelop the arm and finger bones giving the wing a smooth cross section, widest at the leading edge where the arm bones are. This is the classic aerofoil shape, which encourages the air passing over it to move more quickly than the air moving under it. This means there is lower air pressure above the moving wing relative than there is under it, creating lift.

Wing shape and size affect the bat's flight speed, power, and maneuverability. Those with shorter, wider wings are more agile, able to make quick changes of direction while moving at lower speeds. This is perfect for species that forage by picking insects from tree foliage while flying through the enclosed environment of a forest canopy. Bats with longer, narrower wings are faster, more efficient fliers, adapted to chase down fast-flying insects in the open. Brazilian Free-tailed Bats (*Tadarida brasiliensis*) are the fastest-known fliers, able to top 100 miles (160km) an hour when the wind is in their favor (although these could well be outpaced by lesser-known species). They are also high fliers, reaching altitudes of up to 10,000ft (3,000m) on their long migratory flights.

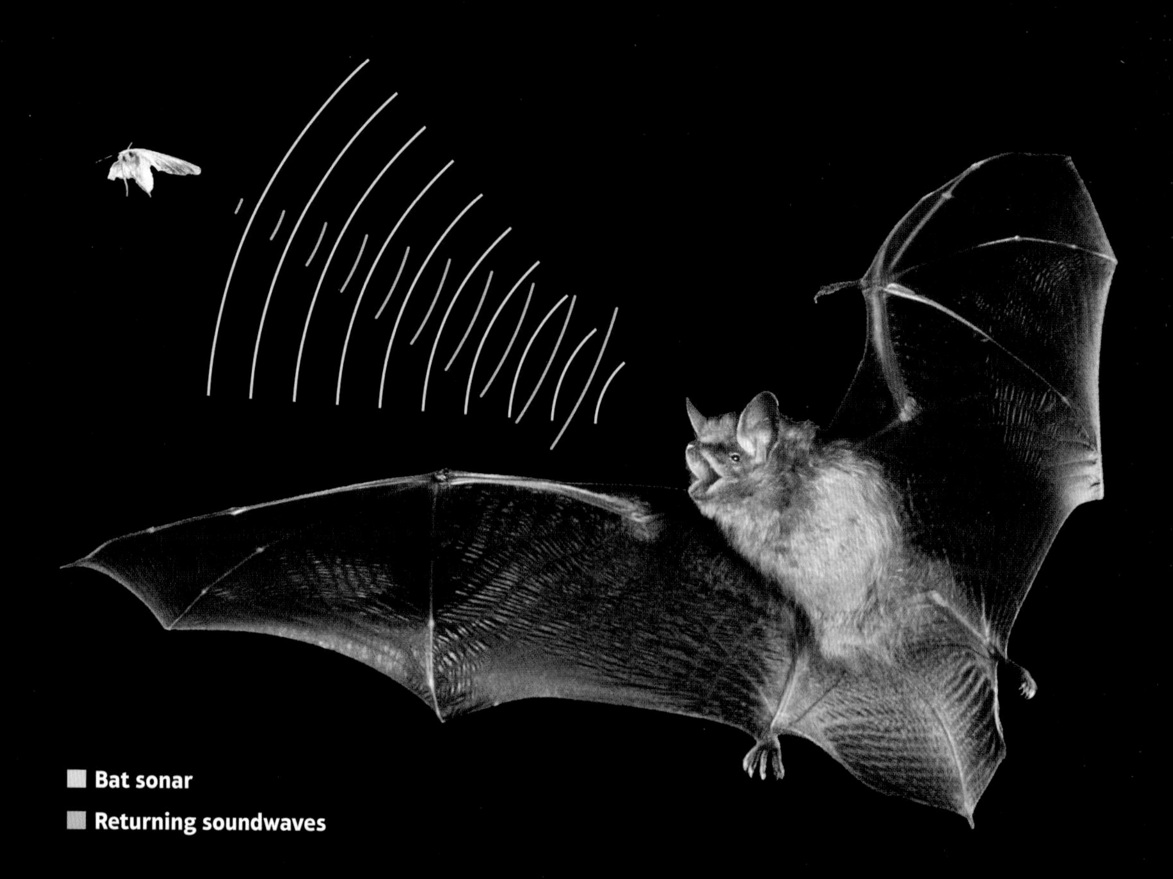


## Echolocation

When soundwaves strike an object, they are bounced back, and this reflected sound is called an echo. Humans do not hear echoes very often, but for some animals the ability to hear and interpret them has become highly developed, and vital to their survival. Among mammals, the cetaceans (whales, porpoises, and dolphins) use echolocation to find food, and a few shrews and tenrecs (curious insectivorous mammals native to Madagascar) do as well. However, the echolocation abilities of the microbats are far more sophisticated than those of any other animals.

A bat in flight produces constant ultrasonic sounds. The resultant echoes that it hears when those sounds are reflected off solid objects or surfaces give it a map of its environment—all the information it needs to navigate a path around obstacles. If you have ever been in the same room as a trapped bat, you will have noticed that it does not fly into the walls, or your face. This is because it is interpreting a constant stream of sound that humans cannot hear. Sound travels through air at about 1,125ft (343m) a second, with slight variations according to temperature and altitude. It is therefore possible to work out how far away an object is by discovering how long it takes for an echo to return. The bat brain swiftly and instinctively carries out these calculations on the fly (literally)—a feat no human brain could hope to emulate.

Bats perceive sound in the same way that humans do. Soundwaves are transmitted as vibrations in the air. These pass via the outer ear, shaped to funnel air efficiently into the eardrum, and make it vibrate. The vibration passes via the middle ear bones to the fluid-filled spiral of the cochlea in the inner ear, lined with sensitive hair cells. As these cells are moved by the pressure waves of sound passing through the fluid, they translate the stimulus from physical movement to electrical signal. Thus the air vibrations are transformed into nervous impulses, which travel via the auditory nerve to the hearing center of the brain.

There are two different systems of echolocation used by microbats, known as low and high duty-cycle. The former involves less frequent calls, which do not overlap with resultant echoes. With the latter, calls are frequent enough to allow overlap, but the bat is able to tell call from echo by using a region of the brain called the acoustic fovea, finely tuned to the "target" frequency range. This system is used by species seeking prey in more "cluttered" environments, such as among foliage.

### A STARTLING DISCOVERY

The fact that bats could navigate primarily by sound rather than sight was discovered in 1793, by an Italian biologist, Lazaro Spallanzani. His experiments involved blocking bats' ears, which left them unable to dodge obstacles in the dark. It took nearly 150 more years before the bats' use of ultrasound was discovered, and the real magic of echolocation revealed. The growing understanding of bat echolocation has helped us to develop our own versions of the system—sonar and radar. It has also allowed researchers to make great strides in discovering and identifying bats in the wild. Many species have their own distinctive pitch and pattern of calls, and these can be picked up by a bat detector, an electronic device that converts ultrasonic bat calls into sounds audible to the human ear.

### BEATING THE SYSTEM

Echolocation is sufficiently sophisticated to enable bats not only to detect small flying insects, but also to distinguish different types of prey by the ways in which their bodies and fluttering wings bounce back echoes. However, not all insects are helpless in the face of this finely tuned hunting technique. Some species of hawkmoths have been found to produce ultrasound of their own. This likely evolved as a means of communication initially, but has a significant side benefit: it confuses bats, reducing their chances of success. In turn, some bats have adapted to produce their calls at a frequency that moths cannot detect, so the moth has no warning that it needs to activate its "jamming signal" to avoid attack. This is an example of an evolutionary "arms race." Selective pressure drives both predator and prey to make more and more extreme adaptations, to stay one step ahead of the other.

A Rafinesque's Big-eared Bat (*Corynorhinus rafinesquii*) uses the thumbs on its wings to help it climb and cling.

## Anatomy

The overall anatomy of a bat is clearly mammalian, but with a few striking differences from typical mammals. The most obvious feature of a bat skeleton, of course, is the great length and delicacy of the wing bones; these are analogous to our arm and hand, and form the framework to which the web of the patagium is attached. Stripped bare of flesh, the wing does look disconcertingly like a skeletal human arm and hand, with enormously elongated fingers. The legs and toes are proportionately much shorter. The ribcage and collarbones are broad and the pelvis tiny. All of the bones in a bat's body are very delicate and finely built. This makes for a low body mass relative to its size, essential to permit flight. Unlike birds, bats do not have pneumatized bones (containing air spaces).

Many bats have a spur of cartilage called the calcar, or calcaneum, on the hind limb, near the ankle. This provides support for the part of the patagium that stretches between the legs (the uropatagium). There is also a tiny extra bone at the elbow, the ulnar sesamoid. It is formed by ossification of part of the triceps muscle, and functions as a sort of "elbow kneecap," strengthening the joint. This is a useful adaptation for these "arm-driven" animals—even when they move on the ground, they are powered by their arms rather than legs. Another difference is that the legs of bats are rotated outward so their knees flex in the opposite direction to other four-limbed mammals.

A bat's flight is powered by relatively large arm, back, and chest muscles, and the exact shape of the wing can be tweaked by tiny muscles within the skin of the patagium. When a bat hangs upside-down to sleep, its feet retain a strong grip because the arrangement of its foot and leg tendons is such that its own weight keeps the toes curled tight. Disk-winged Bats (genus *Thyroptera*) have adhesive disks on their wrists and ankles, enabling them to cling securely to the smooth inner surfaces of large, unfurled leaves where they live.

Most bats, like this California Myotis (*Myotis californicus*), roost head-down, hanging by their feet.

The physical demands of flight mean that bats require a highly efficient circulatory system, to send oxygen quickly to where it is needed. The lungs are proportionately large, with a better blood supply than in similar-sized mammals. Bats' veins contract rhythmically to pump blood back to the heart; in other mammals only the arteries, carrying blood from the heart, contract in this way. Bats can also greatly reduce blood flow to the patagium through a series of sphincter muscles in the main arteries. This allows them to conserve heat when not flying. Bats also need a fast and efficient digestive system, as a gut full of food would impede their flying ability. A fruit bat can consume and digest a meal completely in thirty minutes or less.

The most remarkable feature of bat circulation is that they can exchange gas (expel carbon dioxide and take in oxygen) through their skin. The only other mammals known to be able to do this "skin-breathing" are certain newborn marsupials, which have extremely thin hairless skin and a large surface area relative to body size. A study of Wahlberg's Epauletted Fruit Bat (*Epomophorus wahlbergi*) has revealed that up to 10 percent of the bat's total oxygen intake and carbon dioxide release took place directly via blood vessels running through the thin, bare skin of the patagium.

Spix's Disk-winged Bats (*Thyroptera tricolor*) use adhesive pads on their ankles and wrists to cling to leaf surfaces when they roost.

Flying-foxes such as this Marianas Flying-fox (*Pteropus mariannus*) have large, sharp teeth to grip and eat firm-fleshed fruits.

Jaw shape and the number, size, and shape of a bat's teeth vary according to diet. Depending on genus, insectivorous bats typically have between twenty-eight and thirty-six teeth, which closely resemble those of other insectivorous or carnivorous mammals. Nectar-feeding and vampire bats, which enjoy a liquid diet (though up to 50 percent of the diet of some nectivorous bats is still composed of insects), have the fewest teeth (just twenty in the case of vampire bats). However, nectar-feeders have slender, long jaws, while vampire bats have very short, robust jaws, needing a strong bite force to drive two stout-based, sharp-tipped, top incisors through mammalian skin. Among the fruit-eaters, bats that eat harder fruits have stronger grinding molars, set in shorter, deeper jaws, and more developed molars than bats that take soft, pulpy fruits. The bats that specialize in harder fruits also have large canines and powerful jaws to grip big fruits and pull them free from the plant. Such prominent, large canines account for the vampiric theme of several fruit-eating bats' genus names, such as *Vampyressa, Vampyriscus,* and *Vampyrodes*. The deciduous or milk teeth of baby bats are quite different, having a spiked structure that helps them grip tightly onto their mother's nipple even when she is in flight.

## A TRICK OF THE TAIL

Insectivorous microbats hunt primarily using echolocation. When close enough, they grab at the prey with their mouths, but often simultaneously curve their wings and uropatagium inward to trap the insect in a sort of net made by their own bodies, so they can try again if the first snap of the jaws misses the mark. The entire action is over in a split second, but slow-motion video footage or a well-timed photograph reveals the rapid acrobatic contortion.

The Greater Bulldog Bat (*Noctilio leporinus*) is a fishing specialist, with long but narrow, large-clawed toes to rake through water and seize prey.

frequencies and repetition rates, call from protected locations (as in a thorny plant), or develop varied kinds of armor, sharp spines, and painful bites. Some bats have evolved vocal strategies to fool prey in return. *Lophostoma silvicolum* mimics the mating calls of katydids to locate them. Some toxic-bodied moths actually communicate with bats, giving their own ultrasound calls. A bat that has previously caught one of these moths will remember its call and avoid it in future.

A few bats even catch prey on the ground. Among them is the Fringe-lipped Bat (*Trachops cirrhosus*), a specialist hunter of frogs that can identify and avoid toxic-bodied species. It locates its prey by listening for the frogs' mating calls. Bats of the family Noctilionidae are specialist fish-hunters, and snatch their prey from the water's surface with their oversized feet. Spectral Bats (*Vampyrum spectrum*) often catch rats and small opossums. The Pallid Bat (*Antrozous pallidus*) captures and eats scorpions, and is unaffected by their stings. Some of the largest microbats hunt birds, most notably the rare Greater Noctule (*Nyctalus lasiopterus*). This bat catches small birds in flight, and up to 80 percent of its diet may be made up of birds.

The megabats (Pteropodidae) mainly feed on fruit, and several members of one family of microbats (Phyllostomidae) are also specialist frugivores. All of the fruit-eating bats are of great ecological importance for their role in spreading tree seeds to new locations, because they tend to take a fruit and carry it away to eat elsewhere. Several fruit-bearing trees produce fruit that is especially tempting and accessible for bats: large (but not too large), and hanging on long stems in openings between parallel branches, allowing easy

In Kenya, the Ethiopian Epauletted Fruit Bat (*Epomophorus labiatus*) is a key seed disperser.

access for rapid removal. The fruits also have skin soft enough for bat teeth to grip, plenty of sweet pulpy flesh, and either large seeds, from which the flesh is easily stripped away, or very small seeds that the bat will swallow and later distribute via its droppings. Bats are more even-handed seed dispersers than birds as they defecate in flight, whereas birds tend mostly to defecate from a perch.

Both Pteropodidae and Phyllostomidae also include a number of nectar- and pollen-feeding species. These flower-visiting bats often have long, slim snouts, and very long tongues to access the flowers' nectaries. The tongue of a nectar-feeding bat is also covered in tiny protrusions, to mop up nectar. Bat-pollinated flowers, like bat-attracting fruits, are adapted to be attractive and easily accessed by bats; often the bats can feed on nectar while hovering, rather than needing to land. Some flowers are structured in such a way that the bat's arrival and contact triggers the anthers to spring forward, firing pollen onto the bat's body. Bats are as important as pollinators for some plants as they are seed dispersers for others. Both flower-visiting and fruit-eating bats learn the whereabouts of their favorite food sources, and may undertake long-distance migrations to follow fruiting cycles.

The three species of vampire bats are in Phyllostomidae, and represent the extreme end of bat feeding diversity. The best-known Common Vampire Bat (*Desmodus rotundus*) feeds mostly on the blood of relatively large mammals, less frequently on birds. These bats are seldom encountered in native forests, away from human-introduced livestock. Instead, this extremely sophisticated species has become a costly pest where domestic animals provide easy dining. Common Vampires are skilled at stalking sleeping animals. They land near their victim, approach walking on extra-large thumbs and feet with wings folded. They rely on heat-sensing facial pits to find areas of rich capillaries where a painless incision can be made. Then they lap the flowing blood, aided by anticoagulants in their saliva. The total blood loss is normally not significant, but may look alarming as the animal's skin will be stained with dried blood that flowed after the bat's departure. The other two vampire bats (the Hairy-legged Vampire Bat, *Diphylla ecaudata*, and the White-winged Vampire Bat, *Diaemus youngi*) prefer to take their blood meals from forest birds.

The need to drink water varies on how moisture-rich a bat's diet is. Bats drink in the same way that highly aerial birds such as swallows do; they swoop down over open water and scoop up a mouthful. Many bats require open swoop-zones at least 13ft (4m) in length, some as much as 100ft (30m). Loss of such open water can be a major cause of decline in some species.

The Pallid Bat (*Antrozous pallidus*) is primarily insectivorous but was recently discovered as an important pollinator for desert plants such as the giant cactus *Pachycereus pringlei*.

Numbers of Dawn Bats (*Eonycteris spelaea*) roosting in Khao Chong Phran Cave in Thailand appear to have declined steeply since this photo was taken in 1982.

## Social structure

When we picture bats at rest, we often think of a large cave, its roof densely packed with bats hanging upside-down. Many bats are highly social, using communal roosts both as sleeping spaces and as nurseries, in which they bear and rear their young. Bat roosts may hold just one species or several; they may be used regularly year after year, or temporarily. The largest-known bat roost, in Bracken Cave, Texas, is the summer home of more than ten million Brazilian Free-tailed Bats (*Tadarida brasiliensis*).

Large, safe, and bat-accessible caves are not common in nature, which explains why caves that do meet these requirements attract bats in huge numbers. Some of the slower-flying, most maneuverable species can also inhabit smaller caves. Bats additionally roost in a wide variety of other locations, depending on the special adaptations of different species. Many live in hollow trees, in crevices beneath tree bark, inside and under buildings, and in any other safe, sheltered structure they can find—including purpose-built "bat boxes," designed to be accessible from below. More unusual roost sites include inside termite nests, within colonial spider webs, inside pitcher plants that are adapted to accommodate them, and inside crocodile burrows. Some sites may be suitable only for part of the year, meaning the bats have to move around from roost to roost. A few species even make their own roosts. The aptly named Tent-making Bat (*Uroderma bilobatum*) is one of several species that creates shelters from large leaves, by biting through the midrib so the leaf folds over.

Although the majority of bats are fur-covered, most species are small-bodied, and vulnerable to the vagaries of temperature. So they need protection from the elements, which a covered shelter provides. The option of roosting with other bats also helps keep them warm enough (especially important for

**RULING THE ROOST**

Social behavior is more than friendly cooperation, of course. Within a roost, some spots are better than others, and these will be taken by the most dominant individuals. Communication and recognition between roost-mates is largely vocal, but bats also recognize each other by odor.

These Honduran White Bats (*Ectophylla alba*) are roosting in a *Heliconia* leaf that they have cut to form a tent.

temperate-living species that need warmth for rearing young). A shelter also provides some degree of protection from predators. However, the biggest bat roosts will have resident predators that eat little else besides bats. In such situations, another benefit of roosting en masse is safety in numbers. With bats in their thousands or millions using the same roost, the chances of any one individual falling victim are much reduced, and if a predator makes a surprise attack, the alarm will be quickly raised.

Whether drawn together for protection or social contact, research has revealed that bats are capable of forming lasting, stable relationships that transcend changes in roost location, and are bolstered by time spent foraging together, as well as within the roost. Bats are unusually long-lived for mammals of their size. A study of Bechstein's Bat (*Myotis bechsteinii*) in their summer roosts revealed that it is the senior females (some as old as twenty) who take the lead in group dynamics, keeping family groups together, but ensuring their continued contact with the wider colony.

Bats that roost together in smallish groups regularly break up, mix with other groups, and move to new roosts, in different combinations of individuals than before. Known as fission–fusion sociality, this practice occurs in many wild animals. It means that bats within the wider population encounter many other individuals over time, but also meet the same individuals periodically. They learn from one another, about places to forage, new ways to evade predators, and other survival skills. Over time, bonds of friendship appear to form between certain individuals, which then stick together as they move to new roosts.

Vampire bats are famous for their altruistic behavior of sharing meals with their fellows. A night's searching for prey does not necessarily end with a feast for a Common Vampire Bat (*Desmodus rotundus*), but those that return to the roosts with empty stomachs may be fed blood (by regurgitation) by other bats. A bat is most likely to share food with a close relative—usually a mother giving a meal to one of her offspring. But sharing between unrelated individuals takes place too; the recipient remembers the exchange and is more likely to donate to the giver at some future point. Bats that engage in reciprocal food-sharing are also likely to spend time grooming one another, and this tactile contact may help a bat tell whether its friend is in need of a feed.

**Reproduction & life cycle**

Nearly all bats give birth to a single, well-developed youngster once a year. Pups born in temperate areas in particular must mature rapidly. Some have less than four months in which to reach adult size, learn to fly, and be able to hunt on their own, and to store fat supplies sufficient to sustain them through six to seven months of hibernation. Just one pup annually is the norm in bats, so the parental investment in individual offspring is high. These factors guide all stages of the bat's reproductive process.

There are exceptions to the norm. The bat genus *Lasiurus* is unusual in that many of its members have twin babies rather than singletons. Eastern Red Bats (*Lasiurus borealis)* commonly have quadruplets. Unlike most other bats they have a second pair of teats. Another unusual case is the Big Brown Bat (*Eptesicus fuscus*) of North America, which in the west of its range has a single pup, but in the east frequently produces twins. Several species produce multiple embryos, resorbing those not likely to be supported in years when food is scarce.

As with other mammals, bats feed their young on milk. Only females can provide this with the result that they tend to form single-parent families. Or perhaps *commune* is a better word, as most female bats have their young within colonies, and caring for them is in some ways a shared responsibility. The male's contribution, though, is of sperm only, in the vast majority of species. This sets up a mating system in which males compete to attract as many females as they can, and females are choosy about the males they mate with, selecting only those that are strong, healthy, and whose courtship behavior is most impressive. Such males will father healthy offspring, and their sons will also be successful at attracting mates. In some species, dominant males assemble a harem of females and prevent other males from accessing them; in others males gather at display grounds called leks, where they parade their attributes to visiting females. In a few species, a lasting, monogamous pair bond is formed. Interestingly, in such bats the brain is proportionately larger than in the promiscuous breeders, although the opposite applies to their testicle size.

Courtship behavior is varied, and may include visual, olfactory, and auditory signals. Male epauletted fruit bats (genus *Epomophorus*) have patches of white fur on their shoulders that they display in courtship. They also produce secretions from glands located in shoulder pouches and flap their wings to waft

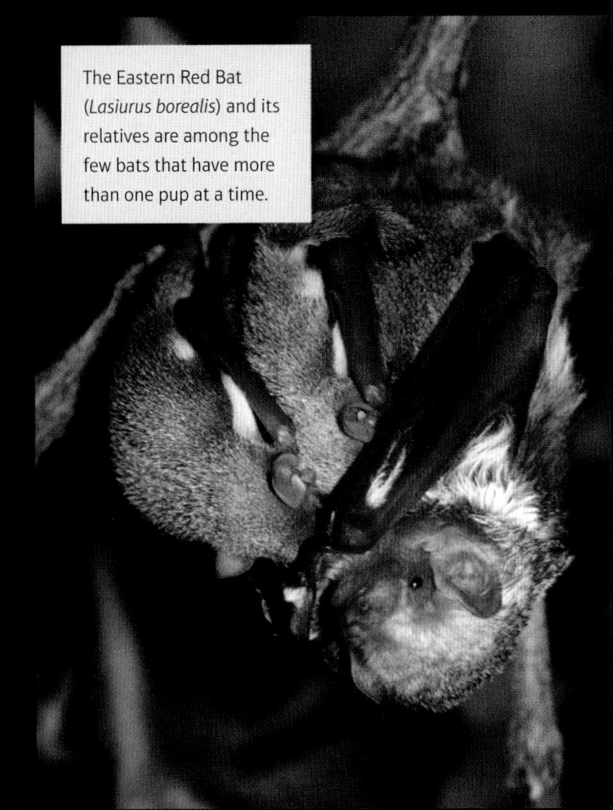

The Eastern Red Bat (*Lasiurus borealis*) and its relatives are among the few bats that have more than one pup at a time.

The cheek pouches of male Ethiopian Epauletted Fruit Bats (*Epomophorus labiatus*) can hold food, but also amplify their courtship calls.

the enticing smells toward an interested female, while using their inflatable cheek pouches to amplify their courtship songs. The champion smell-producers are male Greater Sac-winged Bats (*Saccopteryx bilineata*) and related species. A male sac-winged bat has pouches or sacs on his wings. He fills these with a pungent cocktail of urine and genital and glandular secretions, transferring the substances via his mouth. He then "sings" to attract females, while hovering in front of them to disperse his unique signature scent. Once two bats have "agreed" their mutual interest, they will often hang close together and groom one another before copulation takes place. Male bats may also mate with females that have already entered a torpid state at the start of their hibernation, and are thus unable to reject the male.

Most bats that hibernate over winter will mate in fall, while they are well-fed. In some cases, copulation triggers ovulation, so the female becomes pregnant soon after mating. The embryo may either develop very slowly through winter, or its development may be suspended until the following spring. In many species, females store sperm through winter, with fertilization not occurring until spring. Tropical species, which do not hibernate, may still have a definite breeding season, linked to the best seasons for foraging. Some of the fruit-eating Phyllostomidae have two reproductive cycles a year, the second pregnancy beginning right after the first pup is born. There is a short break from breeding during the peak of the rainy season.

Female bats give birth in their summer roosts, by which time males have often left to form smaller bachelor groups. Although well-developed, the newborn bat pups are often hairless and more or less helpless, though they are able to clamber down the mother's body (most give birth in the usual, upside-down position) and latch onto one of her two nipples, which are pectorally located, as in primates. This led Carl Linnaeus, the father of modern taxonomy, initially to lump bats with primates.

Bats that must hunt insects on the wing normally leave pups at the roost while hunting. In contrast, fruit-eaters—who do not need to be so maneuverable—often carry newborn pups with them when leaving to feed. Cave-dwelling bats leave their pups clustered with other young, sometimes packed in at densities of up to 465 per square foot (5,000 per square meter). When a mother returns, her pup recognizes her voice and calls to her. Amazingly, each recognizes the other's voice. Each mother further recognizes her pup's unique odor and nurses only her own, ignoring or swatting away the attempts of others to latch onto her. This is a remarkable feat, given that thousands of mothers and young are often calling at the same time. However, mothers whose pup dies often adopt another.

The smaller bat species are generally able to start flying when they are three to four weeks old, and quickly learn to feed themselves, though achieving full independence takes much longer in some species. Baby bats making their way in the world face predators of all kinds, as well as the risks of being caught in bad weather or failing to find enough food. Mortality in their first year is high. However, once that period is survived bats may live many years. Brandt's Bats (*Myotis brandtii*) in the wild have been known to live into their forties, and it is likely that related species can do just as well; bats in general are much longer-lived than other mammals of similar size.

Bats appear to be unusually resistant to disease, which seems to be relatively insignificant as a cause of bat mortality. Despite often living in large, densely clustering groups, die-offs from disease are rare—with one notable exception. A fungus, *Pseudogymnoascus destructans*, first recorded in North America in New York in 2006, causes a disease in bats known as white-nose syndrome (WNS). Symptoms include whitening around the muzzle, as well as damage to the wings of bats hibernating in caves. White-nose syndrome has now been identified in thirty-three US states, and is estimated to have killed up to seven million bats, including fifteen species. Some species are minimally affected, but others have experienced a population decline greater than 90 percent. The fungus is common and widespread in Eurasia, where most bats are resistant to its effects. It kills by causing costly arousals from hibernation, forcing bats to starve before spring. It is anticipated to spread throughout North America. Where the disease has already passed in the northeast of the continent, gradual recovery based on

A newborn Brazilian Free-tailed Bat (*Tadarida brasiliensis*) has bare skin but is otherwise well-developed.

Wrinkle-lipped Free-tailed Bats (*Chaerephon plicatus*) emerging from Vihear Luong Cave in Cambodia.

resistant survivors appears to be occurring. These bats are best helped by protecting them from human disturbance—particularly in winter, but also in summer. A team carrying out long-term monitoring of white-nose syndrome in Pennsylvania warns: "All survivors still become infected annually. It is likely that these few survivors are existing on limited fat reserves, and every disturbance is an additional cost on those reserves. The effect of this disturbance may directly cause mortality later in the hibernation season to adults or juveniles fighting infection, or it may lower the fitness of adult females enough to inhibit their ability to successfully reproduce."

Like all other mammals, bats can contract rabies. However, it is not an important source of mortality and, contrary to media emphasis, it is rarely transmitted to humans. Often exaggerated publicity has led to a disproportionate fear of bats. However, for anyone who does not attempt to handle bats, the risk of contracting any disease from them is close to zero (see page 49).

## Migration

Insect-eating bats that breed in temperate climates have a problem when winter comes: the insects disappear. Of course, there still *are* insects, but most overwinter in an early, non-flying life-stage, and are underground, underwater, or otherwise tucked away from the cold. Some insectivores are still able to find enough of them to survive, but not bats, which rely primarily on flying insects. The problem is solved in one of two ways. In most cases, it is by hibernation—dropping body metabolism as close to zero as possible, and surviving the winter on fat stores. The alternative is migration. Even hibernating bats often must migrate to find the relatively few caves that provide sufficiently low, stable temperatures. Some bats migrate to warmer climates, where insects remain plentiful, rather than hibernate.

The best-known migratory bat is the Brazilian Free-tailed Bat (*Tadarida brasiliensis*), a strong-flying insectivorous species that breeds mostly in North America. Not all populations migrate, but some of those that breed in the south of the United States travel as far as 1,000 miles (1,700km) to wintering grounds in Mexico. They travel at altitudes as high as almost 10,000ft (3,000m), and may cover up to 310 miles (500km) in a single night, making use of temporary but traditional roosts on the way. These bats therefore need to cope both with cool, high-altitude air and the stifling heat of roosts that may hold thousands of individuals. Compared to bats of similar

Deep in hibernation, the Tricolored Bat (*Perimyotis subflavus*) is protected from dehydration by moisture condensed on its body.

size but with less extreme lifestyles, they show exceptional thermal efficiency, able to retain and lose heat quickly as the situation demands.

Besides the long-distance migrants, some bats also undertake much shorter but equally essential regular journeys. Examples include montane species that move to lower ground at certain times of year, and fruit- and nectar-eating bats that travel long distances following fruiting and flowering seasons.

The term *vagrancy* describes "migration gone wrong"—a situation in which a migrant becomes lost or disorientated on its journey. This may be through inexperience or being caught in bad weather, and results in the bat ending up far from its intended destination. They can also be accidentally displaced if they take shelter on a ship or other moving vehicle. Kuhl's Pipistrelle (*Pipistrellus kuhlii*) lives in the south of Europe, but has been recorded in Britain a few times: One arrived courtesy of a vacationer who visited Cyprus, and unwittingly brought the bat home in a suitcase. The Particolored Bat (*Vespertilio murinus*), found across central Europe and Asia, is a long-distance migrant whose regular occurrences in Britain are almost certainly powered by its own wings, albeit sending it the wrong way. It has also been found on North Sea oil rigs on several occasions.

## Hibernation/torpor

Many animals living in areas with distinct seasons become less active in the leaner months, (including hot, dry periods), conserving energy. Often they will store supplies to help them survive these times, whether in the form of actual food, or in a blanket of body fat, and they will spend much more time asleep. However, true hibernation takes things a stage further. Here the state of unconsciousness can continue for weeks or months at a time—more like a coma than sleep, with all vital functions slowed down as much as possible, and body temperature far below normal; it can even drop below freezing point in some cases. Some mammals that are often thought of as true hibernators, such as bears, do not really approach this level of metabolic suppression, but bats do.

The advantages of hibernation are obvious: It is a very low-effort way of getting through winter, free from the many hazards of a long migration. There are different hazards though. Bats in hibernation are essentially defenseless, and if a hibernation roost (hibernaculum) is discovered by predators, it can be completely wiped out. Even the most unlikely predators may take advantage of a deeply sleeping bat. In Hungary, the Great Tit—a small songbird that many Europeans enjoy watching on the garden bird table—is one such predator. The birds have been observed entering pipistrelle roosts and killing the hibernating bats in the most gruesome manner, pecking through their skulls and eating their brains.

A similar fate befell hibernating bats of several species in the Netherlands, but this time the killers were, if anything, even more innocuous: wood mice. These charming-looking rodents broke open the bats' skulls to access the brain in the same way that they would nibble into a hazelnut, as well as eating other parts of the body. Both bird and rodent appear to seek out bat hibernacula deliberately. In Wales, United Kingdom, an analysis of otter droppings proved that these aquatic, primarily fish-eating mammals were preying on rare Lesser Horseshoe Bats (*Rhinolophus hipposideros*) that they took from roosts in abandoned mines. In North America, raccoons can prey heavily on hibernating bats.

The other main hazard of hibernation comes from inadequate preparation. Even with all body processes on their lowest possible setting, energy is still consumed by the hibernating bat's body and supplies can run out, starving the bat or leaving it too weakened to emerge from hibernation in a fit enough state to fly and forage. So bats spend the weeks before hibernation feeding frantically, to store as much body fat as they can (while still being capable of flight). Hibernating in dense clusters, as some bats do, helps to protect against sudden temperature changes or dehydration, and enables them to share the considerable cost of arousal from hibernation.

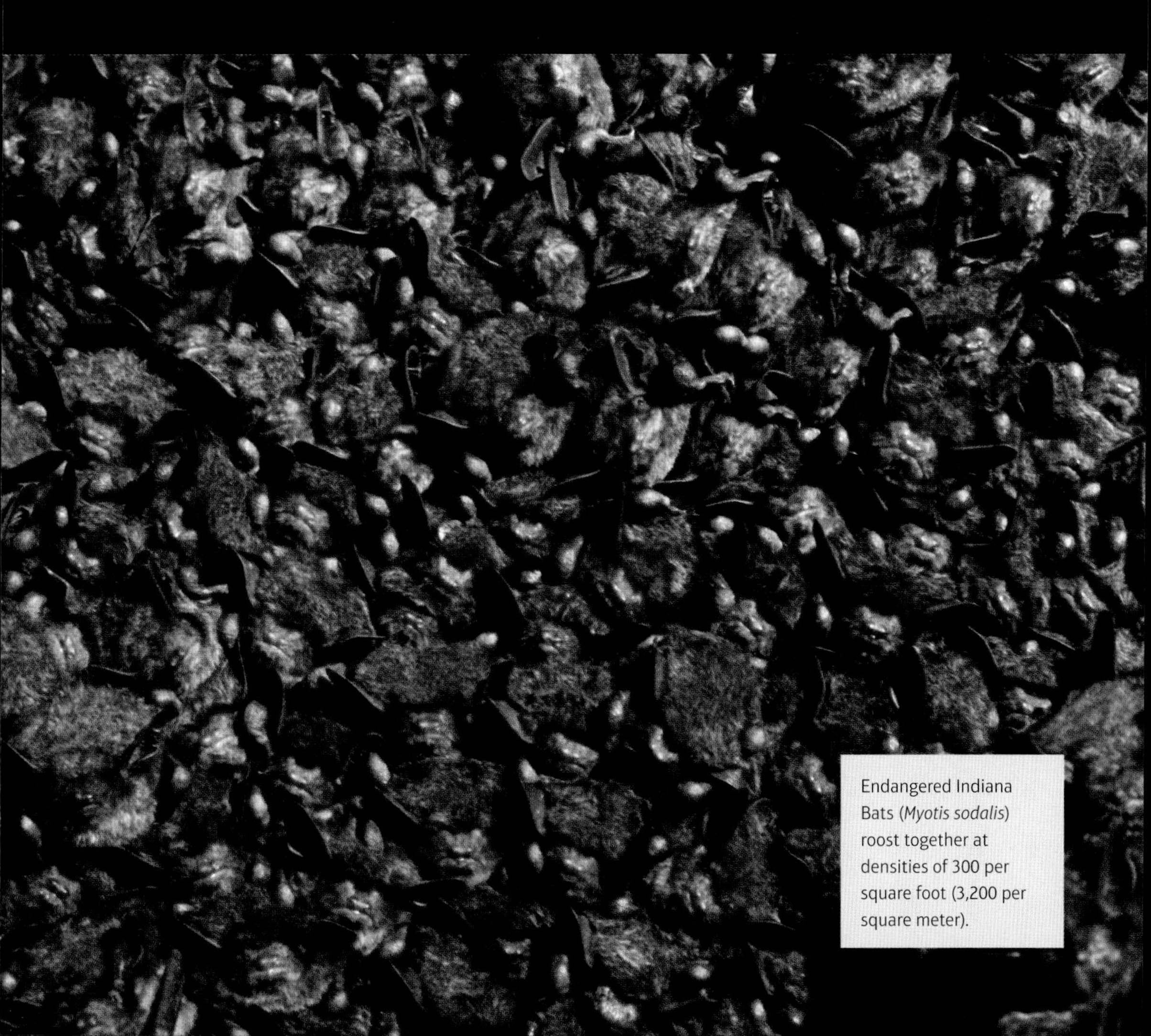

Endangered Indiana Bats (*Myotis sodalis*) roost together at densities of 300 per square foot (3,200 per square meter).

# ECOLOGY

Every organism on Earth has its own part to play in the wider system. The fortunes of each individual bat species affect those of many other organisms, whether directly or indirectly. Those organisms may include its prey and predators, its parasites, its competitors, and even those that clean up after it has defecated, and ultimately after it has died. As children we are taught about "food chains," but this is an oversimplification. The links are more than linear, as suggested by the word *chain*; in reality they are a web. Ecology is the study of the web of life—how organisms and life processes are interconnected, and how any change affects the overall structure.

## Predators & prey

Most bats are predators and also prey. They hunt and are hunted by a great variety of other animals, placing them very centrally in their ecological webs. Different bats have different diets; there are species that feed on nectar, fruit, insects, frogs, fish, scorpions, birds, rodents, and blood. The bats that take a vegetarian diet—the fruit-eaters and nectar-lappers—disperse plants' seeds and spread around their pollen, allowing new plants to grow and even colonize new areas. They also compete with many other organisms, perhaps most obviously the birds, which share their mastery of the air and ability to exploit the

Raptors like the Swainson's Hawk will wait near roosts of Brazilian Free-tailed Bats (*Tadarida brasiliensis*) to catch the bats as they emerge.

vast soup of flying protein in the form of winged insects.

The quantity of insects eaten by bats is quite staggering. Even a tiny microbat can get through 3,000 or more in a night—often half its own body weight. A female bat with a pup to feed consumes more—up to her own weight—to enable her to produce the high-protein, high-fat milk that her pup requires. Fruit- and nectar-eating bats also have tremendous appetites, eating up to twice their body weight in a single night.

The range of animals that prey on bats is extensive. However, bats are hard to catch when flying in the open air. Few birds of prey can match them for agility, and while many raptors might catch the odd bat, only a handful could be considered to be specialized predators of bats. One such specialist is the Bat Hawk—a raptor that has adapted to hunt bats at dusk, relying on its broad wings to make the quick turns necessary to chase down its prey. It is found widely across much of sub-Saharan Africa and also parts of southeast Asia and Australasia. In Mexico and Central and South America is found another raptor, the Bat Falcon, which, like the Bat Hawk, is most active at dusk and takes bats as they leave their roosts. Owls, the archetypal night predators, are among the most important predators of bats. They capture far more rodents, but due to the much lower reproductive rates of bats, their impact on bats can be surprisingly large.

Predators that attack bats at their roosts include snakes, climbing mammals, such as raccoons, martens, and domestic cats, and various opportunistic birds, from crows to grackles and jays. Bats are also vulnerable when leaving their roosts, particularly when they have to funnel through a small access point. In Florida, Gray Rat Snakes climb vines and ledges near the entrances of bat caves and hang down in the bats' flight path. They do not attempt to strike moving bats, but if one makes even the slightest wingtip contact with the snake's body as it passes, the snake grabs it with lightning speed.

A Gray Rat Snake catches a Southeastern Myotis (*Myotis austroriparius*) at the entrance to a cave roost.

Large caves used by many bats for roosting develop their own special ecosystems, of which the bats themselves are, of course, the cornerstone. Entering such a cave is a challenge for any naturalist: the heat can be stifling, and the atmosphere heavy with ammonia, produced by dermestid beetle larva digesting vast quantities of bat guano. But other animals are quite comfortable in this environment. A camera crew filming for the television series *Planet Earth* explored a vast cave system in Borneo, where an estimated three million Wrinkle-lipped Free-tailed Bats (*Chaerephon plicatus*) roost, and revealed the extraordinary wealth of life within. The cave floor is thickly layered with guano, on top of which is an equally continuous layer of cockroaches, busily consuming the guano, as well as any dead bats that fall from the roof. In turn the cockroaches are hunted by centipedes, while crabs scavenge for whatever they can find. These animals derive all of their energy from what the bats bring in from the outside world.

## Habitat

There are few habitats on Earth that do not support at least some bat populations. Woodland, at all latitudes and altitudes, is home to a greater variety of species than more open country, but plains, grasslands, and even some deserts also have their own special bat species. Wherever bats can find food and shelter, there they will be. Some megabats live on tiny tropical islands and commute between them to find enough food, crossing open seas. The only land mammals naturally present in the Maldives, for example, are a couple of species of megabat.

A suitable habitat for bats must include places to forage, places to drink, and somewhere to roost. Nearly all habitats found on Earth offer something for some kind of bat to eat. Insects and other invertebrates live everywhere, from Arctic tundra to deep tropical jungle and the edges of open desert. Insect-eating bats occur in all of these habitats, their abundance and diversity increasing in concert with that of their prey. Fruit-eating and nectivorous bats are more restricted, occurring mainly in tropical and subtropical climates.

In the tropics, bats only require a day roost and perhaps a night roost for short visits, during which they will eat large items they have caught or collected. However, those in temperate climates that hibernate will also need a winter roost, which is likely to be very different in character. Especially in temperate areas, only a few caves may meet bats' year-round needs, forcing them to move seasonally between exceptionally warm, heat-trapping caves for rearing young and caves that trap and store large volumes of cool air for hibernation. A single cave can be critical to the survival of hibernating bat species from thousands of square miles around. Often such caves are also the most used by humans for shelter or religious rituals, explaining how some formerly abundant bats have now become endangered. Long-term use by bats frequently stains ceilings and walls of caves, making them readily identifiable for future conservation protection.

Summertime day roosts are more varied. While the most typical are holes in trees, spaces behind tree bark, caves, rock crevices, and inside buildings; other more unusual roosts include crocodile burrows, termite nests, pitcher plants, and even colonial spider webs. Tent-making bats construct their own shelters by biting through a leaf's midrib, causing it to fold over and form a tented roof. Some bats require a variety of roosts, switching frequently among them. Such moves may be driven by weather changes, parasite build-up, predator discovery, or other factors. This is why conservationists have discovered that providing multiple bat houses, preferably exposed to varied levels of solar heating, aids in a scheme's success.

## Distribution

On a global scale, the tropics support the greatest number of bat species, with South and Central America best of all and equatorial Africa and southeast Asia not far behind. Diversity gradually declines with distance from the equator. The bat found farthest north is (appropriately) the Northern Bat (*Eptesicus nilssonii*), which occurs in Norway at 69°N, well inside the Arctic Circle.

The pattern is interrupted by desert regions and high mountain ranges such as the Himalayas, where diversity is low. The extreme

These White-throated Round-eared Bats (*Lophostoma silvicolum*) have cut into the bottom of a termite nest to make a safe sheltered roosting site.

The durian fruit industry in southeast Asia is worth billions of dollars each year. The Dawn Bat (*Eonycteris spelaea*) is a key pollinator for durians.

## BATS & PEOPLE

Humans are indebted to bats in many ways, and we have much to learn from them. Consider a forty-year-old wild bat, small enough to hide in your hand, yet still able to chase down flying insects for dinner—the equivalent of a hundred-year-old human being able to sprint through an obstacle course. Bats can perform myriad physical feats. Their flight is exceptionally agile, more so than any bird, but some can also fly faster than most birds—free-tailed bats can use tail winds to achieve speeds of close to 100 miles (160km) per hour in level flight. Their physical strength is impressive as well. For example, a Veldkamp's Dwarf Epauletted Fruit Bat (*Nanonycteris veldkampii*), weighing ¾oz (22g), can fly comfortably while carrying a fruit weighing ½oz (15g).

As far as direct value goes, bats benefit us enormously by keeping insect numbers in balance, protecting our crops, and reducing disease-carrying pests such as mosquitoes. Pest control alone has been estimated to save farmers in the United States nearly 23 billion dollars annually. Bats that feed on nectar are key pollinators for many commercially important crops—without bats, we might not have a wide variety of tropical products, from bananas and mangoes to tequila, timber, and nuts. In southeast Asia, the most prized and famous durian fruit still requires bat pollination, even when grown in orchards. Fruit-eating bats are key seed dispersers, and bat guano is a valuable fertilizer. There is even a place in medical research laboratories for bat-derived biochemicals.

Yet despite these wonderful benefits, bats often remain disliked and feared in many cultures. Fortunately the tide is turning, as people become more and more aware of how valuable these animals are to us as a society. So are there any ways in which bats can be harmful to people? The answer is very few. In some areas, admittedly, certain species of fruit-eating bats can damage crops. However, this can usually be managed with non-lethal measures, and bats as a group are overwhelmingly more a friend than a foe to the farmer. One other worry people have about bats is their potential to harbor and spread disease. While this is possible, it is rare, has been greatly exaggerated, and is easily avoided, mostly by simply not handling bats. Researchers and veterinarians who handle unfamiliar animals, including bats, should be vaccinated against rabies (see Myth and truth).

In short, the value of bats to our economies and (arguably no less important) to our ability to connect with and appreciate nature in all its wondrous variety of form and function far outweighs any negative impact bats have upon humankind. Many bat species are needlessly declining and threatened with extinction because of human activity and unwarranted persecution. The next chapter of the story of bats and people has to be one that tells how science, imagination, and committed hard work are being used to help reverse all the harm that has been done.

## Myth & truth

Humans are naturally drawn to animals—on some level we recognize our own emotions and motivations in them, even as they are exploited for our own ends. Some animals, though, are much easier to relate to than others. Bats are so unusual and enigmatic that they almost seem like aliens to us. Although our mammalian cousins by lineage, in every obvious way they are as different from us as can be—a difference that can inspire fear and confusion. Bats, and vampire bats in particular have long been associated with death, ghosts, and general sinister otherworldliness.

In various folk stories, bats are birds that lost their feathers, or mice that learned to fly—and the transformation cost them dearly. Fijian legend tells that bats were once rats that stole their wings from the birds, and then hid in the dark to avoid reprisals. In an Indian legend, bats were birds that prayed to the gods to be made human, but instead were turned into a sort of half-bird, half-human hybrid; they were so ashamed of their strange appearance that they would only venture out at night. A story told with minor variations in ancient Rome, Nigeria, and Australia casts the bat as a turncoat: in a great battle between birds and beasts, the bat switched sides whenever necessary to stay on the winning team. When the war was over and the combatants reconciled, they all agreed that the bat should be banished to a solitary life in the darkness, as penance for its lack of loyalty. Other cultures look more kindly upon bats, however. In China they represent joy and good health; babies wear jade buttons shaped like bats as an amulet to bring long life. There are Native American bat deities, as well as folktales in which bats play the hero, for example by helping a team of mammals defeat a team of birds in a ball game.

Some strange notions about bats persist today, including the idea that they attack humans and tangle themselves in our hair (they do not), that they are blind (they are not), and that they are exceptionally dangerous carriers of disease. This last point warrants more careful consideration, because bats can transmit rabies to humans. However, this is rare and extremely unlikely for anyone who simply does not handle bats. Rabies transmission from the Common Vampire Bat (*Desmodus rotundus*) of Latin America, poses a risk only to people who sleep outside, unprotected by mosquito netting.

Another disease that can pass from bats to humans is the fungal infection histoplasmosis. This can be contracted through inhaling spores released from accumulations of animal droppings. Most human infections are mild and come from birds. Infection is easily avoided simply by not inhaling dust from bird or bat droppings. If you drink raw palm juice in Bangladesh or India, you also might contract Nipah virus. A variety of Megachiropteran bats raid the containers hung to collect sap referred to as palm juice. Occasionally, the juice is contaminated by a bat as it drinks. But it only infects those who fail to head warnings not to drink this juice raw.

Unfortunately, premature headlines speculating that bats are dangerous sources of several of the world's rarest, but scary diseases have proven exceptionally profitable. Despite seemingly endless claims, Ebola and SARS outbreaks still have not been traced to bats. In fact, mortality records show that bats have one of our planet's finest records of living safely with people. Huge bat colonies coinhabit major cities from America to Africa and Australia and have caused no harm. Forecasts of pending bat-caused pandemics are also without foundation. Bat researchers and the millions of humans who eat bats and extract guano from their caves remain healthy.

## Economic importance

As noted, insect-eating bats deserve our endless gratitude for their work as pest controllers. A single Brazilian Free-tailed Bat (*Tadarida brasiliensis*) can catch enough migrating corn earworm or armyworm moths to prevent them laying 20,000 or more eggs in a single night, saving a farmer from spraying at least two acres of land (0.8ha) with pesticides. Reduced pesticide use greatly benefits all wildlife. The bats that roost in Bracken Cave, Texas, can catch 100 tons (90 metric tons) of insects, or even more, in a single night; most are crop pests. The bats from Khao Chong Phran Cave in Thailand are estimated to save Thai farmers some $300,000 annually through control of rice pests. By strategically placing small bat houses around rice crops in the Mediterranean, scientists have been able to eliminate the need for pesticides, by attracting enough bats to take care of pests. Additionally, insectivorous bats consume numerous mosquitoes and other blood-suckers. Bats around the world are not only saving us from being bitten, but also helping to prevent the spread of serious, insect-borne diseases such as malaria.

The presence (or absence) of insectivorous bats in an environment is a good way to assess the general health of the ecosystem. As predators, their falling numbers are an early warning that action must be taken to prevent wider ecological collapse. In some cultures, bats themselves are a source of food for people, although bats' slow reproductive rate makes them unsuited to sustainable hunting; overhunting has seriously harmed many bat populations.

Nectar-feeding bats are extremely efficient pollinators, and hard-working too; they carry more pollen over longer distances than any other pollinating animals. Hundreds of species of plants are adapted to attract pollinating bats—without bats, they might not survive. Many of these plants are grown commercially and are of great economic importance. Among them is durian fruit, which requires bat pollination whether in the wild or in orchards, and supports an economy worth billions each year. Mexico's tequila and mescal industries are of similar economic value, and rely on bat-pollinated agaves. All the world's commercial bananas have been developed from ancestral wild plants that are bat-pollinated, and mangoes, guavas, and hundreds of other fruit-bearing trees are at least partly pollinated or dispersed by bats. Seed dispersal is the vital "job" of fruit-eating bats

**BATS & ECOTOURISM**

The Ann W. Richards Congress Avenue Bridge in Austin, Texas, attracts large crowds on summer evenings. They come to witness the nighttime emergence of up to 1.5 million Brazilian Free-tailed Bats (*Tadarida brasiliensis*), which roost under the bridge and emerge at dusk. Up to 100,000 tourists come to Austin each year to see the bats, making a real contribution to the town's economy. Other large bat gatherings can also entice tourists, while on a much smaller scale bat walks are popular activities. Many people return from such trips with a new respect for bats, and head for the nearest wildlife supplies store to buy their own books, bat detector, infrared torch, and even put up bat houses to attract bats at home.

In Africa, bats can account for up to 95 percent of initial seed rain required to begin reforestation of abandoned clearings.

Bat guano may not make for a pleasant environment, but it does make a superb fertilizer, and in nineteenth-century America was used to manufacture gunpowder. The fact that such vast accumulations can be found in relatively small spaces means that harvesting bat guano from roosting caves can be profitable. A single, large bat cave can produce guano valued at more than $100,000 annually; it is a sustainable resource as long as the bats themselves are protected. Harvesting must be timed and carried out carefully, with an understanding of the bats' behavior, as disturbance can be harmful (especially for mothers with young pups). Over-exploitation has wiped out bat colonies in the past, but if managed properly this is a sustainable resource. Today important bat caves are often under strict protection: many other associated animals and insects benefit from this as well. Some caves have lost their once great populations, but could still be profitably restored.

The Common Vampire Bat (*Desmodus rotundus*) is known and rather feared for its blood-drinking ways, though people are seldom at risk. Vampire saliva contains an anticlotting agent that has been isolated. It is a complex glycoprotein, which researchers have christened "Draculin," and its action on blood has been explored in the lab with a view to medical application, especially to help stroke patients. It may also have a role in the prevention of heart attacks, by reducing the risk of clots.

The benefits that bats bring to people, ranging from financial reward to enhancing the enjoyment of time spent in wild places, are tremendous. In comparison the problems that they can cause, such as transferring disease, raiding crops, or roosting in buildings, are minuscule. Bats roosting in buildings seldom cause structural damage. They may

The first game warden at Khao Chong Phran Cave in Thailand, hired at Merlin Tuttle's recommendation in 1982, to protect the then declining colony of Wrinkle-lipped Free-tailed Bats (*Chaerephon plicatus*) from poachers.

decorate the floor somewhat, but the pleasure and privilege of sharing space with such charismatic wild animals makes the arrangement for many people quite worthwhile. In some countries, such as the UK, bats of all species are strictly protected. Local bat groups in the UK can help safely to relocate bat colonies where absolutely necessary.

## Research & conservation

The study of wild bats is fascinating but challenging. They fly at night, and they fly quickly and erratically: tracking a small bat in the air is often very difficult, and obtaining photos or video footage takes specialist gear and great skill and patience. Even with a bat detector in use, a positive identification of a bat seen in flight is often impossible. Happily, bat roosts are much easier to find and study. It is also now possible to fit tiny dataloggers to bats, enabling scientists to collect information on the bat's changing spatial position over a long time span. This in turn makes it possible to work out flight paths, preferred foraging grounds, and migratory routes.

Bats can be caught by placing "mist-nets" along their flyways—these are too fine-meshed to be easily detected by the bats' echolocation, and they will hold the caught bat unharmed in a pocket of mesh until it can be retrieved. An alternative is the Tuttle trap, named after bat researcher Merlin Tuttle. The original trap employed two vertical frames of monofilament fish line, resembling a harp. A canvas bag hung beneath to catch falling bats, and plastic flaps prevented escape. Some more recent designs include three or even four sets of vertical lines. By catching bats, biologists can assess their size, body condition, stage in their breeding cycle, their parasite load, and much more. Individual bats can be marked with identifying tags before release, and their "now and then" data compared if they are ever caught again.

Mist-netting is a safe, widely used way to catch bats for research purposes. This is a Spotted Bat (*Euderma maculatum*).

Bat roosts are, in the main, scarce and heavily used resources, which means that identifying and protecting them is key to bat study and conservation. Scientists can count the number of bats using a roost from season to season and year to year. It is also possible to determine historical use and importance of a roost by examining the amount of staining and etching on limestone cave ceilings. It is important that cave roosts are protected from accidental and deliberate disturbance. One way to accomplish this is to fit a bat gate across the entrance—a sturdy, metal gate that prevents inappropriate human entry at times when bats are hibernating or rearing their young. The gaps in the gate are large enough for the bats to easily navigate their way through. Tolerance of gates varies among species and locations, so must be planned carefully to avoid harm.

## Habitat disruption & creation

Lots of the human-caused problems faced by bats today are obvious because of their sheer scale. Deforestation and other habitat destruction eliminates the places where they live and breed, while insecticides used over wide areas wipe out their food supply and can

Volunteers build a huge steel gate to protect the endangered Gray Myotis (*Myotis grisescens*) that hibernate in Hubbards Cave, Tennessee.

**BATS & THE LAW**

CITES (Convention on International Trade in Endangered Species of Wild Fauna and Flora), an international treaty intended to protect endangered animals and plants around the world from exploitation, is helping to protect a small but growing number of bat species, although enforcement can be difficult. In Britain and most European countries, all bat species and their roosts are strictly protected by law. In the United States, however, most bats have little federal protection, and protection at state level is highly variable, though endangered species have some protection from disturbance through the Endangered Species Act of the US Fish and Wildlife Service. Elsewhere in the world, bat protection is variable and subject to local vagaries. In most developing countries even endangered species that inhabit designated national parks receive scant protection.

Only by thorough studies of bat populations, and their ecological values and needs, together with an understanding of how they use their habitats, can effective conservation strategies for individual species be implemented. This is a problem in itself, as many species are near-impossible to study properly. This may be because they are already so rare, their correct identification is difficult, or their habitat too dangerous to access.

poison the bats. Climate change forces them to shift their distributions, and renders some habitats unusable. Modern buildings rarely offer suitable roosting spaces, unlike the older structures they replace. Even wind turbines can kill many bats if carelessly operated, as the bats' echolocation systems cannot detect the fast-moving blades. This problem can be reduced by switching off turbines when nighttime wind speeds are low in late summer and fall. In some countries, introductions of non-native animals have harmed bat populations, and domestic cats allowed to roam free can do great damage to any bat roosts they find.

Roads are vital to our way of life, but can literally cut animal populations in half, with devastating effects. The challenge of helping animals navigate our infrastructure has led to some innovative solutions. Where roads pass through woodland, overhead rope bridges can help climbing mammals such as squirrels, dormice, and primates to cross over safely. In open countryside, such as in the Sierra du Andújar Park in Spain, underground pathways

## GONE FOR GOOD

Extinction is a full stop on any species' existence: it can never return, though something like it may evolve again someday. Since the year 1500, a dozen bat extinctions have been documented. The majority were tropical species restricted to a single, small island, and many were declared extinct before any sort of conservation effort could be implemented to protect them. For example, it was decided in 2009 to capture any remaining Christmas Island Pipistrelles (*Pipistrellus murrayi*) for captive breeding (a costly last resort that may or may not save an otherwise doomed species). However, the search that took place in August 2009 found just one bat, which the conservationists were not able to catch, and none have been seen since.

Bat species that form huge colonies in only a few locations, often in caves with a specific temperature range, are especially vulnerable to extinction. The first two eastern North American bat species to become endangered were (formerly) the most abundant. The emergence of white-nose syndrome is a problem that could not have been predicted, and disproportionately affects those species that gather in large roosts at a limited number of sites. Climate change particularly threatens such species. The world's largest remaining bat colony (in Bracken Cave, Texas) appears to be affected by overheating, with many bats forced to seek shelter elsewhere in the hottest weather. Those lacking suitable alternatives are unnaturally exposed to predators. On the other hand, if the size of a nursery colony falls too low, the group may no longer generate sufficient body heat for pups to survive and grow.

It is vital that more effort is put on identifying and studying the needs of key populations of bats of all species. Loss of huge colonies in particular could result in serious and irreversible harm to ecosystems and our own economic interests.

ave helped reduce roadkill in species such as he endangered Iberian Lynx. For bats, there s the "bat bridge." Bats, of course, can safely verfly a road, but a carefully landscaped and lanted bat bridge provides an obvious flight ath for bat species that avoid crossing open reas. If placed carefully, they can ensure that ats make their way from roosts to feeding rounds in the safest, most efficient way.

Light pollution is an issue for many octurnal animals, including bats. While raditional, incandescent lights could attract nsects that in turn helped to feed bats, modern ellow lights do not aid in attracting insects. nstead, they increase bats' vulnerability to owl redation. Many bats avoid such lit areas. In district of the town of Zuidhoek Nieuwkoop n the Netherlands, special LED street lighting vas installed in 2018 to mitigate this problem. The bats living and foraging in the area perceive he red lighting as darkness, but it provides ufficient illumination for the human residents o go about their business at night.

There is still a great deal that can be done o help bats, though—by people as individuals by governments, and at every level in between. Roosts can be protected, and potential new roosts created. Good habitats can be protected, and better connections established between them and other local patches of habitat, through corridors of wildness that traverse farm fields or urban areas. Anyone who owns land can make it more useful for bats by encouraging native plants and reducing or eliminating the use of pesticides. Our lives can be made more environmentally friendly, from reducing air travel to increasing the amount we reuse, and recycling the things we buy. More direct action can also be taken by volunteering at local nature reserves, joining local bat groups, and supporting green initiatives whenever the chance arises.

The dire straits in which bats find themselves today are reflected across the natural world. Biologists now generally agree that we are in the midst of the world's sixth mass extinction event. The fifth such event was the one that ended the reign of the dinosaurs, some sixty-six million years ago, most likely in the aftermath of a major meteor strike.

Wrinkle-lipped Free-tailed Bats (*Chaerephon plicatus*) emerging from their roost cave at dusk.

What is happening to wildlife today, though, is a catastrophe of our own making. It is not just that species are disappearing: entire ecosystems are being lost. Often interventive action is taken in time to save a species from complete destruction, but the remnant preserved is so tiny that it no longer can have ecological or economic relevance, and will remain vulnerable for centuries to come.

A committed, united, global action is needed immediately, to halt the devastation and save what we can. Humans themselves cannot survive without wildlife, even if such a life were worth living. It has been shown how much bats contribute to our lives in practical ways, but their importance for our minds and imagination should not be underestimated. Few wildlife-watching experiences can compare with witnessing a million or more bats streaming out of their roost into a skyscape flushed with sunset colors. And it is no less inspiring simply to lie outside in the evening and watch local bats flitting through the darkening sky, engaged in their age-old dance of survival. With determination, optimism, and hard work, we can help to ensure that the sheer magic of bats, and of all wildlife, can be enjoyed by future generations.

# ABOUT THIS BOOK

In the following section, this book looks in more detail at the 1,384 bat species of the world. The number and names of species can vary from source to source, and new species continue to be described on a regular basis. For this book, the species selection follows a comprehensive review of bat taxonomy by Simmons and Cirranello (2018). The families are arranged in taxonomic order. An overview of each family is presented as a whole with closer details on some of the better-known species within it, exploring distribution, habitat, behavior, and conservation status. Individual species accounts are arranged one or two to a page, each illustrated by a photograph. The book concludes with a full species index.

YANGOCHIROPTERA

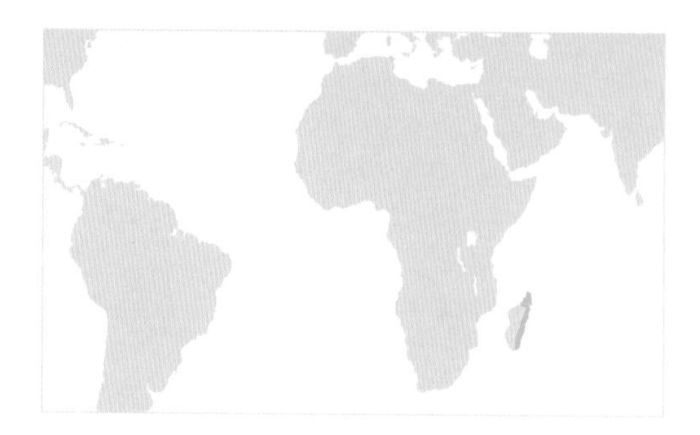

Myzopoda

## MYZOPODIDAE

The family Myzopodidae is endemic to Madagascar, and is of uncertain taxonomic affinity. In Simmons and Cirranello's taxonomy it is the only bat family to be classed as *incertae sedis* ("of uncertain placement") at the superfamily level; other taxonomies have placed it either in superfamily Noctilionoidea or in Vespertilionoidea. In any case, the family appears to originate close to the point where these two superfamilies shared a common ancestor.

The family contains just two species, and until recently only one. *Myzopoda aurita* (described opposite) was the sole member until a 2007 study determined that the *Myzopoda* bats in northwest Madagascar were sufficiently different to those on the east side of the island to warrant a species split. The bats on the northwest side were given the name *Myzopoda schliemanni*.

The two species are nearly identical in appearance, and share the distinctive trait that inspires the family's English name: the Old World Sucker-footed bats. The bats roost head-upward on vertical palm leaves, attached by the skin pads on their ankles and wrists. The attachment is not, however, done through true suction but with wet adhesion, the pads secreting a fluid sticky enough to hold the animal in place. In this respect Myzopodidae differs from the disk-winged bats in the Neotropical family Thyropteridae, which do use true suction on their wing discs to fix themselves to their leaf roosts. The two families are likely to have evolved their sticky abilities independently.

Myzopodidae bats are medium-sized with tawny fur and darker membranes, ears, and face. They have short, pointed snouts and large ears. The long tail appears to help stabilize them as they hang by their pads. They roost in the shelter provided by partly unfurled *Ravenala* leaves, sometimes in groups of more than fifty, though more usually in smaller colonies of fewer than ten. The two species share the remarkable trait of being entirely free of ectoparasites—apparently because crawling arthropods cannot move around on the smooth leaves on which the bats roost. Very few females of *M. aurita* have been observed, with almost all roosting groups being all-male. The females' preferred roost sites are thus still unknown. However, *M. schliemanni* roosts in mixed-sex groups.

Like *M. aurita*, *M. schliemanni* is quite common within its range and seems to have a high tolerance for modified habitats.

***Myzopoda aurita***

## Eastern Sucker-footed Bat

LIKE OTHER *MYZOPODA* SPECIES, this bat is endemic to Madagascar, occurring on the east side of the island. It has dense, red-brown fur and a short, pointed snout, with a receding lower lip and turned-up nose. The eyes are small and the angled-forward ears large, with pronounced inner ridges. The ear shape is distinctive—strongly curved on the leading edge and coming to a backward-angled, pointed tip. Round sucker pads occur on the wrists and ankles. It has a fairly long free tail beyond the uropatagium. It is found in the humid zone, in both primary and disturbed forest, and visits open fields, farmland, and towns to hunt. When roosting it uses its pads to cling, head-upward, to the vertical surfaces of palm leaves. An insectivore, this bat is thought to specialize in hunting small moths.

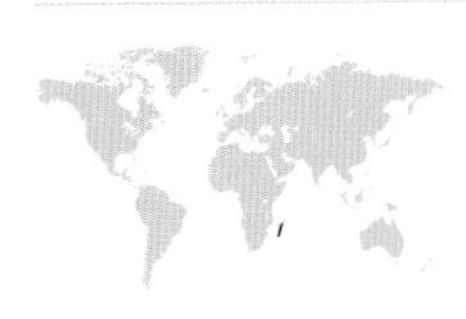

**IUCN STATUS** Least Concern
**LENGTH** 4⅛–5in (10.5–12.5cm)
**WEIGHT** ¼–⅜oz (8–10g)

***Balantiopteryx plicata***

## Gray Sac-winged Bat

THIS BAT RANGES FROM the tip of Baja California and Pacific northwest Mexico through the Pacific seaboard of Central America to Costa Rica. It has light pinkish-gray or brownish fur, and dark membranes. The tail tip just extends beyond the uropatagium. The snout is fairly short with a broad-tipped nose. The eyes are large, and the ears small. It inhabits dry forest and thorn scrub, up to 5,000ft (1,500m) in elevation, and roosts in tree holes or caves. This species is insectivorous, mainly hunting well above the canopy. It flies strongly though slowly, hunting by sight as well as echolocation. Groups of about fifty roost together, and emerge around sunset. Females produce one pup a year—timing varies by region. Populations may be affected by deforestation in some areas, as well as roost disturbance.

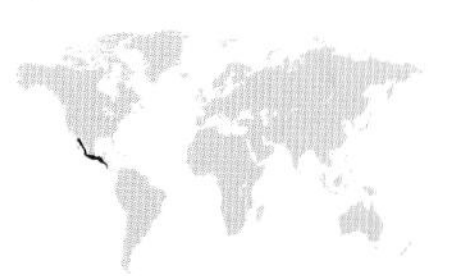

**IUCN STATUS** Least Concern
**LENGTH** 2⅝in (6.6cm)
**WEIGHT** ¼oz (6.1–7.1g)

Balantiopteryx

## EMBALLONURIDAE

The family Emballonuridae contains fifty-four species in fourteen genera. They can be found in South America (with a few species reaching North America), Africa, and across south and southeast Asia to Australia, mainly in tropical and subtropical regions. They are known as the "sac-winged bats," as males of many species have large sacs on the wings, which are used to spread their scent during courtship. Another name for some members of the family is "sheath-tailed bats," because the tail appears to be almost fully enveloped in a sheath of skin in the uropatagium. Most of the species in the family show sexual size dimorphism, with females being a little larger.

Emballonurids are mostly small bats with attractive, pointed faces that lack nose-leaves or other facial ornamentation. They have simple, broad-based, cup-shaped, and often slightly hook-tipped ears, and their eyes are quite large. They prey mainly on insects, which they catch in the open air—some species are fast and powerful high-fliers, while others fly with great agility among forest foliage. They roost in various shelters, but rarely in deep hollows, instead preferring partly sunlit spaces with easy access to open air. They usually roost with their whole bodies in contact with a vertical surface, rather than clinging by the feet to a horizontal ceiling as many other bats do.

The function of the wing sacs, and their link with vocal displays, has been well studied in *Saccopteryx bilineata*. In this species, males attempt to defend a harem of three to five females. However, females frequently leave the group, regularly visiting two or three different harems a day, and the male cannot prevent this. So he is constantly trying to attract "new" females, as well as to retain the ones he already has. Males spend long periods cleaning out their wing sacs using a mixture of saliva and urine. They then use their chins to transfer smelly secretions from glands around the genitals to the wing sacs. The scent can then be wafted toward a female by beating the wings in her direction. The male hovers around females and performs a complex "song." A receptive female responds with a "song" of her own, and the pair then mate. However, in around 80 percent of attempts the female rejects the male by striking him with her wings, and flies away. Females already in the harem will also sometimes attempt to drive away a new female.

### Balantiopteryx

The genus *Balantiopteryx* occurs in Central America and the north of South America. These are medium-sized, dark, sleek-furred emballonurids, with rounded ears. Besides *B. plicata* (see page 61), there are two other species. *B. infusca* is known from just four sites in Ecuador and Colombia; it seems to be very rare, but much potential habitat for this

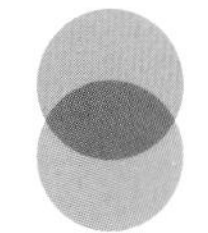

species has not been surveyed. *B. io* is quite widespread, occurring from south Veracruz and Oaxaca in Mexico to Guatemala and Belize. However, it has lost much of its habitat from deforestation, and many of its cave roosts suffer severe disturbance from tourism.

## Diclidurus

The ghost bats of the genus *Diclidurus* are found from Mexico through to South America. They are small, short-faced bats notable for their white fur (except for *D. isabella*, which is pale brown). *D. albus*, the best-known species, is described on page 66. Of the other three species, *D. ingens* is known just from a few specimens from the north of South America. *D. isabella* is an elusive species, present in a relatively small area of south Venezuela to Guyana and northwest Brazil, while *D. scutatus* is found from northeast Brazil, the Guyanas, and east Venezuela across to Ecuador.

## Peropteryx

Bats of the genus *Peropteryx* are known collectively as the "dog-faced bats," because of their strong, slightly snubbed nose and the suggestion of jowls, recalling a short-faced dog. They have wide and round-tipped ears and dark fur. The dark eyes look rather deep-set under the steep, domed forehead. The group occurs in Central and South America, and the far south of North America. *P. macrotis* is described on page 67. The genus includes a further four species: *P. leucoptera* (South America) and *P. kappleri* (Mexico to the north of South America) are widespread, and *P. trinitatis* (north of South America, Trinidad and Tobago, and Antilles) also appears to have an extensive range. *P. pallidoptera*, found in the west Amazon region, was only described to science in 2010. It is still virtually unknown.

***Peropteryx kappleri***

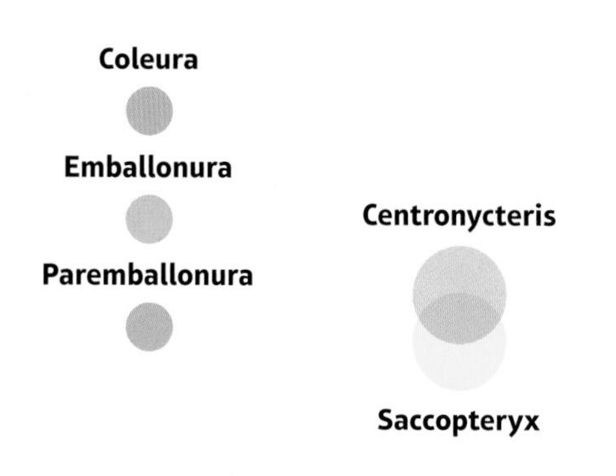

## Centronycteris

The small bats of the genus *Centronycteris* are known as the "shaggy bats" because of their thick, slightly unkempt-looking coats. They also have very short noses for emballonurids. The ears are rather narrow, with delicately pointed, back-curved tips. These bats have a slow, fluttering flight and catch insects in the air. The two known species, *C. centralis* and *C. maximiliani*, are both found in South America; *C. centralis* is also present in Central America and Mexico.

## Saccopteryx

Three species of the widespread American genus *Saccopteryx* are described on pages 68–70 (*S. bilineata, S. canescens*, and *S. leptura*). The remaining two are *S. antioquensis*, an endangered species known only from a tiny area of Colombia and not observed since 1996, and *S. gymnura*, which occurs sparsely in the Guianas and north Brazil. All *Saccopteryx* species are forest bats, adapted to catching flying insects in spaces within the tree canopy.

## Coleura

The African sheath-tailed bats of the genus *Coleura* include the widespread *C. afra*, described on page 70. Two other species are known—the newly described *C. kibomalandy*, present in lowland west Madagascar and *C. seychellensis*. This species is native to the Seychelles, occurring on three islands (but formerly on at least two more). It declined catastrophically following forest clearance in the early twentieth century, and is further pressured by predation from non-native Barn Owls and feral cats. Fewer than one hundred now survive.

## Emballonura & Paremballonura

Bats of the genus *Emballonura* are also known as "sheath-tailed bats." This book recognizes seven species: *E. alecto*, *E. beccarii*, *E. dianae*, *E. furax*, *E. monticola*, *E. raffrayana*, *E. semicaudata*, and *E. serii*. All are found in southeast Asia, New Guinea, and certain Pacific Islands. *E. semicaudata* occurs (in four subspecies) on various Pacific Island groups; it has declined severely since the 1940s. Threats include invasive non-native species, disturbance at its cave roosts, and severe

***Paraemballonura tiavato***

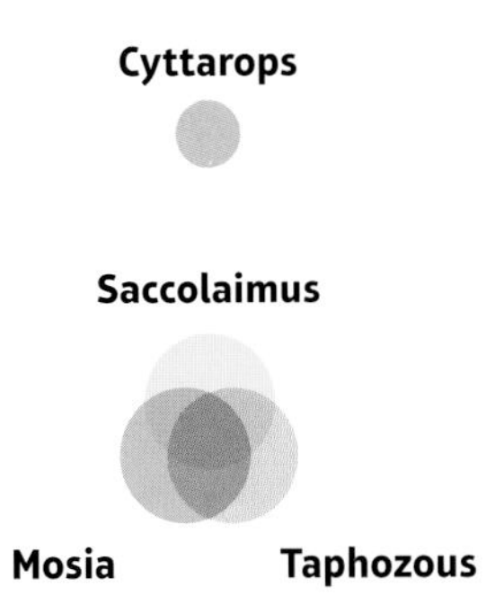

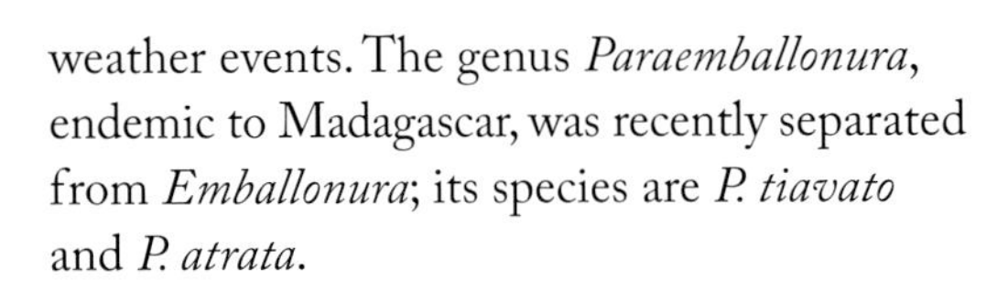

weather events. The genus *Paraemballonura*, endemic to Madagascar, was recently separated from *Emballonura*; its species are *P. tiavato* and *P. atrata*.

## Saccolaimus

The sturdily built, sometimes boldly patterned, pouched bats of the genus *Saccolaimus* occur from Africa across Asia to Australia. There are four species: *S. flaviventris*, *S. mixtus*, *S. peli*, and *S. saccolaimus*. *S. mixtus* is threatened—due to loss of the tall *Eucalyptus tetrodonta* forest that forms its main habitat—in both Australia and New Guinea.

## Taphozous

The largest genus within Emballonuridae is *Taphozous*, although it is smaller than it used to be as it formerly included *Saccolaimus*. Three species are described: *T. theobaldi* and *T. mauritianus* (see page 71), and *T. melanopogon* (page 72). There are a further eleven species, namely *T. achates* (Indonesia), *T. australis* (north Australia), *T. georgianus* (north and west Australia), *T. hamiltoni* (central and east Africa), *T. hildegardeae* (east Africa), *T. hilli* (west and south Australia), *T. kapalgensis* (Northern Territory, Australia), *T. longimanus* (south and southeast Asia), *T. nudiventris* (west Africa across into Asia), *T. perforatus* (west Africa to south Asia), and *T. troughtoni* (Queensland, Australia). Variously known as "sheath-tailed bats" or "tomb bats," these bats are sleek, strong-flying, and larger than most emballonurids.

## Monotypic genera

Emballonuridae also contains four monotypic (single-species) genera. *Cormura brevirostris* is described on page 73 and *Rhynchonycteris naso* on page 75. The other two are *Mosia nigrescens*—a common species of Indonesia and New Guinea, formerly classed within *Emballonura*—and *Cyttarops alecto*, an exceptionally small-eared and short-nosed species. The latter is very widespread (but also apparently very rare) in South and Central America and the Caribbean.

***Diclidurus albus***

# Northern Ghost Bat

**IUCN STATUS** Least Concern
**LENGTH** 3⅜in–4⅛ (8.6–10.3cm)
**WEIGHT** ¾oz (20g)

**THIS STRIKING SPECIES RANGES** from central Mexico (Pacific seaboard) to north Peru and east Brazil; it also occurs in Trinidad. Despite its white fluffy-looking fur and translucent, pale-pink wing membranes, it is not albino; the thumb is very small. This bat is mainly found in humid tropical forests and plantations, and also visits urban areas at times. By day it roosts (usually alone) in palm tree foliage, where its white fur provides camouflage against the bright sun shining through fronds. It hunts insects through echolocation, flying fast in the open air. Adults form small groups in the breeding season (winter), with one male attracting usually two or three females—probably through the scent of secretions from a two-horned gland on the uropatagium. Each female gives birth to a single pup in May or June.

***Peropteryx macrotis***

# Lesser Dog-like Bat

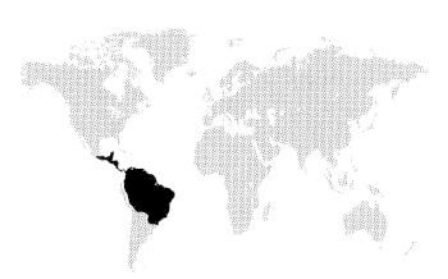

**IUCN STATUS** Least Concern
**LENGTH** 2⅜in (6cm)
**WEIGHT** ⅛–⅜oz (3–9g)

**THIS SPECIES RANGES FROM** south Mexico through Central America, and across most of the north of South America—its range broadly overlaps with its close relative, the slightly larger *P. kappleri*. It is found mostly in tropical, deciduous forest in the lowlands, below 3,300ft (1,000m), and also in open countryside in Mexico. Its roosts, mainly in caves and rocky crevices, are small, usually consisting of no more than fifteen individuals, though it may share a roost with other species. It tends to roost by clinging to a vertical rock face rather than the cave roof. An agile insectivore, it hunts within the forest and around artificial lights. Timing of breeding varies by region but occurs once a year and the pups are born after four months of gestation. The mothers shown are nursing their pups in a hollow Kapok tree.

***Saccopteryx bilineata***

# Greater Sac-winged Bat

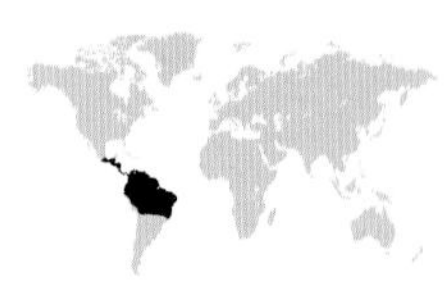

**IUCN STATUS** Least Concern
**LENGTH** 2⅞–3in (7.3–7.6cm)
**WEIGHT** ¼–⅜oz (8.5–9g)

**THIS BAT OCCURS EXTENSIVELY** through Central and South America. It is dark brown or black, with two pale stripes down its back and a grayish underside. It is usually associated with water and is most common in lowland tropical forest, but also occurs in many other habitats, including around human habitation. It roosts in the entrances to caves, in rock crevices, under bridges, and in hollow trees (as seen here), often with other bat species. Dominant males defend harems (although these are unstable, females moving around quite readily); they use a wing-beating display, dispersing scent from their wing sacs, to attract new females. Receptive females respond with a courtship "song." Mating occurs near the end of one rainy season, with pups born at the start of the next.

***Saccopteryx leptura***

# Lesser Sac-winged Bat

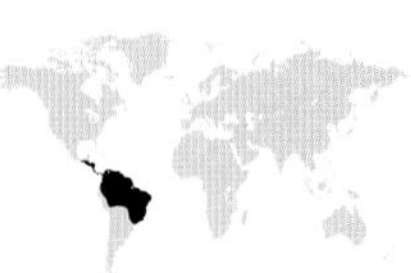

**IUCN STATUS** Least Concern
**LENGTH** 1¾in (4.4–4.5cm)
**WEIGHT** ⅛–¼oz (4.4–5.7g)

**THIS BAT'S DISTRIBUTION BEGINS** in south Mexico and extends through Central America into South America, where it is present in north, northeast, and central regions, and also on Trinidad and Tobago. It is found in varied habitats, especially lowland, evergreen forest, and roosts in crevices and tree hollows. It emerges around sunset to hunt for small flying insects within spaces and clearings in the forest, working around the same "track" in repeated circuits. It is reported to live in small groups of up to nine animals; males and females may form monogamous pairs, rather than the polygynous structure favored by the better-known *S. bilineata*. The species is subject to special protection in Mexico, but elsewhere is common with no special status.

***Saccopteryx canescens***

## Frosted Sac-winged Bat

**THIS SPECIES OCCURS** in South America, in a large area in the north of the continent including Venezuela, the Guianas, most of Ecuador, east Peru, and northwest Brazil, as well as Trinidad and Tobago. In appearance it is rather similar to *S. bilineata*, but the far has a more mottled, silvery look. The broad-based ears have swept-back, hooked tips, and the snout is short, pointed, and blackish. This bat occurs in forest, particularly close to clearings with streams; it hunts flying insects in open patches and on the forest edge, at both high and low levels, and also around artificial lights. It roosts in crevices, in small groups (mixed-sex gatherings of twelve to fifteen have been noted). The species is reported to be quite common in general, but is likely to be affected by deforestation in many areas.

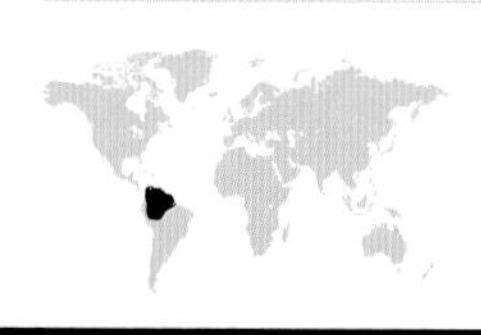

**IUCN STATUS** Least Concern
**LENGTH** 2⅞–3in (7.3–7.6cm)
**WEIGHT** ¼–⅜oz (8.5–9.3g)

***Coleura afra***

## African Sheath-tailed Bat

**THIS BAT HAS A PATCHY** distribution in east Africa and parts of west Africa, and has recently been found in northwest Madagascar (although this could possibly be a separate subspecies). It is light brown with pinkish-gray membranes; the tail tip just projects beyond the uropatagium. It occurs in open, dry, rocky habitats, especially near the coast, including scrub and savanna, and also light woodland, provided suitable roosting places are available. On Madagascar it occurs in dry, deciduous, tropical forest. By night it catches insects in fast, dashing flight. It roosts inside well-lit parts of large caves, which can hold tens of thousands. Mature males assemble a harem of up to twenty females, which give birth in one or both annual rainy seasons. Young males leave their breeding colonies to join "bachelor swarms"; young females stay with their natal roost.

**IUCN STATUS** Least Concern
**LENGTH** unknown
**WEIGHT** ⅜oz (10–12g)

***Taphozous theobaldi***

## Theobald's Tomb Bat

**THIS IS A TYPICAL** *Taphozous* with dense, short, velvet-brown fur and slightly darker brown membranes, largish eyes, rounded ears, a 1in (25mm) free tail, and long, slender wings. The chin has a slight, dark "beard." It appears in south and southeast Asia, from India to south Myanmar, Thailand, Vietnam, Cambodia, and Laos; it also occurs in Indonesia on Java, Sulawesi, and Borneo, and is thought to exist in Yunnan, China. It occurs in forested habitats, forming very large roosts within deeper parts of large cave systems, but also roosts in smaller colonies (ten to forty) in tree hollows. This bat is exploited as a food source and for guano, and suffers some deliberate disturbance and persecution. Its status and population trend require study to determine if protective measures are needed.

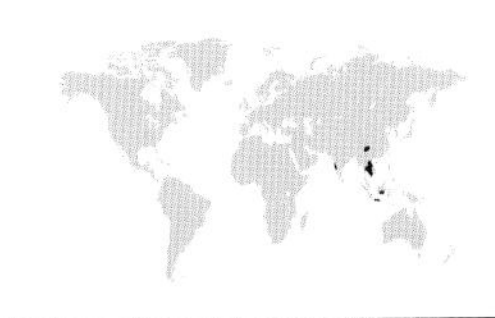

**IUCN STATUS** Least Concern
**LENGTH** unknown
**WEIGHT** unknown

***Taphozous mauritianus***

## Mauritian Tomb Bat

**THIS IS A FAIRLY LARGE** species with short, velvety, silver-gray fur, becoming pure white on its belly. The face, ears, and wings are pinkish-gray. This bat has an extensive distribution across sub-Saharan Africa; it is also present on Madagascar, as well as Mauritius, Réunion, Zanzibar, Bioko, Annobón, and São Tomé and Príncipe islands. It occurs in most habitat types, even desert edges, where its efficient kidneys allow it to survive on minimal water intake. It roosts on tree trunks under loose bark, and also in caves, tree holes, and buildings, with easy, direct access to open air. Roosting colonies are small but highly vocal, their chirps and screeches audible to human ears. It uses echolocation, but may also hunt by sight in broad daylight. This common species is adaptable enough to withstand some habitat modification.

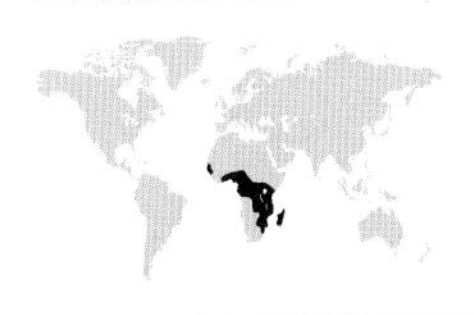

**IUCN STATUS** Least Concern
**LENGTH** 4–4⅝in (10.1–11.6cm)
**WEIGHT** ¾–1¼oz (20–36g)

***Taphozous melanopogon***

# Black-bearded Tomb Bat

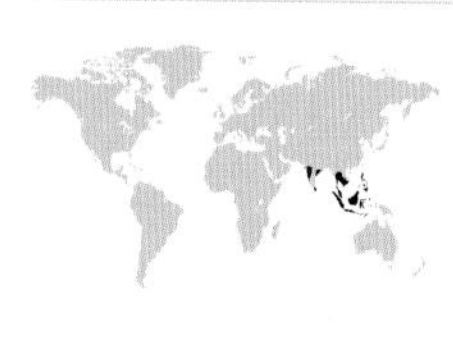

**IUCN STATUS** Least Concern
**LENGTH** 2¾–3⅛in (7–8cm)
**WEIGHT** ⅜–1¾oz (10–50g)

**THIS BAT HAS A PATCHY** but extensive distribution in south and southeast Asia (mainland and insular). It has long, sandy-tinted fur, and a black patch under its chin. It occurs in many habitats and roosts in caves, rock crevices, or buildings. Roosts usually hold 100–400 individuals, sometimes well over 1,000. They cling to vertical walls and can climb up and down with ease. It hawks flying insects at heights sometimes exceeding 300ft (90m)—at closer range its echolocation clicks are usually audible to human ears. Males try to monopolize and defend females during the mating season, after which the sexes separate. Females bear singleton pups after three to four months' gestation. This common bat is an important controller of insect pests in agricultural areas.

***Cormura brevirostris***

# Chestnut Sac-winged Bat

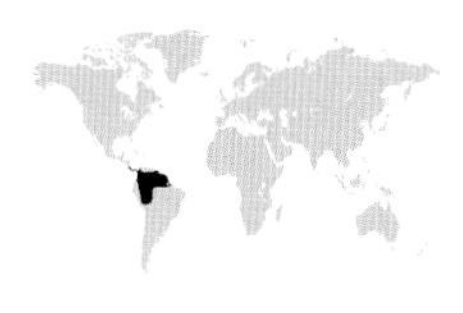

**IUCN STATUS** Least Concern
**LENGTH** 2⅜–2¾in (6–7cm)
**WEIGHT** ¼–⅜oz (7–11g)

**THE SOLE SPECIES IN ITS GENUS**, this is a beautiful bat, with rich, dark chestnut fur and black wings, ears, and face. The male's wing sacs are further from the body than in other sac-winged bats, almost reaching the elbow; those of females are vestigial. The tail just extends beyond the uropatagium. This bat occurs widely across Central and South America, though is absent from the northwest and west coasts of South America. It prefers damp, primary forest with streams or ponds. It hunts in spaces within the forest canopy and roosts in small groups (up to five) inside holes in decaying tree trunks, or hanging from the undersides of branches. Though widespread, the species is at risk from deforestation; its range has contracted, especially in the far south.

***Rhynchonycteris naso***

# Proboscis Bat

**IUCN STATUS** Least Concern
**LENGTH** 1½–1¾in (3.7–4.3cm)
**WEIGHT** ⅛oz (3.8–3.9g)

THIS DISTINCTIVE EMBALLONURID has an extensive range, from southern Mexico through Central America and across much of northern South America, reaching Bolivia and southeast Brazil, as well as east Peru and Ecuador. A very small bat, it has a well-camouflaged gray and brown marbled coat, with regularly spaced white tufts along its forearms. This "disruptive" pattern helps conceal the bat as it rests on lichen-covered tree trunks. The bare parts are blackish, and it has a slim, elongated snout. It lacks wing sacs. This bat occurs in lowland tropical forest, close to slow-flowing streams and rivers, and roosts in tree holes. It preys mainly on flies and caddisflies as they emerge from or swarm over water, but eats other aerial insects too, according to availability. Within the small colonies (up to forty, but usually much smaller), dominant males defend groups of females. Breeding occurs once a year in the rainy season; exact timing varies by region. This species enjoys special protection in Mexico, where it has a restricted distribution; but elsewhere it is common and found in many protected areas.

Nycteris

## NYCTERIDAE

The family Nycteridae comprises just one genus, *Nycteris*, which includes sixteen species. The group is distributed across Africa and south Asia. These are mostly small bats with very distinctive facial anatomy. A long slit runs down the center of the face, evident on close examination but partly concealed by fleshy folds on either side. This pouch is accommodated by a depression in the skull between the orbits. Its function is not known with certainty, but it is likely to be related to echolocation. *Nycteris* bats tend to have light-colored fur, and to have long, flat faces with barely protruding snouts. The eyes are very small, the ears large and oval-shaped with small tragi. The wings are broad, permitting agile flight. Unique among mammals, the last (outermost) tail vertebra of these bats is T-shaped.

*Nycteris* bats are strictly nocturnal and catch their prey mainly by gleaning it from foliage, rocks, or even the ground. Prey is primarily insects and other arthropods, but the large *N. grandis* feeds on a high proportion of vertebrate prey. *Nycteris* bats occur in varied habitats and use all kinds of shelters for roosts. These include buildings and tree hollows, as well as holes in rotting logs and the tunnels of burrowing animals.

See pages 77–9 for detailed accounts on *N. grandis*, *N. macrotis*, *N. thebaica*, and *N. tragata*. Of the other species, four (*N. madagascariensis, N. major, N. parisii,* and *N. vinsoni)* are currently little known. *N. madagascariensis* is known from just two specimens, taken in north Madagascar in 1910. *N. major* has been recorded from several locations in west and central Africa, but only in extremely low numbers. *N. parisii* is known from south Somalia, south Ethiopia, and northern Cameroon, while *N. vinsoni* is known only from its type specimen, caught in Mozambique in 1965.

*N. javanica* depends on fast-disappearing forest habitat on Java and nearby islands. The remaining species, are *N. arge* (lowland west and central Africa), *N. aurita* (east Africa), *N. gambiensis* (west Africa, from Senegal to Nigeria), *N. hispida* (widespread in sub-Saharan Africa), *N. intermedia* (Liberia to Tanzania and Angola), *N. nana* (west, central and east Africa), and *N. woodi* (the south of Africa).

***Nycteris macrotis***

# Large-eared Slit-faced Bat

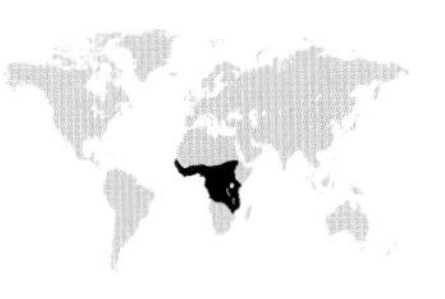

**IUCN STATUS** Least Concern
**LENGTH** unknown
**WEIGHT** unknown

**THIS BAT OCCURS THROUGH** a large area of sub-Saharan Africa, as far south as northern Botswana. Here we see a mother nursing her pup in Kenya. The populations in west Africa are smaller than those elsewhere, and may be a different species. It has rather luxuriant fur, colored light gray-brown or yellow-brown, sometimes bright orange-yellow. Its usual habitat is moist savanna, and it roosts in various shelters including caves, hollow trees, and under bridges—in larger groups than is typical of *Nycteris* bats (up to 600). However, smaller colonies are not unusual; for example, groups of twenty or so may roost within roof thatching. A slow-flying, agile gleaner, this bat feeds on many beetles, termites, and grasshoppers, appearing to migrate or roam nomadically in response to food shortages. Agricultural changes are probably affecting its population, at least in some areas.

***Nycteris thebaica***

# Cape Long-eared Bat

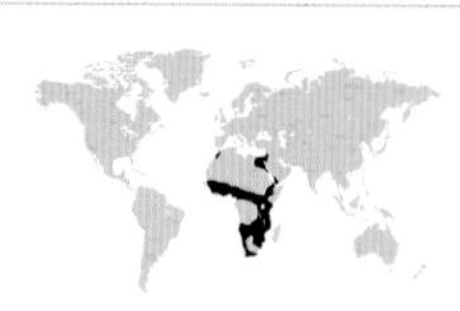

**IUCN STATUS** Least Concern
**LENGTH** 3¼–6⅝in (8.3–16.8cm)
**WEIGHT** ¼–⅝oz (6.5–16g)

**THIS BAT IS PRESENT** over much of Africa, and the west of the Arabian peninsula. It has light gray-brown or golden-brown fur, with grayish ears, face, and membranes. It has a fairly large nose-leaf, split in two by a depression running from the tip of the nose to the forehead. It has broad wings, permitting hovering and other agile maneuvers. This bat occurs in wet and dry savanna and riverine habitats, and uses all kinds of shelters for day roosts, moving around to find optimum cool temperatures. It preys on a variety of insects, including grasshoppers and green stink bugs, which it picks from the ground as well as foliage. The breeding season varies by location. Females produce a single pup per year, which they carry with them while it is small to protect from predators.

***Nycteris grandis***

## Large Slit-faced Bat

**THIS SPECIES IS FOUND** across much of sub-Saharan Africa, from Senegal east to Tanzania and south to Zambia, Zimbabwe, and Mozambique. A large *Nycteris*, it has light ginger-brown fur and broad wings, and a typical long, flat, *Nycteris* face with a central longitudinal slit, small eyes, and large, narrow ears. It is found in swampy forest areas, but also occurs in savanna; it roosts in hollow trees, alone or in small groups. A formidable predator, this bat uses echolocation to hunt frogs and lizards, birds, and other small vertebrates, as well as large arthropods—often flying out from a perch to intercept passing prey, but also actively searching for them. Discarded prey remains reveal nocturnal roosts. This species may breed in spring or in winter. It is scarce but is not considered threatened.

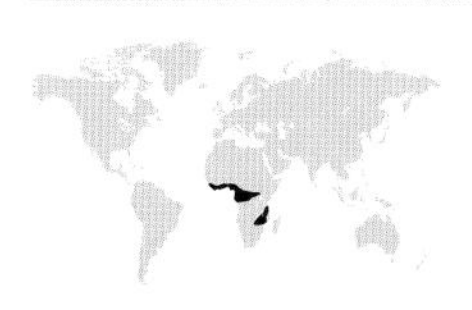

**IUCN STATUS** Least Concern
**LENGTH** 2½–2¾in (6.3–9.3cm)
**WEIGHT** ¾–1¼oz (23–36g)

***Nycteris tragata***

## Malayan Slit-faced Bat

**THIS BAT IS FOUND IN** southeast Asia, from Myanmar and Thailand through peninuslar Malaysia, Sumatra, and Borneo. It is a smallish *Nycteris* with reddish-brown fur, large, broad ears, and a flattened face; a slit runs from between its very small eyes to the nose tip. There are skin folds around the slit, most likely to help direct its echolocation calls. The wings are short and broad. This bat occurs in damp primary forest in both uplands and lowlands, where it roosts in crevices between boulders, in tree hollows, and occasionally in buildings. It hunts mainly by gleaning and can easily take off from the ground while holding even relatively heavy prey. Its Near Threatened status reflects its natural rarity and the rapid rate of deforestation within its range. It is known from several national parks and other protected areas.

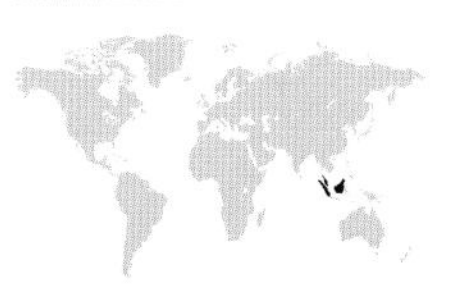

**IUCN STATUS** Near Threatened
**LENGTH** 2–2½in (5–6.5cm)
**WEIGHT** ½–⅝oz (14–19g)

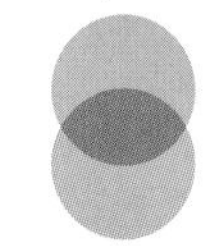


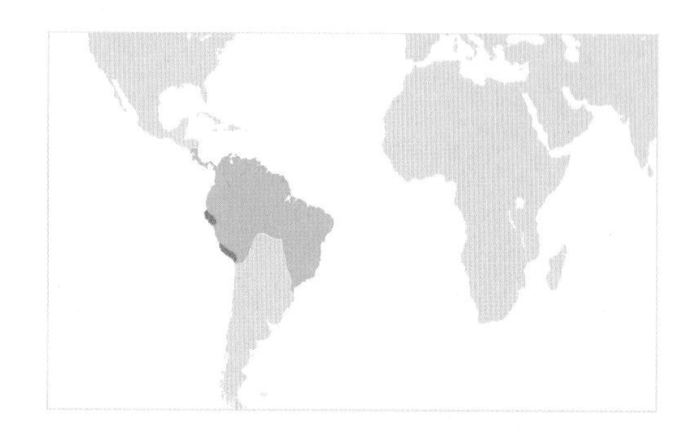

## FURIPTERIDAE

Furipteridae is a small family, containing just two genera (*Furipterus* and *Amorphochilus*), each of which contain a single species. The family is known as the "smoky bats," because of their coloration, or the "thumbless bats," because the thumb digit is extremely small and completely enclosed within the wing membrane, thus seeming at first glance to be absent. They are small, insectivorous bats with very short faces, steep foreheads, and rather small ears. They tend to forage at low to mid-levels, and mainly prey on moths. Roosts are usually fairly small and in caves, with the bats emerging after dark. They navigate and hunt using echolocation; their calls are short, quiet, and at a high frequency. Females of these species are typically larger than males. They have their teats located on the abdomen, in a lower position than in most other bats. This means that when a female is feeding a pup, it hangs head-up on its mother as she hangs head-down.

This family is part of the superfamily Noctilionoidea, along with the families Phyllostomidae, Thyropteridae, Mormoopidae, Mystacinidae, and Natalidae. The grouping as a whole is largely confined to the Neotropics (excepting the small family Mystacinidae, which occurs in New Zealand). Furipteridae is likely to be most closely related to Natalidae and Thyropteridae. However, taxonomic relationships within Noctilionoidea are not yet fully clarified, and further studies are needed.

### Furipterus & Amorphochilus

*Furipterus horrens* is described on page 82. The other species, *Amorphochilus schnablii*, is a rare bat. Its range is strictly coastal and extends from central-west Ecuador south through Peru, reaching to northern Chile; it is also present on Puna island, off Peru. Like *F. horrens*, it has no obvious thumb, and is a small bat with dusky gray fur. Its tail does not reach the edge of the uropatagium. It has a more projecting snout and more prominent ears than *F. horrens*, and lacks the facial bumps that characterize the latter species. Its main threat is loss or disturbance of roosting sites and foraging habitat, but the species is also sometimes used in local religious practices.

## MORMOOPIDAE

This small family occurs in the Americas, from the southwest of the United States through to southeast Brazil, including parts of the Caribbean. It comprises just two genera: *Mormoops* (the "ghost-faced" bats) with two species, and *Pteronotus* (the "mustached," or "naked-backed" bats) with fifteen. Mormoopidae is part of the superfamily Noctilionoidea and is a sister family to Noctilionidae.

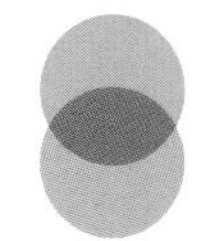

## Mormoops & Pteronotus

Mormoopids have elaborate facial anatomy, especially *Mormoops*—which are quite bizarre with their complex cheek, lip, and chin folds, and almost circular, low-set ears. *Pteronotus* bats are more conventional-looking, but still show skin folds around the lower lips. They also have a bump at the tip of the nose. These structures are presumed to have a role in echolocation; the lip folds give the open mouth a definite funnel shape. Other features of mormoopids include a short length of tail extending beyond the uropatagium. The muzzle has moustache-like bristles, and some *Pteronotus* species appear to have bare skin on the back, but this is in fact the wing membranes, which meet along the midline of the body.

These small to medium-sized bats are insect-eaters that hawk their prey on the wing; they are agile and fast fliers. They mostly roost in caves, sometimes in sizeable colonies—in some cases more than 100,000 strong. Leaving the roost after sunset, they forage in varied habitats, often over water. Most species are quite widespread, and of no immediate conservation concern.

The two *Mormoops* species, *M. blainvillei* and *M. megalophylla*, are described on page 82 and page 83 respectively. There are also accounts for four of the *Pteronotus* species: *P. davyi* and *P. gymnonotus* on page 84, *P. macleayii* on page 86, *P. parnellii* on page 85, and *P. personatus* also on page 85. Several of the other *Pteronotus* species are recent splits and are subject to some taxonomic disagreement. This book recognizes the following—*P. fulvus* (found in Colombia and Venezuela), *P. fuscus* (found in El Salvador, Honduras, and parts of Mexico), *P. portoricensis* (Puerto Rico), *P. pusillus* (Hispaniola), *P. mexicanus* (Mexico), and also *P. psilotis* (much of Central America).

*P. mesoamericanus* is a scarce bat of Central America. The most threatened mormoophoid bat is *P. paraguanensis*, found only on the Paraguaná Peninsula, Venezuela. The final two species are both recent splits from *P. parnellii*. *P. quadridens* is a common species found on Cuba, Jamaica, Hispaniola, and Puerto Rico, while *P. rubiginosus* ranges from Honduras south to Peru and Bolivia.

***Furipterus horrens***

## Thumbless Bat

**THIS TINY BAT RANGES FROM** Costa Rica south through Central America and then across northern and eastern South America. It is also found on Trinidad and Tobago. It has velvety gray or gray-brown fur. The wings are fairly short and broad; the tail membrane is large. It has a short, upturned, pig-like snout; its small ears are almost lost within the dense fur, giving it a round-headed look. The face bears small, warty bumps, which may have a sensory function. The species occurs in a wide variety of both wet and dry tropical forest, roosting near ground in tree hollows and other small crevices, in groups of about four; it also forms larger roosts (up to sixty) inside caves. It hunts around forest foliage, feeding on mainly moths and butterflies. Females give birth to one pup a year.

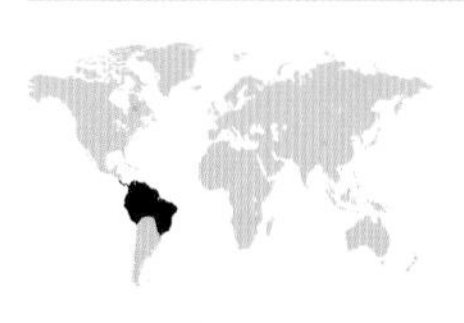

**IUCN STATUS** Least Concern
**LENGTH** 1½–2¼in (3.8–5.8cm)
**WEIGHT** ⅛–¼in (3–5g)

***Mormoops blainvillei***

## Antillean Ghost-faced Bat

**THIS BAT OCCURS IN** the Greater Antilles and other islands nearby. It has a flattened, wide face with tiny eyes, overshot lower jaw, and folds and frills of skin around chin, nose tip, and cheeks. The ears are rounded and low-set. The fur is reddish brown, the membranes dark gray. It roosts in hot caves, hanging from the cave roof. Most roost alone, but may cluster when temperatures are lower. The bats emerge about an hour after sunset and hawk for flying insects on forest edges. The large uropatagium sweeps prey toward the mouth. It flies fast for a mormoopid and its echolocation calls may be audible to humans. It breeds once a year, with singleton pups born in midsummer. Disturbance of its roosts, for guano collection and mining, threatens this species.

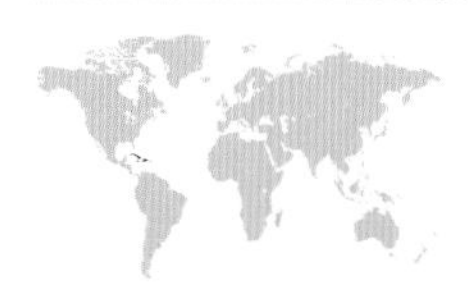

**IUCN STATUS** Least Concern
**LENGTH** 3⅛–3¼in (7.8–8.4cm)
**WEIGHT** ⅜oz (9g)

***Mormoops megalophylla***

# Ghost-faced Bat

**IUCN STATUS** Least Concern
**LENGTH** 2⅝in (6.7cm)
**WEIGHT** ½–⅝oz (13–19g)

**THIS NEW WORLD SPECIES** is present in the southwest of the United States, south to Honduras, and separately from south Panama into the northwest of South America, including Trinidad; it just reaches Peru. Populations in the north appear to be partly migratory. It has warm-toned brown fur, dark membranes, and a distinctive squashed, folded, bulldog-like face, with projecting lower jaw and steep forehead. This is a forest species, but is also common in arid areas of Big Bend, Texas. It requires deep, karstic limestone caves for roosting. Colonies within roosts are large—if numbers fall below a certain threshold, the bats cannot maintain a high enough temperature (104°F–40°C) for pups to survive. It is insectivorous, emerging in a swarm after dark, then dispersing. It often hunts for moths and other insects over water. It is declining, as few suitable roosting caves exist, and some are unprotected.

*Pteronotus davyi*

## Davy's Naked-backed Bat

**THIS BAT OCCURS FROM** Mexico through much of Central America, and northern South America into northwest Peru, as well as Trinidad and the southern Lesser Antilles. Its apparent bare back is the patagia of the two wings, which meet along the midline. Small and broad-winged, it has smallish ears pointing forward; their outer edge extends to frame the small eyes. Its snout is short and pointed, the large lips forming a funnel to direct calls more precisely. The species occurs in rainforest and along watercourses. It roosts deep within hot, humid caves (which suffer little disturbance), also sometimes in buildings. It is not highly colonial, but shares its roosts with other bat species. Strictly insectivorous, it hawks over water and around vegetation, but on cool nights may only leave its roost to drink.

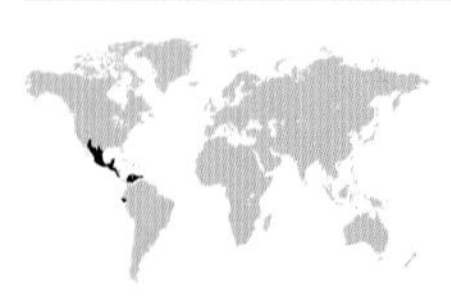

**IUCN STATUS** Least Concern
**LENGTH** 2¾–3⅜in (7.1–8.5cm)
**WEIGHT** ¼–⅜oz (6.5–10g)

*Pteronotus gymnonotus*

## Big Naked-backed Bat

**THIS BAT RANGES FROM** the far south of Mexico through Central America into northwest and central South America, becoming more abundant in the south of its range. It has bright rufous or duller brown fur, with darker grayish membranes and bare parts. The ears are forward-angled with pointed tips and an outer ridge that curves down around the small eyes. The snout is short and upturned. Long wings connect at their bases along the midline of its back, giving a "bald-backed" appearance. Found in dry, deciduous, low-lying forest (below 1,300ft/400m), this bat joins other mormoopid bats in cave roosts. Leaving the roost at dusk, it hunts in the open air, taking various flying insects, particularly beetles. The species is common and widespread, but vulnerable to disturbance of its roosts and to deforestation.

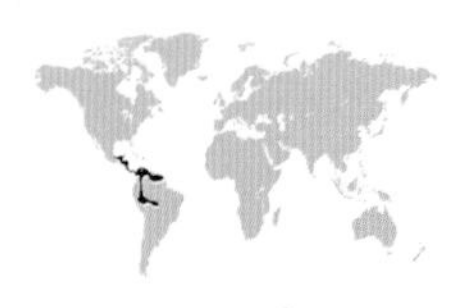

**IUCN STATUS** Least Concern
**LENGTH** 2½in (6.4cm)
**WEIGHT** ⅜–½oz (12.6–13.9g)

***Pteronotus parnellii***

## Common Mustached Bat

**THIS BAT IS FOUND IN CUBA** and other Caribbean islands. It is dark brown or reddish with brown membranes. It lives from the lowlands up to 10.000ft (3,000m) and forages mainly in semi-open and forest-edge habitats. The species roosts inside suitable deep, warm caves, alongside up to five other species. The largest-known roost, on Puerto Rico, holds some 140,000 *P. parnellii*. It is insectivorous and opportunistic, often hunting close to the ground and taking varied prey types; it will feast on ant swarms. Females give birth to a single pup in late spring or early summer, which become independent after about three months. This bat's echolocation abilities allow it to compensate for the rapid and complex changes in soundwave frequencies generated by the beating wings of a flying moth.

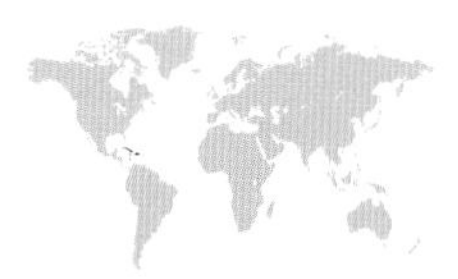

**IUCN STATUS** Least Concern
**LENGTH** 2⅞–4in (7.3–10.2cm)
**WEIGHT** ⅜–¾oz (10–20g)

***Pteronotus personatus***

## Wagner's Mustached Bat

**THIS BAT OCCURS FROM** the south half of Mexico on both coasts, extending through Central America into South America. Here it occurs across the north coast including Trinidad, through parts of Peru and south to Bolivia—but is absent from a broad swathe of the continent's northeast. It is a small bat with warm brown or darker gray-brown fur, rather large eyes for a *Pteronotus*, but a snub-nosed face typical of its genus. It occurs in both humid and dry forest, and hunts over water for insects. Roosts are in deep, warm caves; they may hold up to 10,000 individuals, as well as various other bat species. Like others in its genus, this bat shows Doppler-shift compensatory behavior, changing the frequency of its echolocation pulses to adjust for Doppler shifts in the returning echoes as it flies through the "cluttered" forest environment.

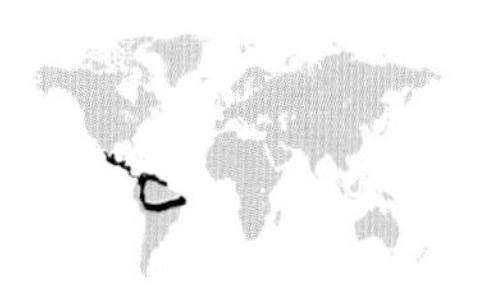

**IUCN STATUS** Least Concern
**LENGTH** 1¾–2⅛in (4.3–5.5cm)
**WEIGHT** ⅛–⅜oz (5–10g)

***Pteronotus macleayii***

# Macleay's Mustached Bat

**IUCN STATUS** Least Concern
**LENGTH** unknown
**WEIGHT** ⅛–¼oz (4–8g)

**THIS POORLY KNOWN** small bat occurs in Cuba (including nearby islets) and Jamaica in two subspecies. It has mid-gray-brown fur and rather long wings. The face is short in profile with an upturned snout and thin but prominent lips. Long hairs on the upper lip form a definite moustache. This bat shares its range with the slightly larger *P. quadridens*. In the hand, *P. macleayii* can be distinguished by the series of parallel, vertical ridges on the underside of its uropatagium. It feeds mainly on winged beetles and flies, and forages in varied habitats, its distribution dictated by the presence of suitable roost caves. Those that best suit it are deep, sometimes warm, but with a consistent temperature. Disturbance, including excessive guano collection, threatens some colonies.

**Mystacina**

## MYSTACINIDAE

Mystacinidae is a very unusual family, with an array of unique anatomical and behavioral traits. Most notably, these bats are highly terrestrial, able to move with ease on the ground and often pursuing prey on foot. The wing membranes can be folded away into a protective sheath to aid with this. As well as catching insects, they feed on fruit, nectar, pollen, and even carrion, and can dig into earth and decaying wood with their large, strong feet. It has long been assumed that the unusual terrestrial traits of mystacinids arose because they evolved in an environment without other land mammals, but an ancestral genus, *Icarops*, showed similar adaptations and was native to Australia, home to an array of land mammals. The family is classed within the superfamily Noctilionoidea, but diverged from the rest of that group some forty to fifty million years ago.

This family contains one genus (*Mystacina*) of just two species, only one of which is certain to be extant, as there have been no confirmed records of *M. robusta* since the 1960s. The family is endemic to New Zealand, and along with the vespertilionid bat *Chalinolobus tuberculatus* these are the only land mammals native to the islands. New Zealand's natural vertebrate fauna is dominated by birds, with many unique species evolving and adapting into ecological roles occupied in other countries by mammals. For example, the large, flightless moas were the New Zealand equivalent of grazing and browsing hoofed mammals, and the kiwis played the part of terrestrial, nocturnal, undergrowth-foragers such as hedgehogs. New Zealand does now have large populations of several non-native mammals, some of which caused serious problems for the native fauna, including the three bat species.

*M. tuberculata* is described on page 90. The other species, *M. robusta*, was last observed in the 1960s and, although the IUCN still class it as Critically Endangered, it is likely to be extinct. Like *M. tuberculata*, it was a medium-sized bat with dense, dark gray-brown fur, proportionately small wings, and large, strong legs. The few recorded observations of the species date back to the early twentieth century, when it was already lost from North and South Island and occurred only on some small islands around Stewart Island, off the southwest tip of South Island. Edgar Stead, visiting Taukihepa (Big South Cape Island) in 1936, noted that the bats foraged with a low, fluttering flight, and when captured would crawl quickly about on the floor of their cage. Introduction of the Polynesian rat brought about the demise of the species—its last outposts are now rat-free, and surveys are still ongoing to search for bats on the islands.

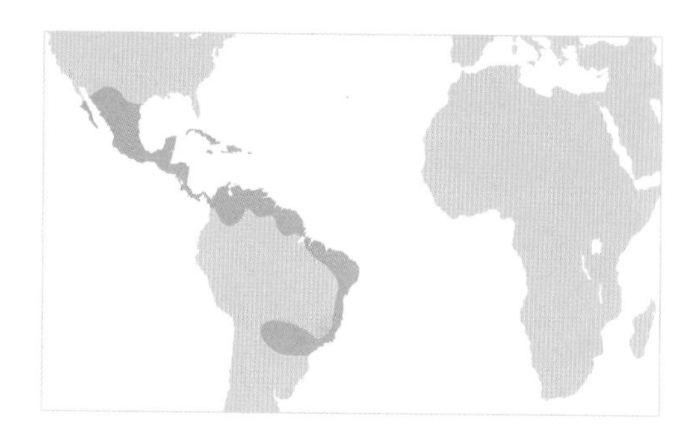

Natalus

## NATALIDAE

The family Natalidae is a distinctive grouping of very small to small bats. They are endemic to the Neotropics (at their most diverse in Central America and the Caribbean), and are sometimes known as "funnel-eared" bats. Their characteristics include a very large uropatagium that contains a long tail (often longer than the combined length of the head and body), a very short, delicate snout, and the distinctive, funnel-shaped, pointed ears; these are forward-angled and set deep within the thick fur on the rounded head. Males possess a facial gland (the natalid organ) on the upperside of the snout. It secretes a fluid that probably has a function in communication.

Natalids roost inside caves, many species only using the deepest parts of very hot and humid caves. It is the presence of suitable caves that tends to limit their distribution, as they are able to forage in all kinds of habitats. These bats hawk for insect prey in flight, typically at rather low levels. They use night roosts in between their foraging bouts, which tend to occur in two main phases—just after sunset and a short time before dawn. Prey is caught in flight, and the bat uses its oversized uropatagium as a scoop that sweeps forward to help contain the insect before the bat can seize it in its mouth. These bats live in single-sex groups, getting together to breed once a year. They have a very long gestation period of eight months or more. The pup is born at an advanced state of development and may be 50 percent of its mother's weight in the smaller species. Nevertheless, she carries the pup with her on foraging flights when it is newly born.

Natalidae belongs to the superfamily Noctilionoidea. Although a well-defined group when considered as a whole, the family has undergone extensive taxonomic revision over the last century, with the number of species recognized ranging from just four in one genus, to thirteen in three. This book recognizes three genera and eleven species.

### Natalus

The largest genus is *Natalus*, which holds seven species. *N. stramineus* and *N. tumidirostris* are described on pages 92 and 90 respectively. The other species in the group are similar in appearance. The largest of all natalids is *N. primus*, which can weigh up to ⅖oz (12.6g). This bat occurs on Cuba and has a population of just a few thousand; all use a single roosting cave. Another relatively large natalid, *N. jamaicensis*, is in a far more precarious position. Its only known roost, St. Clair Cave on Jamaica, holds just fifty individuals; it is unprotected and highly vulnerable to disturbance. The microclimate of hot caves such as this is easily disrupted by outside factors, so climate change is an additional threat to these bats. *N. major* of Hispaniola

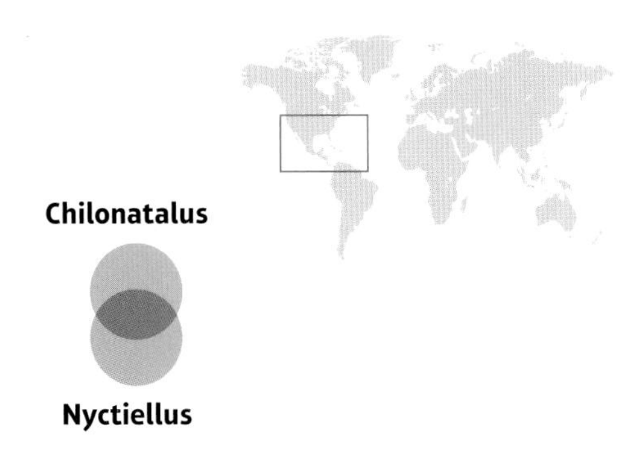

has a small population spread over both islands with several small colonies known. *N. macrourus* (known by the IUCN as *N. espiritosantensis*) is widespread in Brazil, Paraguay, and Bolivia, but, like its congeners, is completely dependent on hot, deep cave types. Suitable caves are both scarce and highly vulnerable to disturbance. *N. mexicanus* occurs in Mexico and much of Central America. It is a bat of lowland forests, roosting in warm moist caves.

## Chilonatalus

Bats of the genus *Chilonatalus* are endemic to the Caribbean. They differ from *Natalus* species in several ways. The snout has small bumps and folds, giving the impression of a "double lip"; in males the natalid organ is larger and more prominent. The ears are also larger, with a pronounced convex curve to the outer edge. The three species this book recognizes are *C. macer*, *C. micropus* (see page 91), and *C. tumidifrons*. *C. macer* is found on Cuba and the Isle of Pines. *C. tumidifrons* is found on the Bahamas where, like other natalids, it roosts only in hot, deep, and damp caves. Its IUCN status of Near Threatened recognizes the vulnerability of its roost habitats.

## Nyctiellus

The third natalid genus is *Nyctiellus*, which contains just one species: *N. lepidus*. This is one of the smallest bat species in the world, with its adults weighing as little as 1/16oz (2g). It is rather dark-furred with a split-lip appearance to its snout, and relatively narrow ears for a natalid. The male's natalid organ is small and positioned near the end of the snout. It is found throughout Cuba and also occurs on the Bahamas. Unlike most other natalid bats, it is very common and widespread throughout its range, and has many large cave roosts. It chooses humid but not necessarily hot caves, and can access small spaces with ease. The sexes roost separately except in winter when they come together to mate. The females give birth to their large, singleton pups in summer.

***Mystacina tuberculata***

## New Zealand Lesser Short-tailed Bat

**THIS BAT OCCURS SPARSELY** on North and South Islands, New Zealand, and on Little Barrier Island and Codfish Island. It is medium-sized, with thick, gray-brown fur and darker ears and membranes. The wings are smallish, the tail free, and the feet and claws large. It has a short, pointed face, long, narrow ears with slender tragi, and small eyes. It occurs in remnant areas of primary forest, using old, decaying trees for roosts, and foraging in deep leaf-litter. It spends up to 40 percent of its active time seeking food on the ground. The diet includes insects, fruit, and carrion. Females form small maternity roosts (about twenty) in tree holes. Eradication of non-native predators is necessary to conserve it— this process took place on Codfish Island, where bat numbers have risen accordingly.

**IUCN STATUS** Vulnerable
**LENGTH** 3⅛–3½in (8–9cm)
**WEIGHT** ¾oz (22g)

***Natalus tumidirostris***

## Trinidadian Funnel-eared Bat

**THIS BAT IS PRESENT** in northern South America, from north Colombia through the Guianas. It is also present on Trinidad and Tobago, and some of the Netherlands Antilles. It has light brown fur with darker wings, and a short, Slightly pointed muzzle. The pale, triangular ears are angled forward and the small eyes set far back, framed by the bases of the ears. It has long wings and a large uropatagium. This species occurs in wet and dry forest, and in suburban and cultivated areas. Roosts, in the deepest parts of damp caves, may hold thousands of individuals. It hunts close to the ground, targeting flying insects by echolocation, and often uses its tail membrane as a scoop to capture them. It is often affected by cave fumigations, intended to control vampire bats.

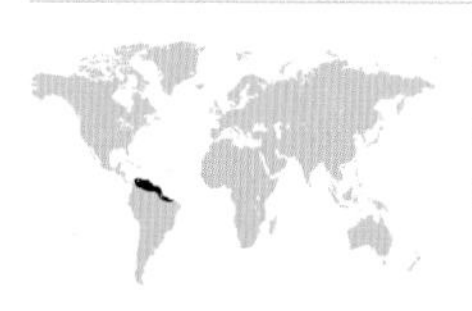

**IUCN STATUS** Least Concern
**LENGTH** 1⅜–1¾in (3.5–4.5cm)
**WEIGHT** ¼oz (6.3g)

***Chilonatalus micropus***

# Cuban Lesser Funnel-eared Bat

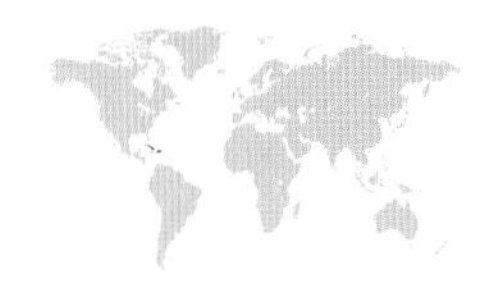

**IUCN STATUS** Near Threatened
**LENGTH** unknown
**WEIGHT** ⅛oz (2–3g)

**THIS TINY BAT OCCURS** in the Caribbean, on Cuba, Jamaica, Hispaniola, and Providencia islands. It has a long, soft coat of light yellowish-brown fur, creamy on the underside; the broad wing and tail membranes are light gray. It has low-set, funnel-shaped ears with very broad bases and pointed tips, and a short, deep-chinned snout tipped with a hairy bump; its eyes are extremely small. The wings are relatively long and slim. The species roosts in loose groups of up to several hundred inside deep, damp cave systems, and forages nearby, hunting tiny flying insects in the open air with an extremely refined echolocation system. Females appear to give birth to their single pups in late summer. It has a relatively restricted range and depends on vulnerable undisturbed caves for roosting.

*Natalus stramineus*

# Mexican Funnel-eared Bat

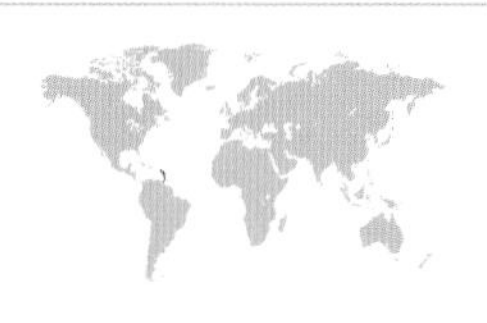

**IUCN STATUS** Least Concern
**LENGTH** 1½–1¾in (3.8–4.6cm)
**WEIGHT** ⅛oz (3–5g)

THIS TINY BAT OCCURS only in the northern Lesser Antilles in the Caribbean; it was split as a separate species following genetic research in 1997. It has tiny eyes, a short snout, and a steep forehead, which in males bears a pronounced, ridge-shaped gland on its center. It uses diverse habitats—from dry, thorny scrub to rainforest—and roosts deep inside dark, humid caves, in groups of up to a few hundred. It specializes in navigating and locating prey through echolocation in cluttered environments. Prey is caught in the air, and also picked from foliage. It breeds once a year, with young born eight to ten months after mating. Females appear to locate their own pup by scent. This bat is vulnerable to roost disturbance and to extreme weather events.

## NOCTILIONIDAE

This very small family, part of the superfamily Noctilionoidea, holds only one genus *Noctilio*), which itself contains just two species. They are sometimes known as the "bulldog" bats, in reference to their short, stout, dog-like snouts, but should not be confused with some of the molossid bats, as these may also be nicknamed "bulldog" bats. Another name for the group is "fishing bats," describing their characteristic feeding behavior.

In new molt, *Noctilio* bats have gray or light brown fur. After molting, depending on where they roost, their fur sometimes bleaches to a bright orange, believed to be due to ammonia bleaching. They are strong and fast fliers,with long, narrow wings. The feet are large with very big, long, and curved claws. They hunt mostly over open water. When hunting, the bat uses its echolocation to detect ripples in the water, then flies low over water with the feet trailing, seizing small fish and any other prey in its feet (*N. albiventris* takes aquatic insects rather than fish). Each trawling flight may cover a distance as long as 10ft (3m). The bats have cheek pouches which are used to store food, while the hunting foray continues. Both species can swim well if necessary.

The two species are described overleaf, *N. albiventris* on page 94 and *N. leporinus* on page 95.

***Noctilio albiventris***

# Lesser Bulldog Bat

**IUCN STATUS** Least Concern
**LENGTH** 2¼–3⅜ in (5.7–8.5cm)
**WEIGHT** ⅝–1½oz (18–44g)

**THIS BAT RANGES FROM** the far south of Mexico, through central America, and as far south as northern Argentina; it may include more than one species. It has bright yellowish fur and pinkish-gray face, ears, and membranes. The ears are long and narrow, the snout pointed, and the cheeks and chin pendulous and jowly (concealing cheek pouches). It roosts in tree hollows and sometimes buildings, and may share the space with other species such as *Molossus molossus*. Its roosts can be located by their distinctive musky scent. It forages in varied habitats, but always near water. It preys mainly on insects, catching them in the air and from the water's surface. It may hunt in small groups of up to fifteen. The species is of conservation concern in Mexico.

***Noctilio leporinus***

# Greater Bulldog Bat

**IUCN STATUS** Least Concern
**LENGTH** 3⅞–5¼in (9.8–13.2cm)
**WEIGHT** 2⅛–3⅛oz (60–78g)

**THIS SPECIES RANGES FROM** the Pacific coast of north Mexico south through Central America to north Argentina. It is a large bat with gray or light-brown fur and darker brown wings, face, and ears. The legs are long and the feet very large. Its hanging jowls conceal internal, elastic cheek pouches. This bat occurs in all sorts of habitats close to water, and is a fish-eater. It uses echolocation to find prey, and can detect hair-fine objects just $1/16$in (2mm) long on the water's surface. It catches fish with its feet and transfers them to the mouth. Breeding males visit maternity roosts (which can hold several hundred females), and may participate in care of the young. Pups hunt independently at about one month old. This species is affected by water pollution and sometimes deliberately persecuted.

## PHYLLOSTOMIDAE

Phyllostomidae is a very large family of bats, making up by far the majority of species in the superfamily Noctilionoidea. This book recognizes 216 species of phyllostomids, in sixty genera. The group is found in the Americas, and its members are known as the "leaf-nosed bats," or "New World leaf-nosed bats," to distinguish them from the Old World family Hipposideridae. The nose-leaf in question—a fleshy leaf- or spear-shaped upright projection at the tip of the nose, often with a rounded or horseshoe-shaped lower part where it attaches to the face—is obvious in nearly all species. Otherwise, this family exhibits remarkable diversity in both appearance and ecology, probably more so than any other bat family.

The phyllostomids are in some ways the New World's answer to the pteropodids, as many of the species are specialist fruit-eaters and some are nectar-feeders. However, the majority of members of this family are omnivorous to some extent, eating insect prey along with plant-derived food. The group also includes some unusual specialists, including the blood-drinking vampire bats, and the largest of all American bats, *Vampyrum spectrum*. This imposing and powerful animal preys almost exclusively on vertebrates and hunts them through stealth, in the manner of an owl or cat.

Many of these bats have strong social structures, with bonds maintained by an array of auditory and olfactory communications. In several species, males in particular possess obvious scent glands around the throat or shoulders, and may apply scent to other members of their group. In some species twenty or more different kinds of social call have been identified. The phyllostomids often show sexual dimorphism; in some cases males are larger than females, in others the reverse is true. In a few species, males have fleshy facial ornamentations that females lack.

Phyllostomids use nasal echolocation, and the nose-leaf is thought to play a role in directing their calls, or amplifying them, or both. Its shape and angle are often helpful for identification, and it is remarkably large in some species, such as those of the genus *Lonchorhina* ("sword-nosed bats"), in which the nose-leaf is as tall as the (very large) ears. These bats are insectivorous, with very accurate and highly developed echolocation. However, fruit-eating and nectar-feeding phyllostomids can also locate and assess their food using echolocation, as well as other cues such as scent. Many plants are adapted to make use of phyllostomid bats as pollinators or seed dispersers, and have evolved fruits and flowers that are easily detected and accessed by the bats.

The fruit-eating phyllostomids have short, strong jaws and are powerful fliers. They often pull fruits from the tree and carry them away to eat elsewhere—assisted by their large and sharp teeth that grip the fruits. Figs and *Piper* (pepper) fruits are particularly popular with these species. In most cases the bats swallow the fruit seeds and excrete them intact. In some cases the journey through a bat's digestive tract improves the seed's chances of successful germination by softening the seed coat (the outer protective layer).

Those bats that visit flowers for pollen and nectar usually have longer and more slender snouts with small nose-leaves, and also long tongues with nectar-mopping, brush-like projections at the tip. These bats can often feed in flight, hovering around flowers and flicking their tongues into them. Flowers that are bat-pollinated usually have large, upright blooms with a strong scent. Several species of nectivorous phyllostomids are of great economic importance as pollinators of commercial crops.

Those phyllostomids that take an entirely or mostly carnivorous diet are usually short-faced and may have very prominent nose-leaves, as well as large, tall, and broad ears. Their hunting behavior is variable but most often involves slow and agile flight within the mid- or understory, with prey gleaned from foliage or the ground as well as captured in flight.

An interesting trait seen within the frugivorous Phyllostomidae is that of tent-making. Finding suitable shelters for day roosts is a key challenge for bats. Many species that are otherwise adaptable in their ways are limited in their distribution because they require a particular kind of cave in which to shelter—one where the size, ceiling height, temperature, and access points are all just right. Roost disturbance is a major conservation issue for bats as a group. Many endangered species are in the situation where the entire population uses only one or a few cave roosts, placing them at very high risk of extinction. Bats that do not roost exclusively in caves still need to find a safe shelter to pass the daylight hours, and this is often difficult. Large, decaying trees with cavities tend to collapse or be felled, and because most bats prefer to roost in groups, other natural hollows are often too small for them. However, tent-making allows bats to create their own roosting spaces. A supply of large, living leaves is all that is needed, and while the roosts will not last forever, they can sometimes be used for up to nine months before they deteriorate—certainly long enough for a group of females to rear their young to independence. The tents are accessed from below; any predator approaching through the trees will cause the roost to shake, and the bats can easily escape the same way if danger does threaten.

Eight different kinds of bat-built tent have been documented, of which one has only been observed in the Old World. This is the work of pteropodids of the genus *Cynopterus*, and is made from palm flower and fruit clusters. All of the other seven types may be made by phyllostomid bats, and at least one bat species has been found roosting in tents of all seven types. Tent styles include shelters in the shape of an upturned boat; a leaf is folded parallel to the midrib by a long cut, making a curved roof. There are also more complex cones and cylinders, perhaps involving multiple leaves collapsed together, as well as the simple apical type; in this the bats chew across a leaf midrib so it folds over to form a peak, like an old-fashioned ridge tent. Dozens of different plant species are used by the tent-makers. The bats that make tents are typically smaller species of forest-dwelling fruit-eaters; they include many species in the genus *Artibeus* and various others, such as *Uroderma* and *Ectophylla alba*. Related species that do not make tents often roost within dense foliage, gaining similar benefits. Some bats may use tents made by other species.

The vampire bats are the most notorious of phyllostomids. Of the three species, only one (*Desmodus rotundus*) is likely to bite humans; the other two mainly take blood from wild and domestic birds. Their bites are usually painless and not noticed by the sleeping host till the next morning. Extremely effective anticoagulants in the bats' saliva means blood continues to flow from the wound for many minutes, so the bat can lap up as much blood as it requires. Its kidneys work rapidly to filter out the nutrients, and the bat then

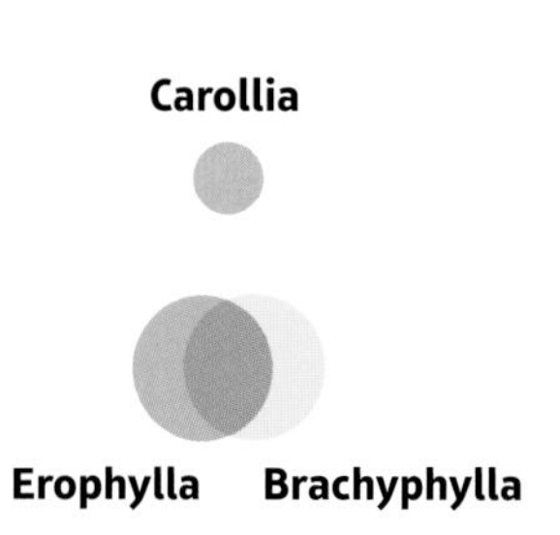

has to urinate continuously while drinking in order to obtain enough protein while still being light enough to fly. *D. rotundus* approaches its prey on the ground, and accordingly is one of the most agile of all bats in "four-footed" locomotion. All vampire bats are unusually large-brained and intelligent; they are among the most socially sophisticated of bats. The reciprocal blood-sharing seen in *D. rotundus* is a well-known example of altruism among wild animals. Unfortunately, vampire bats are persecuted as possible carriers of rabies and due to the introduction of livestock are overpopulated, often requiring control. Control measures are not sophisticated and involve fumigating entire roost caves, with the result that many other bat species are also killed unintentionally. This is a significant cause of population decline among many cave-roosting bat species.

## Carollia

The genus *Carollia*, short-tailed fruit bats, occur in Central and tropical South America; they include the well-studied *C. perspicillata* (see page 102), as well as *C. brevicauda* (page 99), and *C. castanea* (page 100). As their name suggests, these bats have short tails that do not reach the edge of the rather small uropatagium. They are fruit-eaters, specializing in the fruits of *Piper* plants, and roost in caves and tree hollows in small harems; unmated males form bachelor roosts. Most appear to have two breeding seasons each year. The other species include *C. sowelli* (Mexico to Panama), *C. manu* (southeast Peru and north Bolivia), and *C. subrufa* (Mexico to Nicaragua and Costa Rica). *C. benkeithi* (Peru, Bolivia, and Brazil) is a recent split from *C. castanea*. *C. monohernandezi* is found in Panama and northwest Colombia.

## Brachyphylla

The two species in the genus *Brachyphylla* occur in the Caribbean and are both common. They are dark, fruit-eating bats with very small, inconspicuous nose-leaves, and form sizeable roosts in caves. *B. nana* is discussed on page 103. *B. cavernarum* is a common species of Puerto Rico, Virgin Islands, and the Lesser Antilles, then reaching south to St. Vincent and Barbados.

## Erophylla

Another Caribbean genus comprising two species is *Erophylla*. These are adaptable fruit-, insect-, and nectar-feeders, with long noses and tongues. *E. bombifrons* is found on Hispaniola and Puerto Rico. *E. sezekorni* is described on page 104.

***Carollia brevicauda***

# Silky Short-tailed Bat

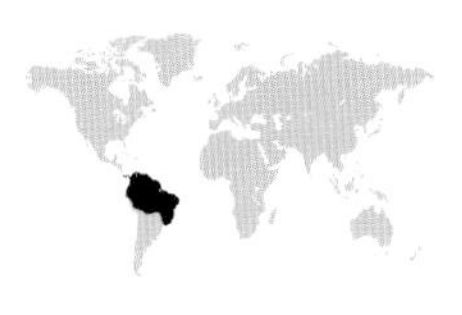

**IUCN STATUS** Least Concern
**LENGTH** 2¼in (5.8cm)
**WEIGHT** ½oz (12.5–13.7g)

**THIS SMALL BAT RANGES** through north and central South America, reaching Bolivia; it also occurs on Trinidad and Tobago. It has light-gray fur and a short tail. The large lower lip bears a V-shaped pattern of warts. It fares best around secondary growth and cultivation, roosting in caves, tree hollows (as seen here), buildings, and on the undersides of banana leaves. This species mostly forages in the understory, and the diet is primarily fruit, but it feeds on insects, nectar, and pollen too, depending on season. Its preferred food is the fruits of wild *Piper* shrubs. *Piper* seeds have been found to germinate more successfully if they have passed through the digestive tract of the bat first. Like other species, they are often mistaken for vampire bats and killed as a result.

***Carollia castanea***

# Chestnut Short-tailed Bat

**IUCN STATUS** Least Concern
**LENGTH** 1⅞–2⅜in (4.8–6cm)
**WEIGHT** ⅜–⅝oz (11–16g)

**ONE OF THE SMALLEST** in its genus, this species ranges from Honduras through Central America and into South America as far as north Ecuador and Peru, Venezuela, the Guianas, and into north Brazil. This is a small, rufous-colored phyllostomid, with a tall nose-leaf and medium-sized pointed ears. It is a species of tropical evergreen forest, including edges and clearings, and usually roosts in small groups in holes or under overhanging roots in earthy banks, or in caves and tree hollows. A specialist feeder on fruits of pepper (*Piper*) shrubs, particularly in the dry season, it finds fruits within the forest understory and carries them to a nearby perch to consume. It also feeds on a few insects. This bat emerges well after dark, and studies indicate that it finds most of its food in the first three hours of foraging. Two breeding peaks occur each year, with pups born in early spring and late summer. The species is widespread with a high tolerance for changes in extent of forest cover.

***Carollia perspicillata***

# Seba's Short-tailed Bat

**IUCN STATUS** Least Concern
**LENGTH** 1⅞–2¾in (4.8–7cm)
**WEIGHT** ½–⅞oz (15–25g)

**THIS BAT RANGES FROM** the south of Mexico through almost all of Brazil, and Paraguay; it also occurs in Trinidad and Tobago. It is sturdy, broad-winged, and short-faced with dark brown fur and wing membranes, and very pink limbs. A forest bat, it prefers lowlands with standing water. It forages mainly at low levels, and roosts in underground caves. It can detect fruit by scent even when odor molecules are sparse, and can distinguish colors. This species prefers *Piper* fruits; just one bat can carry up to 60,000 *Piper* seeds to new locations in a single night. In some areas it is partly migratory, leaving arid forest areas in the dry season. Dominant males attract females to their harem through calls and a hovering display, and repel rivals with a different range of calls and posturing.

***Brachyphylla nana***

# Cuban Fruit-eating Bat

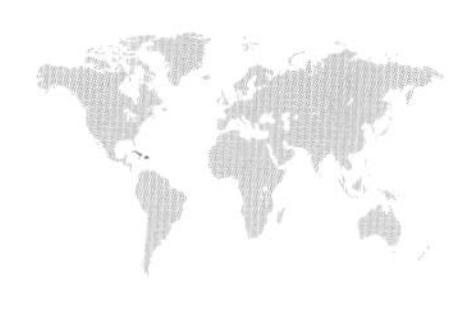

**IUCN STATUS** Least Concern
**LENGTH** 3–3½in (7.6–9cm)
**WEIGHT** unknown

**THIS BAT IS FOUND IN CUBA**, Grand Cayman, Hispaniola, Isle of Pines, and the central Caicos islands. It has quite uniform, light brown fur, and a rather short snout with prominent nostrils. There is no nose-leaf. The eyes are rather small, the ears small and pointed. It forages in many habitat types and roosts in warm, humid caves. Colonies can be large, with dominant males defending small harems of females high in the cave; roosts may also be shared with other fruit-eating bat species. It feeds mainly on fruit, such as *Spondias*, but also hunts insects, and visits flowers to take nectar and pollen. This bat cannot hover to feed, so settles on large flowers with stout stems to feed. Though common on Cuba and the Isle of Pines, it is reportedly rare on Haiti, and extinct in Jamaica.

***Erophylla sezekorni***

# Buffy Flower Bat

**IUCN STATUS** Least Concern
**LENGTH** 3–3⅝in (7.7–9.2cm)
**WEIGHT** ⅝oz (16–18g)

**THIS SPECIES OCCURS IN CUBA,** Jamaica, and some parts of the Bahamas, Caicos, and the Cayman Islands. Large for its genus, it has light gray-brown fur with darker wings, and prominent, dark eyes; it has a long muzzle with a small, upright nose-leaf. It forages in forests, and in caves forms large roosts, with roughly equal numbers of males and females. In December, when the mating season begins, adult males establish a territory within the roost, giving ultrasonic calls and performing a wing-flapping display to attract females. The most successful mature males assemble harems, although males that do not display have some mating success too. The following summer, each female produces a single pup. The diverse diet includes fruit, nectar, and insects, including beetles and moths.

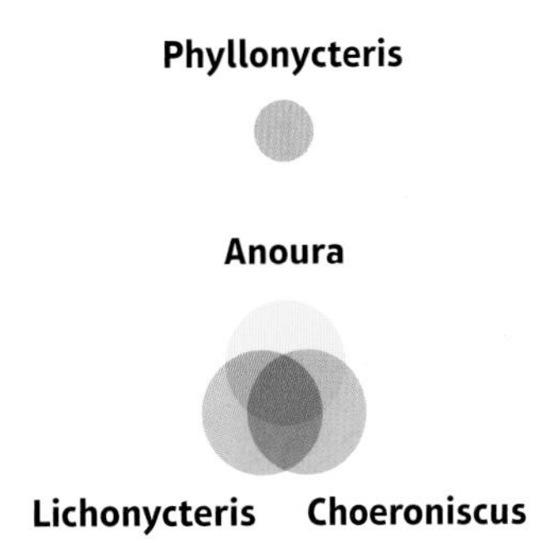


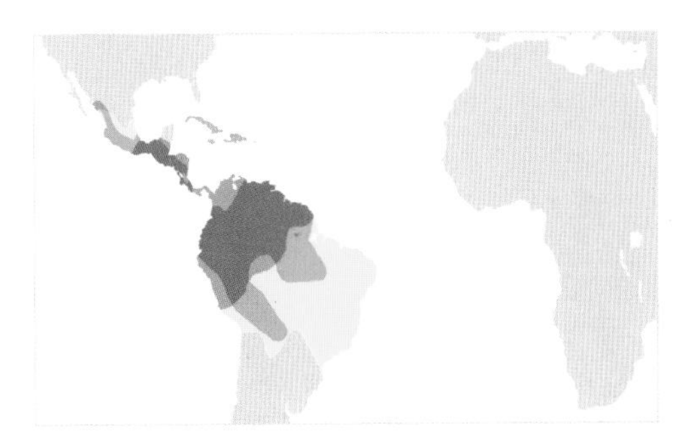

## Phyllonycteris

*Phyllonycteris* is similar, in ecological terms, to *Erophylla*. It is also confined to the Caribbean, and has two species. *P. aphylla* occurs only on Jamaica. It has just two known roosts, between them holding no more than 250 individuals. The other species, *P. poeyi*, is described on page 107.

## Anoura

One of the larger phyllostomid genera, *Anoura* holds ten species. These bats are mostly found in high-altitude forest habitats. Most lack tails (and so the genus is known in English as "tailless bats"). They feed on nectar and insects, and have slim snouts, very small eyes, and rather small nose-leaves. *A. geoffroyi* is described on page 108. *A. caudifer*—a species which occurs in north, northwest, and central South America—does possess a very short tail, and so has the tautological English name Tailed Tailless Bat. It is known to have an unusually high metabolic rate, allowing it to maintain its body temperature in its cool cave roosts; to do so it requires a high calorific intake. *A. cadenai*, which occurs in the southwestern Colombian Andes and into Ecuador, and *A. fistulata*, present on both Andean slopes in Ecuador, are both little known. *A. cultrata* is an upland and montane forest species of Costa Rica, Panama, Venezuela, Colombia, Ecuador, Peru, and Bolivia. *A. latidens* is patchily distributed in northwestern South America, while *A. luismanueli* occurs in the Andes in Venezuela, Colombia, and Bolivia.

The IUCN still recognizes *A. aequatoris* as a subspecies of *A. caudifer* and *A. peruana* as a subspecies of *A. geoffroyi*; neither are thus assessed as yet. *A. carishina* is a new species, described in 2010 from Colombia, and not yet assessed by the IUCN.

## Choeroniscus & Lichonycteris

The small genus *Choeroniscus* comprises three species of nectar-feeding, long-nosed bats with small nose-leaves. They are known as the "long-tailed bats," but in fact the tail does not extend beyond the uropatagium. *C. periosus* has a limited distribution in Colombia and north Ecuador. *C. godmani* is more widespread, ranging from south Mexico through much of Central America to the north of South America. *C. minor* is covered on page 116. The IUCN treats the similar genus *Lichonycteris* as monotypic, but this book recognizes two species (*L. obscura* and *L. degener*). The former ranges from Mexico through to west Colombia; the latter replaces it south and east of the Andes.

## Glossophaga

*Glossophaga* are long-tongued, nectar-feeding phyllostomids, similar in appearance to *Choeroniscus* and *Lichonycteris*. *G. commissarisi* and *G. soricina* are described on page 110 and page 109, respectively. A further three species are recognized: *G. longirostris* (widespread in the north of South America and the Caribbean, but little studied), *G. leachii* (Mexico to Costa Rica), and *G. morenoi* (south Mexico).

## Leptonycteris

Yet another genus of long-tongued bats is *Leptonycteris*. *L. curasoae* is a species found in the north of South America. *L. yerbabuenae* ranges from Arizona and New Mexico south to El Salvador. The other species, *L. nivalis*, is described on page 115. These species are vital pollinators of agaves and cactus in Mexico.

***Leptonycteris yerbabuenae***

## Monophyllus

The name *Monophyllus* means "one leaf," and the two species in this genus are known as the "single-leaf bats," though their nose-leaves are not noticeably different to the other mainly nectar-feeding phyllostomids. They are close relatives of *Glossophaga*. There are two species, both of which occur in the Antilles. *M. redmani* is described on page 112. *M. plethodon*, a less widespread species, occurs in the Lesser Antilles; it is quite common, though some of its cave roosts are vulnerable to disturbance.

## Glyphonycteris

The bats of the genus *Glyphonycteris* are distinctive, short-faced animals with large, broad ears that taper to pointed tips; they are known as the "big-eared bats." They are insect-eaters and also take some small fruits. Of the three species, two (*G. daviesi* and *G. sylvestris*) are described in detail, on page 113. The third, *G. behnii*, is a poorly known species from the west of South America. It appears to be very rare. Surveys are needed to determine what threats it faces and what conservation action, if any, should be taken.

***Phyllonycteris poeyi***

# Cuban Flower Bat

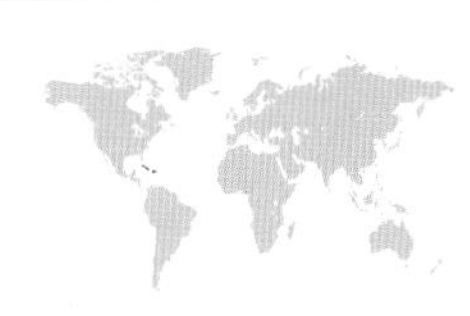

**IUCN STATUS** Least Concern

**LENGTH** 2½in (6.4cm)

**WEIGHT** ½–1oz (15–29g)

**THIS BAT OCCURS ON CUBA,** including the Isle of Pines, and in the Dominican Republic. It is an attractive phyllostomid with bright orange-yellow fur and brown membranes. It has a long snout, with a tiny nose-leaf at the top, small eyes, and small, pointed ears. It occurs in forest, forest edges, and scrubland, and feeds from large, open-structured flowers able to support its weight. The short, dense fur collects pollen as the bat feeds. Its roosts, in large caves, can be huge. Cueva de los Majaes, an enormous cave in east Cuba, hosts as many as a million individuals, alongside thousands or millions more bats of another dozen or so species. Their presence heats the cave to 109.9°F (43.3°C). Each night they pollinate tens of millions of flowers, both wild and cultivated.

***Anoura geoffroyi***

# Geoffroy's Tailless Bat

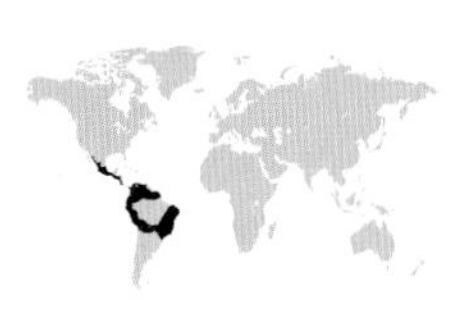

**IUCN STATUS** Least Concern
**LENGTH** 2¼–2⅞in (5.8–7.3cm)
**WEIGHT** ½–⅝oz (13–18g)

**THIS WIDESPREAD SPECIES** ranges from the north of Mexico through much of Central America to north, northwest, and central South America (not the Amazon basin). It also occurs in Trinidad and Tobago. A medium-sized bat, it has mid-brown fur (gray on the underside), long, narrow wings with dark membranes, and a long snout, with projecting lower jaw and an upright nose-leaf. It has hairy legs and no tail. It occurs in forests, farmland, and near water, up to 8,200ft (2,500m) elevation, and roosts in well-lit caves or crevices. Insects comprise up to 90 percent of its dry season diet, but can consist of more fruit and nectar at other times. It roosts alone or in small groups (up to seventy-five recorded), sometimes alongside other *Anoura* species. Most females produce a single pup, late in the rainy season.

***Glossophaga soricina***

# Pallas's Long-tongued Bat

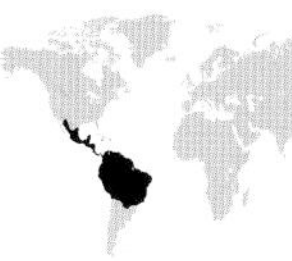

**IUCN STATUS** Least Concern
**LENGTH** 2–2¾in (5–6.9cm)
**WEIGHT** ¼–⅜oz (7–12g)

**THIS BAT OCCURS THROUGH** most of Mexico through Central America and South America to north Argentina. It has relatively large eyes, round-tipped ears, and a delicate snout. Roosts are usually in caves or buildings, and may exceed 2,000 individuals. It forages in all kinds of habitat, from undisturbed forest to suburban areas. An important pollinator, it may hover or settle when taking nectar. Plant species it visits include the rare Ecuadorean cactus *Espostoa frutescens*, which has specialized furry material around its flowers to absorb echolocation signals and help the bat locate the flowers more easily. Its high-sugar diet fuels the fastest metabolic rate ever observed in a mammal, depleting more than half of the bat's body fat stores every day; it also hunts small insects and feeds on some fruit.

***Glossophaga commissarisi***

# Commissaris's Long-tongued Bat

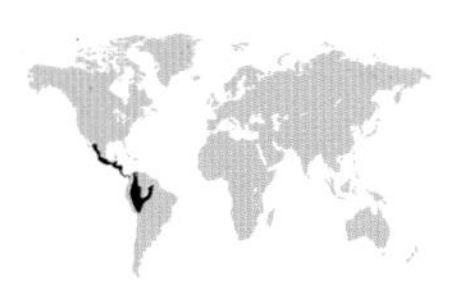

**IUCN STATUS** Vulnerable
**LENGTH** 1¾–2⅞in (4.6–7.2cm)
**WEIGHT** ¾oz (23g)

**THIS SPECIES IS FOUND** from central, Pacific-coast Mexico south through much of Central America, and in Peru, Ecuador, parts of Colombia and Venezuela, Guyana, and northwest Brazil. It is a small, rotund bat with a proportionately big head. The snout is shortish and delicate, with a small, slender nose-leaf. It has mid-brown fur, slightly paler on the underside, with broad but short wings, which allow it to maneuver skillfully in flight. It occurs in varied habitats, from thorn forest and savanna to rainforest, and roosts in tree holes or caves; it can accept some habitat disturbance. Primarily a nectar- and pollen-feeder, this bat feeds on nectar from flowers while hovering. One of the plants it visits, seen here, is the sea bean flower (*Mucuna holtonii*). It is a legume that grows along streams, rivers, and other rain forest edges from southern Mexico and Belize throughout Central America. It is highly specialized for exclusive bat pollination. The bat's diet also includes soft fruit and small insects. Females produce a single pup once or twice a year. Its large range and adaptable habits mean this species is not currently of conservation concern.

***Monophyllus redmani***

# Leach's Single-leaf Bat

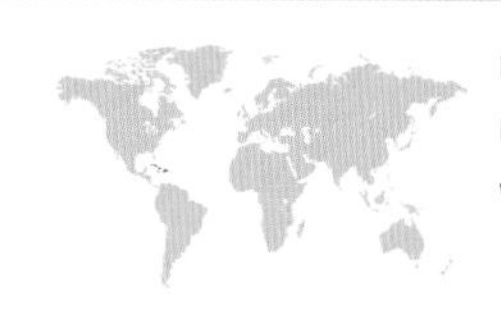

**IUCN STATUS** Least Concern
**LENGTH** unknown
**WEIGHT** ⅜oz (8.7g)

**THIS BAT IS FOUND** widely across the Caribbean. It has a dull brown coat and a rather round head with a short muzzle. Roosts of this species can be very large, with more than 100,000 individuals recorded in some cases; the roosts may also be shared with other species. It prefers warm caves. Emerging after dark, it forages in woodland and more open, dry countryside. It feeds mainly on nectar, but also eats some fruit and insects. This bat is an important pollinator for a variety of Cuban plants, including *Marcgravia evenia*, an endemic Cuban vine that is highly adapted to attract bats. Its bowl-shaped leaves act as reflectors for the bats' echolocation calls, helping to guide bats to the flowers that hang below. The bat is seen here pollinating a relative of the Organ Pipe Cactus.

***Glyphonycteris daviesi***

## Davies's Big-eared Bat

**THIS BAT, THE LARGEST** *Glyphonycteris* species, ranges from Colombia and Venezuela south to northern Brazil and Bolivia. It is dark brown with a dense, long coat, a furry face, broad ears with pointed tips, a wide nose-leaf, and a short, robust snout, with a prominent, puffy-looking lower lip. Its front top incisors are as long as its canine teeth, and its bite is very powerful—captured specimens have been known to chew their way out of cloth bags. It occurs in primary evergreen forest, and more sparsely in adjacent secondary growth. It roosts in groups inside tree hollows, in small, mixed-sex groups. This species feeds on large insects, small vertebrates, and some fruit. It is quite rare, and its strong preference for primary forest makes it vulnerable to habitat loss and population fragmentation.

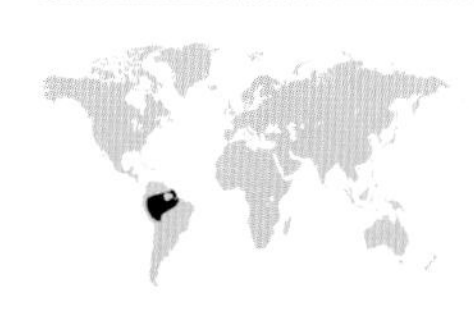

**IUCN STATUS** Least Concern
**LENGTH** 2½–3⅜in (6.3–8.4cm)
**WEIGHT** ⅝–1oz (19–30g)

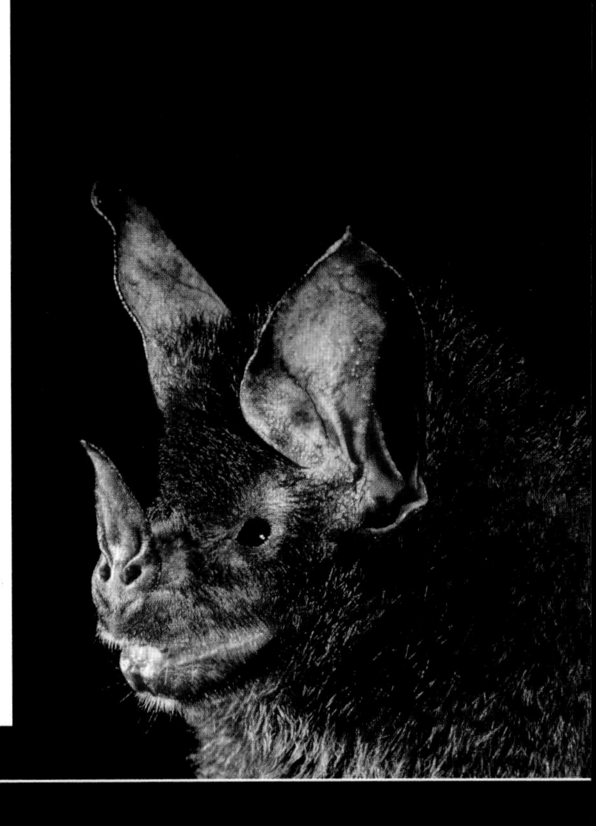

***Glyphonycteris sylvestris***

## Tri-colored Big-eared Bat

**THIS BAT OCCURS** in parts of south Mexico, Central America, and South America in at least four clearly separated populations. It has gray-brown fur, blacker around the eyes and whitish on the belly, and a blunt, mastiff-like face, with a long, pointed nose-leaf and large, triangular ears. It lives in lowland forest and roosts in caves and hollow trees, in groups of up to seventy-five; it also sometimes lives alongside other bat species. The diet includes large insects picked from foliage, but also some fruit; this bat carries food items back to a favorite night roost to consume them. Although it has been recorded over a wide area, the species seems to be rare across its whole range; it is almost certainly suffering declines because of deforestation in South America.

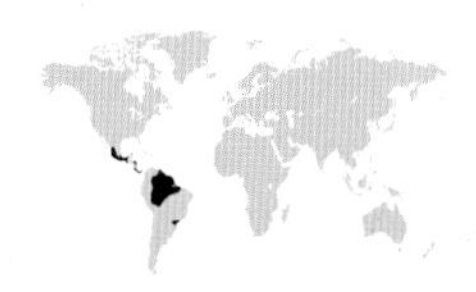

**IUCN STATUS** Least Concern
**LENGTH** 2½–3⅜in (6.3–8.5cm)
**WEIGHT** ¼–⅜oz (7–11g)

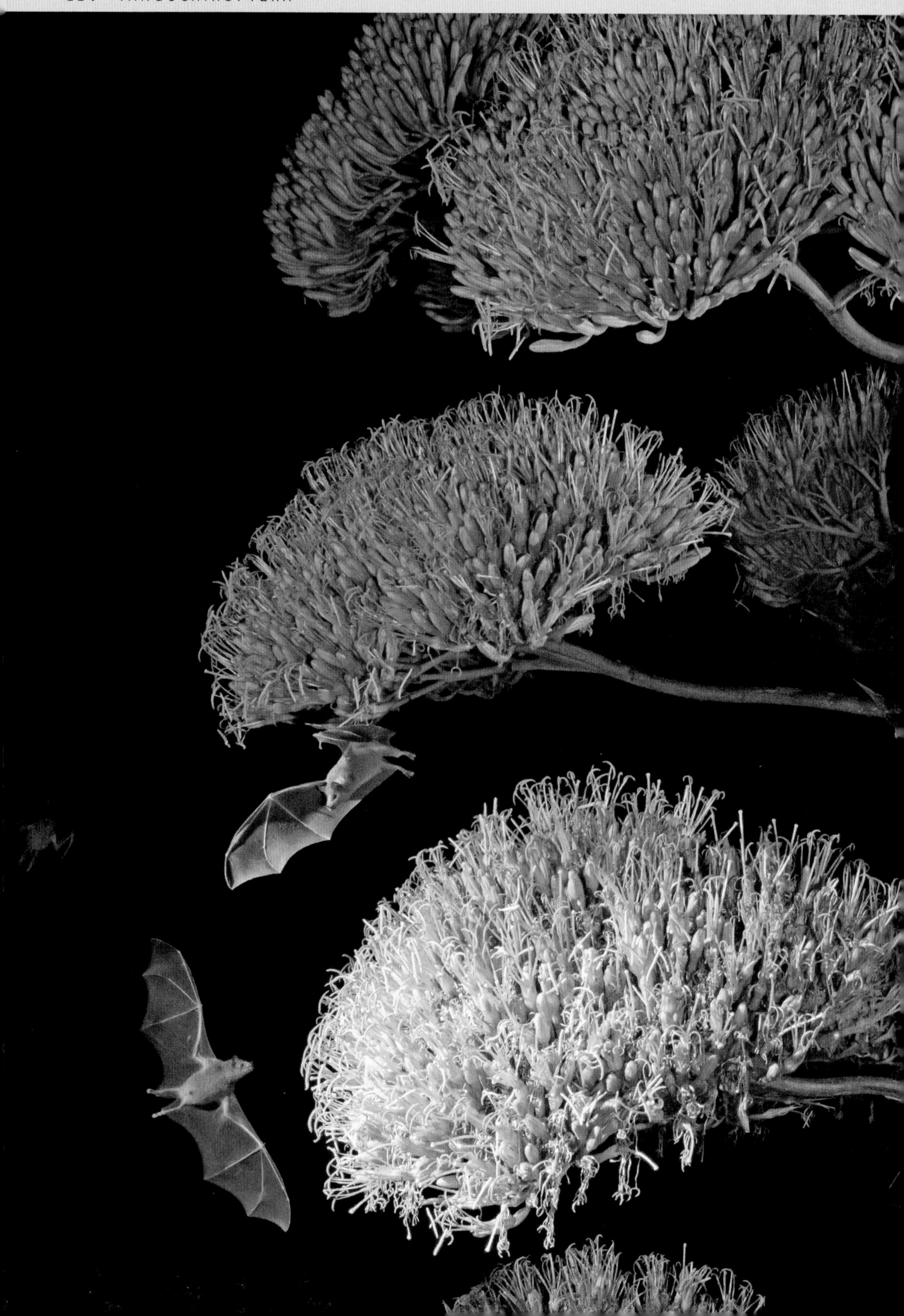

*Leptonycteris nivalis*

# Greater Long-nosed Bat

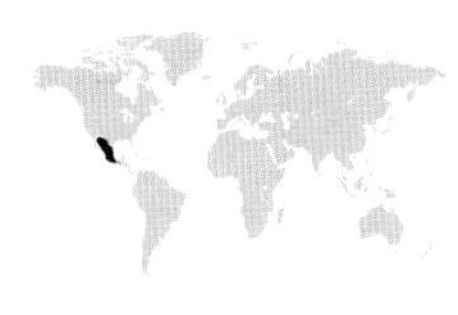

**IUCN STATUS** Endangered
**LENGTH** 2¾–3½in (7–9cm)
**WEIGHT** ⅜–1oz (18–30g)

**THIS SPECIES IS FOUND** in Mexico (except the northwest) and just reaches west Texas; populations in the north undertake short migrations south in winter, following the flowering seasons of the plants they feed on. It is dusky gray-brown, a shade paler on the underside, with gray wing membranes. The ears are small with pointed tips, the snout long, with a protruding lower lip, and a small, triangular nose-leaf. It has a long tongue with papillae at the tip for "mopping up" nectar. This bat is found in dry areas, from desert edges to dry woodland (pine and oak), wherever agave plants (seen here) and cacti grow. It roosts in large colonies in caves, and also sometimes in buildings and tree hollows. The diet is principally nectar and pollen, but it also feeds on fruit and insects. Females give birth to a single pup annually, in May or June. This bat is endangered because of a declining population. Disturbance and destruction of roosts is a main factor, along with habitat loss; it requires large, protected areas of wild agave.

***Choeroniscus minor***

## Lesser Long-tailed Bat

**THIS SPECIES IS PRESENT** in much of northern, eastern, and western South America, though not in the Amazon basin, and on Trinidad and Tobago. It is a mid- to dark brown bat with dense fur and a relatively long tail. The snout is long and slim, with a small nose-leaf; the eyes are largish, and the ears small and rounded. The long, brush-tipped tongue can be extended to an impressive length without the bat needing to open its jaws (via a space between the teeth). It occurs in tropical forest, including secondary growth, especially around water, and will visit plantations. Eight or so roost together in tree hollows. It visits flowers for nectar and pollen, probably also feeding on some insects. It is rare and difficult to observe, but has been recorded from several protected areas.

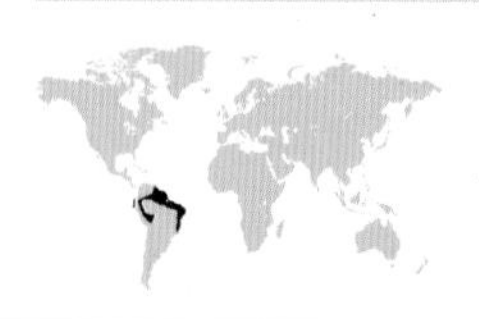

**IUCN STATUS** Least Concern
**LENGTH** 2⅜–2¾in (6–7cm)
**WEIGHT** ⅜oz (10g)

***Hsunycteris thomasi***

## Thomas's Nectar Bat

**THIS BAT IS FOUND** in northern, western, and central South America, reaching Ecuador and northwest Brazil. It is a small, nectivorous phyllostomid with a fairly long, slender snout, tipped with a small, spear-shaped nose-leaf. The ears are small and round, the eyes smallish and low-set; the chin protrudes. It has light red-brown or mid-brown fur on the upperside, with a paler belly. This bat occurs in low-lying tropical evergreen forest near rivers and streams, and is more sparse in clearings and forest edges; it usually roosts inside tree hollows, in small groups. The diet is mainly nectar, taken with the long, brush-tipped tongue from banana flowers and similar species; it also includes pollen and small insects. Although poorly studied, it appears to be common over much of its range.

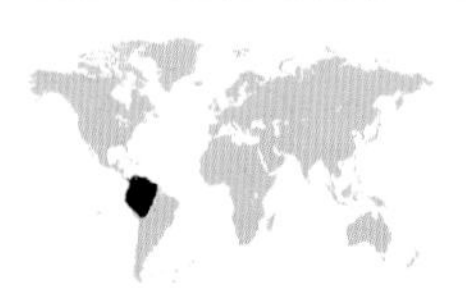

**IUCN STATUS** Least Concern
**LENGTH** 1¾–2⅜in (4.5–6cm)
**WEIGHT** ¼–½oz (6–14g)

## Hsunycteris

The genus *Hsunycteris* contains four species of small, nectar-feeding phyllostomid. These are closely related to the larger species in *Lonchophylla*, and only recently separated from that genus. Only one, *H. thomasi*, is currently recognized by the IUCN, although their taxonomy still places it in *Lonchophylla. H. thomasi* is described opposite. The other three species are all relatively new to science, all being described since the turn of the century. *H. dashe* and *H. cadenai* are both from the north of Peru, while *H. pattoni* has been found to occur in both Peru and Ecuador.

## Lonchophylla

*Lonchophylla* is a large genus of thirteen species. These are medium-sized, nectar-feeding bats, with long, slim snouts, large domed heads, and small but obvious pointed nose-leaves. The eyes are small and set rather far forward on the face, the ears are also rather small and triangular, and the lower jaw sticks out further than the upper. The fur tends to be light sandy-brown or ginger, with a paler underside. These bats occur in forest, but also in more open habitats, and are important pollinators, though they do take some fruit and insects as well as nectar. *L. robusta* is described on page 120.

Two species are little known, these are *L. chocoana* (known from just four sites in Ecuador and Colombia) and *L. orcesi*, described from a single specimen taken in northwest Ecuador in 2005. *L. bokermanni* and *L. dekeyseri* are both found in the *cerrado* (savanna) in the east of Brazil, a fragile and degraded habitat. Both species are rare. A further two species are considered threatened because their habitat is disappearing and their known ranges are rather small: *L. hesperia* is found in Peru and Ecuador, while *L. mordax* occurs in east Brazil.

*L. concava* is present in Central America and parts of northwestern South America. *L. handleyi* has a rather small range in south Ecuador and north Peru, while *L. peracchii* is confined to a small area of Atlantic forest in southeast Brazil, but is present in several protected areas. The final three species are all rather new to science. *L. fornicata* is a newly described species from west Colombia and Ecuador. *L. inexpectata* is a recent discovery from Brazil, and *L. orienticollina* has been found in the east of the Andes in Venezuela, Colombia, and Ecuador.

## Lonchorhina

*Lonchorhina*, the "sword-nosed bats," are six species that have very large ears with prominent, pointed tragi, and tremendously long, spear-like nose-leaves. Their short faces bear elaborate skin folds. These are insectivores and highly colonial. Their extreme powers of echolocation often enable them to detect and avoid mist-nets, so they are a difficult group to study and survey. Additionally, most species have small

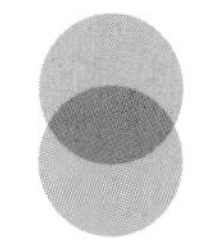

Micronycteris

geographical ranges. *L. aurita* is described on page 121. Of the remainder, *L. fernandezi* of southern Venezuela is endangered, with no observations since a few dozen individuals were found in a cave roost in 1988. It has suffered considerable habitat loss since then, with cattle ranching a fast-expanding industry in its range. *L. marinkellei* of southeast Colombia remains very rare, with only two known locations. This species too is affected by habitat loss due to cattle ranching. *L. orinocensis* is a species that occurs in south Venezuela and southeast Colombia. *L. inusitata* is little known, having only recently been split from *L. aurita.* It occurs in Venezuela, the Guianas, and north Brazil. *L. mankomara* is a newly discovered species from Colombia.

## Macrotus

*Macrotus* is a small genus of appealing-looking, insectivorous phyllostomids with very large, wide ears, large eyes, and short noses with small, delicate nose-leaves. They are broad-winged with a slow, agile, and fluttering flight, often gleaning prey from vegetation while hovering. By day they roost in well-lit caves. Both species are described in detail: *M. californicus* on page 122, *M. waterhousii* on page 123.

## Micronycteris

*Micronycteris* is closely related to *Lonchophylla.* They are known as the "little big-eared bats," and there are eleven species. They have large, wide, and round-ended ears, short sparsely furred faces with a prominent, overshot lower jaw, small eyes, and quite long nose-leaves. They are insectivorous. *M. hirsuta* is described on page 125, *M. megalotis* on page 124, and *M. minuta* also on page 125.

No fewer than six of the remaining eight species are quite new to science and reflect increased interest and improved methods of surveying for more elusive bats in forest habitats. One such is *M. brosseti*, described in 1998 from Brazil, French Guiana, Guyana, and Peru. *M. buriri*, discovered in 2005, occurs on St. Vincent island in the Lesser Antilles, where it was found to be quite common in forests. *M. giovanniae* was described in 2007. It is so far known only from a single specimen, caught in west Ecuador; the same goes for *M. matses*, found in Peru. *M. sanborni,* described in 1996, has three known localities, all in northeast Brazil in dry *caatinga* (desert thorn-scrub) habitat. *M. yatesi* was described in 2013 and occurs in Bolivia.

The remaining two species are *M. microtis,* a common and widespread bat found in Mexico; its range extends through Central America and much of the north of South America. *M. schmidtorum* has a similar distribution, though it is much rarer.

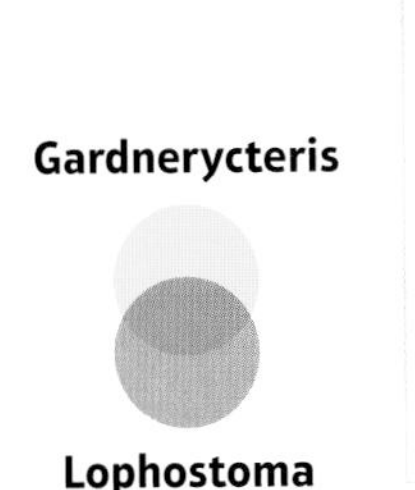

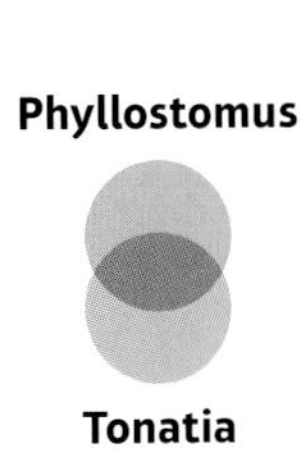

## Gardnerycteris

*Gardnerycteris* has been split from the genus *Mimon*. It comprises two species of large-eared, insectivorous phyllostomids, which sport very long, broad-based nose-leaves. *G. crenulatum* is a dark-furred bat with a white stripe running from its forehead down its back. It is a widespread species of the Neotropics, occurring from Mexico to Brazil and Bolivia. *G. koepckeae* is an apparently uncommon bat of Peru.

## Lophostoma

*Lophostoma*, the round-eared bats, are another group of insectivorous species, with very large ears that broaden toward their rounded tips. They have bulging chins and quite large nose-leaves; their faces are almost bare. *L. brasiliense* is described on page 130, *L. schulzi* on page 129, and *L. silvicolum* on page 130. This book recognizes a further four species, two of which are *L. carrikeri* and *L. evotis*. The former is widespread across tropical South America, while the latter occurs in Mexico and Central America. *L. kalkoae*, a new species described in 2012 in Panama, is little known. *L. occidentalis* completes the genus; it is a rare and declining bat of west Ecuador.

## Phyllostomus

The four species of the genus *Phyllostomus* are known as the "spear-nosed bats," because of their sharply pointed nose-leaves. They are rather large phyllostomids with short, strong jaws and pointed ears, and take a varied diet including pollen, fruit and insects. *P. discolor* and *P. elongatus* are described on page 128, and *P. hastatus* on page 126. The fourth species, *P. latifolius*, occurs in northwest Brazil and adjacent Guyana, French Guiana, and Suriname, Venezuela, and Colombia.

## Tonatia

*Tonatia* is a small genus of just two species. They are rather plain, dark brown phyllostomids with blackish membranes and large, dark ears with rounded tops. The head is acutely domed, the snout medium-length. It has a tall nose-leaf and a prominent, warty chin. *T. saurophila* is described in detail on page 134. The other species, the larger *T. bidens*, is a little-known bat found in South America, from eastern and southern Brazil west through Paraguay to northern Argentina and eastern Bolivia.

*Lonchophylla robusta*

# Orange Nectar Bat

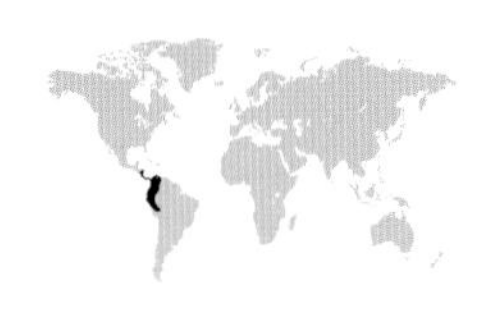

**IUCN STATUS** Least Concern
**LENGTH** 2⁷⁄₁₆in (6.2cm)
**WEIGHT** ½oz (17g)

**THIS FAIRLY COMMON BAT** is found from Nicaragua south to Venezuela, Ecuador, and Peru. It occurs in varied habitats, from forests to gardens, parks, and plantations. It roosts in small groups in small caves, among boulders, and in rocky crevices. Its diet is primarily nectar, but also includes some fruit and insects. Studies of its tongue anatomy and preferred feeding method reveal that this species is highly adapted to nectar-feeding. The tongue has pronounced lateral grooves that open up when the bat is feeding; they show a peristaltic muscular action to pump nectar into the mouth, without needing to draw the tongue in and out. This feeding method is quantifiably more efficient than other bats' "lapping" action. The bat hovers steadily as it feeds.

*Lonchorhina aurita*

# Common Sword-nosed Bat

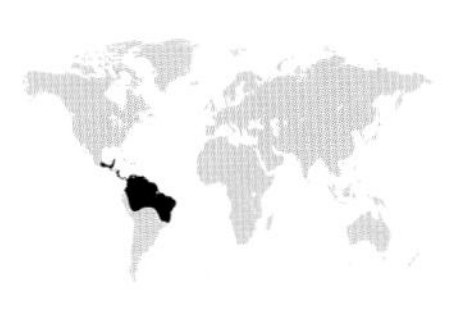

**IUCN STATUS** Least Concern
**LENGTH** 2⅛–2⅝in (5.3–6.7cm)
**WEIGHT** ⅜–⅝oz (10–16g)

THIS BAT'S RANGE EXTENDS from south Mexico through parts of Central America and Trinidad; it also occurs in north, central, and eastern South America, reaching Bolivia and south Brazil. It has an extraordinarily long, tall, nose-leaf, and long, broad ears with pointed tips. Most frequently found in dense, wet forest, it roosts in caves. It is an insectivore with incredibly precise echolocation, used to target insects in the air and on foliage; this bat can often even detect mist-net mesh, so is difficult for researchers to capture. The diet may also include fruit. Cave roosts can hold several hundred individuals—they will roost with other species, but are invariably the last to leave the roost, well after sunset. Females give birth once a year, at the start of the rainy season.

***Macrotus californicus***

# California Leaf-nosed Bat

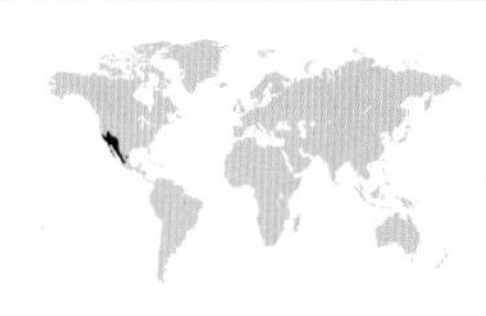

**IUCN STATUS** Least Concern
**LENGTH** 3¾in (9.4cm)
**WEIGHT** ⅜–¾oz (12–22g)

**THIS SPECIES OCCURS** in the far southwest of the United States, and south into northwest Mexico; it also occurs separately in northeast Mexico. It has short, glossy brown fur, a short snout with prominent nose-leaf, and long, pointed ears. Its short, broad wings permit great agility in flight. It occurs in scrubland and along rivers through deserts, roosting in well-lit, cool caves and old mines in groups of 100 to 200. This bat feeds on a variety of insects including moths, beetles, and crickets from foliage or the ground. It detects prey by sound, often from footsteps or fluttering wings and traps it within its uropatagium. It mates in the fall, with females giving birth, usually to twins, the following May or June. Disturbance of maternity roosts is a major problem, and reopening mines can displace colonies.

***Macrotus waterhousii***

# Waterhouse's Leaf-nosed Bat

**IUCN STATUS** Least Concern
**LENGTH** 3¼–4¼in (8.3–10.7cm)
**WEIGHT** ⅜–⅝oz (12–19g)

THIS BAT OCCURS on the Pacific side of Mexico and into central Guatemala—and also parts of the Caribbean, where fossil remains indicate that it was formerly more widespread. It has fluffy, gray-brown fur and broad wings with gray membranes. The large ears have obvious pointed tragi, the eyes are prominent, and there is a small, triangular nose-leaf. It prefers arid, thorny habitats, and roosts near the entrances of large caves and mineshafts, often in large groups. It leaves the roost an hour or two after sundown to forage, making a second flight a couple of hours before sunrise. Its prey is insects, such as katydids, and other arthropods, often picked from the ground or from foliage, then eaten at a night roost. Mating occurs in the fall, and the young are born in June.

***Micronycteris megalotis***

# Little Big-eared Bat

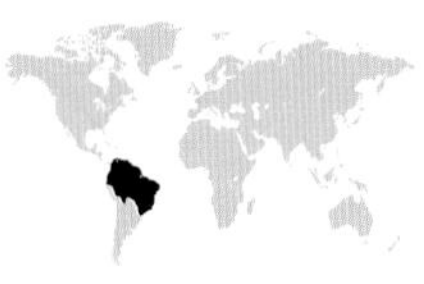

**IUCN STATUS** Least Concern
**LENGTH** unknown
**WEIGHT** unknown

**THIS BAT APPEARS** to be widespread across north and central South America, including Trinidad. It is a small phyllostomid with broad wings, large ears, a spear-shaped nose-leaf, a steep forehead, and relatively large eyes. It is found in both evergreen forest and dry thorn forest, and roosts in tree holes, rock crevices, and buildings, in groups of up to twelve. Primarily an insectivore, it prefers to hunt near streams and rivers. It probably also feeds on fruit; its tooth arrangement suggests an unspecialized diet. The bat's slow and maneuverable flight allows it to glean prey from foliage, and it appears quick to learn how to exploit new feeding opportunities. Prey is eaten at a night roost. Despite its large range, observations are infrequent due to its ability to detect mist nets.

***Micronycteris hirsuta***

## Hairy Big-eared Bat

**THIS SPECIES IS FOUND** in the southeast of Central America and on across a large part of northeastern South America; a separate population also occurs in east Brazil and Trinidad. It is a dark, robust phyllostomid with dense fur. The male's head bears a crest of long hairs, with a scent-producing gland at its base. The crest is presumably fanned out during courtship to disperse the scent to prospective mates. This bat is found in primary forest, roosting in hollow trees but also around habitation, where it roosts in buildings. Studies of diet suggest it is mainly insectivorous, and it is attracted to katydid calls. It eats more fruit in the dry season. It is common in parts of Costa Rica but rarer elsewhere; it is probably suffering the consequences of deforestation.

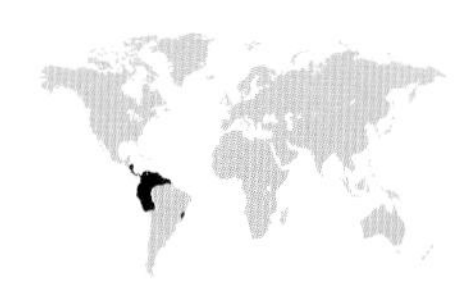

**IUCN STATUS** Least Concern
**LENGTH** unknown
**WEIGHT** unknown

***Micronycteris minuta***

## White-bellied Big-eared Bat

**THIS TINY BAT IS FOUND** across much of the north of South America, including Trinidad, and the southern half of Central America. It has light brown fur, shading to almost pure white on the belly. The ears are large and rounded, the snout rather short, and the chin large. It has small eyes and a medium-sized nose-leaf. Females average a little larger and heavier. It is adaptable, most often occurring in lowland forest (both deciduous and evergreen), but also in farmland with scattered trees. It roosts in small groups in tree holes, sometimes in caves, usually alongside several bat species. It is mainly insectivorous, but also takes some fruit, the proportion probably higher in the dry season. Its tolerance of disturbed habitats means this species is under no immediate threat.

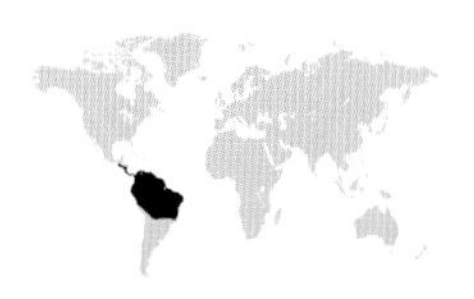

**IUCN STATUS** Least Concern
**LENGTH** 1⅝–2⅛in (4–5.5cm)
**WEIGHT** ⅛–¼oz (4–8g)

*Phyllostomus hastatus*

# Greater Spear-nosed Bat

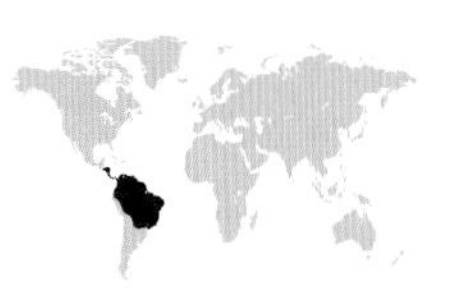

**IUCN STATUS** Least Concern
**LENGTH** 4⅝–5½in (11.8–13.8cm)
**WEIGHT** 2⅞–4¼oz (80.5–120.4g)

THIS BAT'S RANGE EXTENDS from Guatemala south through Central America and across the north half of South America; it is also present on Trinidad. It is rather a large species with glossy brown fur and gray-brown membranes. It has pointed-tipped ears and a short snout with a pointed, upright nose-leaf. The species occurs in forested habitats in both lowlands and uplands, and roosts in caves and tree cavities. The diet includes insects, nectar, fruit, pollen, and occasional vertebrate prey. Adult males, substantially the larger sex, defend a harem of a dozen or more females; the female groups stay together when the dominant male is replaced, which may happen frequently. Males without a harem live in bachelor groups. Populations in central America produce one pup a year, toward the end of the dry season; populations further south may breed twice a year. If a pup falls from the cave roost, other females in the group will "guard" it until the mother retrieves her young—one of the reasons why females living in larger harems have more breeding success.

***Phyllostomus discolor***

## Pale Spear-nosed Bat

**THIS SPECIES IS WIDESPREAD** and often common in the Neotropics, from southern Mexico to Paraguay. It also occurs in Trinidad and Tobago. The fur may be light gingery brown or grayer, with a paler underside; the bat seen here is covered in pollen. This species occurs in forests, farmland, gardens, and plantations. Roosts of up to 400 individuals gather in caves; it may also roost in hollow trees. The varied diet includes pollen, nectar, fruit, and (particularly in the rainy season) insects. Males attract a harem of females, and there is apparently no fixed breeding season. Some twenty different call types have been documented in this species, indicating sophisticated social behavior. Scent-marking allows individual recognition and builds close social bonds; females—and even the harem male—may care for a pup when its mother goes foraging.

**IUCN STATUS** Least Concern
**LENGTH** 3¼in (8.4cm)
**WEIGHT** 1¼oz (35–38g)

***Phyllostomus elongatus***

## Lesser Spear-nosed Bat

**THIS BAT OCCURS** over much of north and central South America. It is large and uniformly dark, with a long, sharply pointed nose-leaf atop a block-shaped snout, and pointed ears. There is a white spot at the wingtip. Most records are from dense, humid, tropical forest in the lowlands, but it also occurs around streams in drier forest. The roost sites are mainly tree hollows, with larger aggregations sometimes forming in caves. Within the roost, the bats live in small harem groups of about ten individuals; these remain stable over months or years. Males without harems form bachelor groups of up to fifty individuals. This species takes fruit within the forest, but also hawks for insects over the canopy, visiting flowers for pollen or nectar. Plant-derived foods are more important in the dry season.

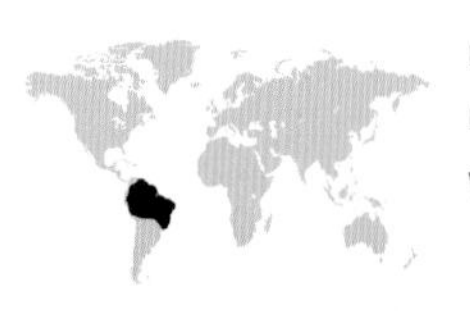

**IUCN STATUS** Least Concern
**LENGTH** 3⅞–4½in (9.9–11.5cm)
**WEIGHT** 1–1½oz (30–44g)

***Gardnerycteris crenulatum***

## Striped Hairy-nosed Bat

**THIS BAT'S MOST STRIKING** feature is a very elongated spear-shaped nose-leaf, covered in hair—a feature unique to this species; the ears are large and broad. Its dense fur is variable in color, with a blackish face and brown body that includes some yellow or orange tones. There are usually pale patches behind the ears, and a pale stripe along the length of the back. This species occurs in south Mexico and the Caribbean side of Central America, extending to Panama and through the north and central regions of South America; it is also found on Trinidad. It lives in lowland forest, roosting in tree hollows in small groups, and hunts by picking insects, and occasionally small invertebrates, off foliage. There is some evidence that pairs hunt together, and that parental care may be protracted (up to nine months).

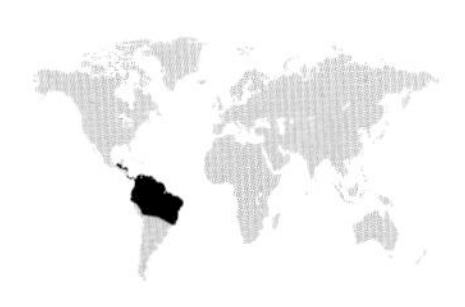

**IUCN STATUS** Least Concern
**LENGTH** 2–3in (5–7.5cm)
**WEIGHT** ⅜–½oz (12–12.8g)

***Lophostoma schulzi***

## Schultz's Round-eared Bat

**THIS BAT OCCURS** from the west half of Guyana through Suriname and French Guiana into north Brazil. It is a dark-furred, fairly large *Lophostoma*. Its bare parts are dark brown; uniquely among bats, they have a rough, grainy texture, as if covered with numerous tiny warts. It is present in primary forest, both dry and wet, and also in secondary growth. An insectivore, it hunts by gleaning prey from foliage. The species is very little known. It seems to be rather rare and is probably declining, as deforestation is occurring in its range. The Guiana Shield region is also affected by invasive, non-native plants. As the largest area of undisturbed tropical rainforest in the world, this region has exceptionally high biodiversity, and requires more extensive and rigorous protection.

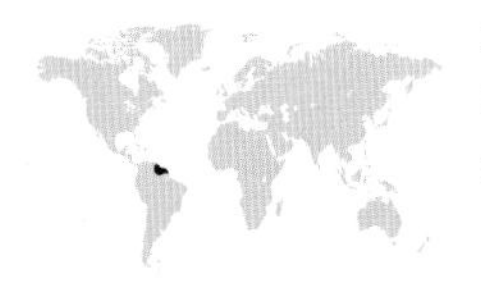

**IUCN STATUS** Least Concern
**LENGTH** unknown
**WEIGHT** unknown

***Lophostoma brasiliense***

## Pygmy Round-eared Bat

**THIS IS THE SMALLEST** *Lophostoma* species. Its wide range runs from the south of Mexico through Central America to north, central, and eastern South America, and also Trinidad. It has light brown fur with a slightly darker brown face, ears, and membranes. The snout is short with a sharply pointed, wide nose-leaf. The ears are large and broad, the lower jaw protrudes, and the eyes are small. It occurs around streams and in wet areas of primary and secondary forest, and also around cultivation. This bat roosts in small groups; roosting sites include both abandoned and occupied termite nests in tree branches, which it accesses by chewing a round hole in the base. It is insectivorous with a slow, agile flight, emerging just after sunset to hunt, and mainly taking its prey by gleaning.

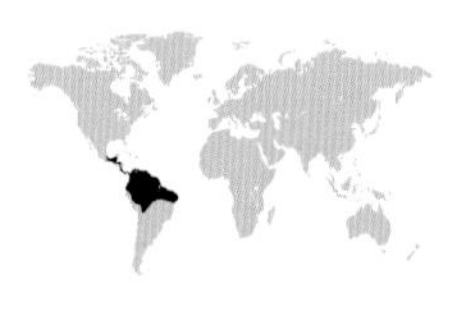

**IUCN STATUS** Least Concern
**LENGTH** 1⅞–2¾in (4.7–7.1cm)
**WEIGHT** ⅜oz (7–13g)

***Lophostoma silvicolum***

## White-throated Round-eared Bat

**THIS BAT IS FOUND** from Honduras and Guatemala through Central America, and broadly through tropical northern and eastern South America down to Paraguay and the east tip of Brazil. A medium-sized *Lophostoma*, it has light brown fur, whitish on the throat, and a pinkish face, with dark ears and membranes. It occurs in primary and secondary forest, also visiting cultivated land. Like some other *Lophostoma* species, it uses arboreal termite nests as roosts, accessing them by chewing a hole at the base. It hunts mainly at low levels within the forest's understory, picking prey from foliage; it is attracted by insects' sounds, such as the courtship calls of katydids. Some fruit and pollen are taken in the dry season. It is often common, though some populations may warrant separation as full species.

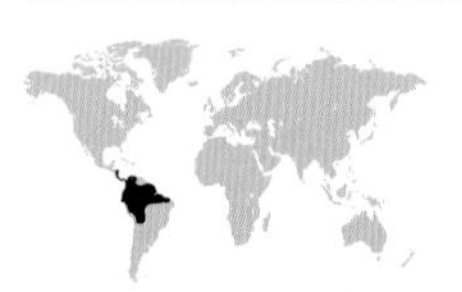

**IUCN STATUS** Least Concern
**LENGTH** 2–2¾in (5–7cm)
**WEIGHT** ½–¾oz (15–23g)

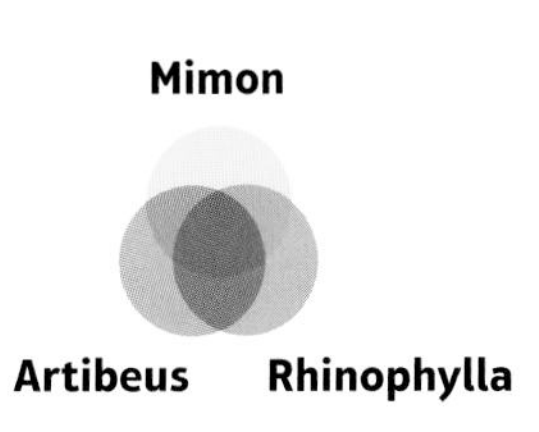

## Mimon

Until recently, there were four species within the genus *Mimon*, but two have been separated into the genus *Gardnerycteris*, leaving just *M. bennettii* and *M. cozumelae*. These two species, the "golden bats," are very similar, and were formerly considered conspecific. They have richly colored chestnut-golden fur, large, pointed-tipped ears, and short snouts with very long, slender, upright nose-leaves. *M. cozumelae* ranges from Mexico south to Colombia, while *M. bennettii* replaces it further south, reaching from Colombia across north Venezuela to the Guianas, and down to southeast Brazil. While their biology is poorly known, their anatomy suggests they are gleaning insectivores.

## Rhinophylla

The *Rhinophylla* bats are known as the "little fruit bats." They are indeed small bats, with the simple ear structure and broad, medium-length snouts that characterize the frugivorous phyllostomids. They take a little pollen as well as a wide range of fruits. There are just three species, of which *Rhinophylla pumilio* (see page 134) is best known. The others are *R. alethina*, a rather rare species confined to coastal Pacific Colombia and adjacent Ecuador, and *R. fischerae*, which is widespread in tropical South America.

## Artibeus

With 23 species, *Artibeus* is one of the largest genera within Phyllostomidae. They can be found in Mexico, Central and South America, and parts of the Caribbean. These are small to medium-sized, fruit-eating bats, which typically have short faces, largish pointed ears, and a white line below and above each eye. However, there is considerable variation in how prominent these markings are. Several of the *Artibeus* species construct leaf-tent roosts; they tend to live in harems of one male with several females. The species described in detail are: *A. amplus* (see page 135), *A. cinereus* (page 136), *A. hirsutus* (page 137), *A. jamaicensis* (page 138), *A. lituratus* (page 139), *A. obscurus* (page 135), *A. phaeotis* (page 142) and *A. toltecus* (page 141) (both formerly of the *Dermanura* genus), and *A. watsoni* (page 140).

Of the other recognized species, *A. aequatorialis* is of South America and *A. ravus* found in Colombia and northwest Ecuador. *A. concolor*, a distinctive species, is sometimes separated into its own genus (*Koopmania*). It has light brown fur without obvious facial stripes, and black wing membranes with contrasting pinkish digits. It has a large range in tropical South America. *A. fimbriatus* is a common but poorly known species in south Brazil, Paraguay, and Argentina. *A. fraterculus* occurs in west Ecuador and northwest Peru, where it is common and sometimes considered a crop pest, visiting plantations in open, dry

countryside. *A. planirostris* is common across most of South America, reaching the north of Argentina. *A. schwartzi* and *A. inopinatus* are both little known; the former is present in the Lesser Antilles and Trinidad, while the latter is found on the Pacific side of El Salvador, Honduras, and Nicaragua. There are a further six small *Artibeus* species, which until recently were classed as part of the *Dermanura* genus. These are *A. anderseni* (found in the north of South America), *A. aztecus* (central Mexico to Panama), *A. bogotensis* (the north of South America), *A. glaucus* (Colombia south to Bolivia), *A. gnomus* (Peru to French Guiana), and *A. rosenbergi* (west Colombia to northwest Ecuador).

## Chiroderma

The six *Chiroderma* bats are rather similar to the *Artibeus* species, being short-faced fruit-eaters with often prominent white facial stripes. The broad, fleshy nose-leaf is often pale-edged. These pale markings may provide disruptive camouflage—making the outline of the head less obvious while the bat is roosting in patchy sunlight. Unlike *Artibeus*, *Chiroderma* species are not known to construct tent roosts; instead they roost in clusters within foliage high in the forest trees where they live.

The newly described Brazilian species *C. vizottoi* is poorly known, while *C. improvisum* of Guadeloupe and Montserrat, in the Lesser Antilles, is endangered. This species was presumed extinct after a nineteen-year period of no observations, which followed volcanic activity as well as some severe weather events. However, it was rediscovered in 2005, though no conservation measures have yet been implemented to protect the small surviving population. The other *Chiroderma* species are: *C. doriae*, which occurs in southeast Brazil and Paraguay, and *C. trinitatum*, which is found in lowland forests from Costa Rica south to Amazonian Brazil, the Guianas, Bolivia, Ecuador, Peru, and also Trinidad. The widespread *C. villosum* is described on page 144 as is *C. salvini*.

## Platyrrhinus

*Platyrrhinus* is a large genus, holding twenty-one species. These forest-dwelling bats are similar to *Artibeus* in appearance, with white facial stripes and sometimes an additional white stripe, running from the crown of the head down toward the back. They are sometimes known as the "broad-nosed bats," reflecting both their short, robust snout shape and their wide nose-leaves. As a genus they are very widespread in South America, and some are present in Central America.

Three species are described in detail: *P. fusciventris* and *P. lineatus* on page 145, and *P. helleri* on page 146. The remaining species are *P. albericoi* (the east Andes, in Ecuador, Peru, and Bolivia), *P. angustirostris* (Colombia, Venezuela, east Ecuador, and north Peru),

**Uroderma**

*P. aurarius* (east Venezuela, Guyana, and Suriname), *P. brachycephalus* (widespread in Amazonian South America), *P. dorsalis* (Andean Colombia and Ecuador), *P. incarum* (the north of South America), *P. infuscus* (Colombia, Peru, Bolivia, and northwest Brazil), *P. masu* (east Andes, in Peru and Bolivia), *P. nigellus* (Andes, from Venezuela south to Bolivia), *P. recifinus* (south and east Brazil), and *P. vittatus* (from Costa Rica south to north Colombia and Venezuela).

*P. aquilus* (Panama), as well as *P. nitelinea* and *P. umbratus* (both found in west Colombia and Ecuador), are all recently described species and little known. *P. guianensis* of the Guyanas was only described in 2014, and has no IUCN assessment as yet. *P. chocoensis* has a broad distribution from Panama south along the west of South America to Ecuador. It is thought to have declined steeply because of habitat loss. *P. ismaeli* (Andean Colombia, Ecuador, and Peru) and *P. matapalensis* (northwest Peru and west Ecuador) are considered threatened because of their reliance on the rapidly disappearing primary forest in their ranges.

***Platyrrhinus vittatus***

## Uroderma

The bats of the genus *Uroderma* include the best-known tent-making species, *U. bilobatum*, described on page 147. Like *Artibeus* and *Platyrrhinus* they are "white-lined bats," with pale or white facial stripes to provide disruptive camouflage; they have short, blunt snouts with broad nose-leaves. Many authorities, including the IUCN, recognize only two species: *U. bilobatum* and *U. magnirostrum*, which occur from Mexico through Central America and across South America, reaching as far as south Bolivia and south Brazil. This book recognizes a further three species: *U. bakeri*, described in 2014 from Colombia and northwest Venezuela, *U. convexum*, and *U. davisi*—recently split from *U. bilobatum* and occurring in Panama and further north respectively.

***Tonatia saurophilia***

## Stripe-headed Round-eared Bat

**THIS BAT OCCURS IN** Central America through to northern Brazil, reaching eastern Peru and northern Bolivia in the west. It has dusky gray-brown fur with a distinct pale forehead stripe. The face is blunt with small eyes, a pink-tinted nose-leaf, and pronounced warty lower lip. The ears are large, broad and pointed with delicate tragi. It occurs in forest (primary and secondary) and forages in the mid-story and often over streams; it also hunts over pastures. It is an insectivore, preying on beetles, katydids and bugs, and also takes some small verterbrates, as well as some fruit. Two birth peaks occur each year, and females can bear twins. Much of its range is affected by deforestation.

**IUCN STATUS** Least Concern
**LENGTH** unknown
**WEIGHT** unknown

***Rhinophylla pumilio***

## Dwarf Little Fruit Bat

**THIS SMALL, DAINTY BAT** is widespread in tropical South America. It has thick, uniform, golden-brown fur. The snout is slim, with a broad-based nose-leaf and a warty bottom lip; the ears are delicate-looking and round-topped. It occurs in and around dense evergreen forest, but rarely in deciduous forest. Its tent roosts, usually well above ground level, are occupied by one male with a harem of two or three females. They return to roosts to eat during the night, as well as sleep by day, switching roosts every few days. The diet mainly comprises small fruits from lower storys of the forest; they also feed on pollen and nectar. This common species is an important seed disperser. Squirrel monkeys prey on these bats, attempting to approach the leaf tents without causing vibrations.

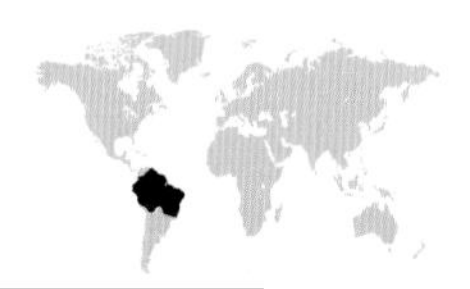

**IUCN STATUS** Least Concern
**LENGTH** 1¾–2¼in (4.3–5.8cm)
**WEIGHT** ¼–½oz (7–13g)

***Artibeus amplus***

## Large Fruit-eating Bat

**THIS BAT, DESCRIBED** to science in the 1980s, occurs across northern South America. One of the largest *Artibeus* bats, it has dark brown fur with pale stripes above and below the eye, giving it a dark-masked appearance. It is present in both lowlands and uplands, and roosts in caves, preferring smaller ones. It forages in evergreen forest habitats, especially around streams, where it feeds mainly on fruit, especially figs. It is also likely to prey on insects and to feed on pollen and nectar. A little-known species, it closely resembles *A. jamaicensis* (also present in Venezuela), but can be distinguished from it by various anatomical features, including a longer, narrower skull and different nose-leaf shape. Deforestation and disturbance of cave roosts are active threats in its range, but the extent to which they are affecting the species is not known.

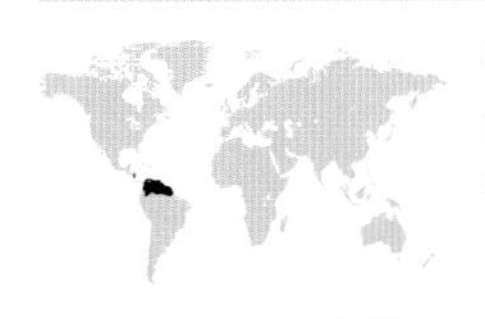

**IUCN STATUS** Least Concern
**LENGTH** 3½–3⅝in (9–9.1cm)
**WEIGHT** 2⅛oz (60g)

***Artibeus obscurus***

## Dark Fruit-eating Bat

**THIS BAT IS PRESENT** over much of north, east, and central South America. It is a large and uniformly dark *Artibeus* with a short, broad snout. The wing membranes are pink-tinted, and it has large, prominent teeth to deal with hard fruits. It occurs primarily in lowland, moist, dense forest, but is adaptable and can thrive in modified habitats including plantations and gardens. The species may also be encountered in more open and dry habitats such as farmland and meadows at times. A strong flier, its large home range makes it an important seed disperser. It picks and carries fruits (especially figs) to a low-level night roost to consume them. It probably feeds on some insects in addition to fruit. Day roosts include spaces beneath loose bark on tree trunks, well above ground level.

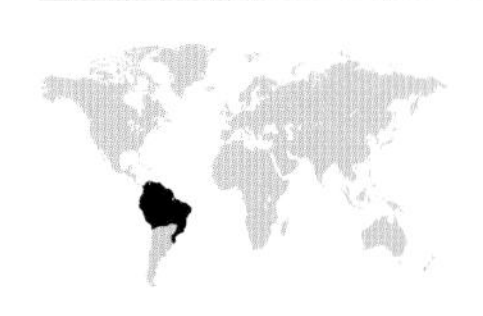

**IUCN STATUS** Least Concern
**LENGTH** 2¾–2⅞in (7–7.3cm)
**WEIGHT** 1–1¼oz (29–35g)

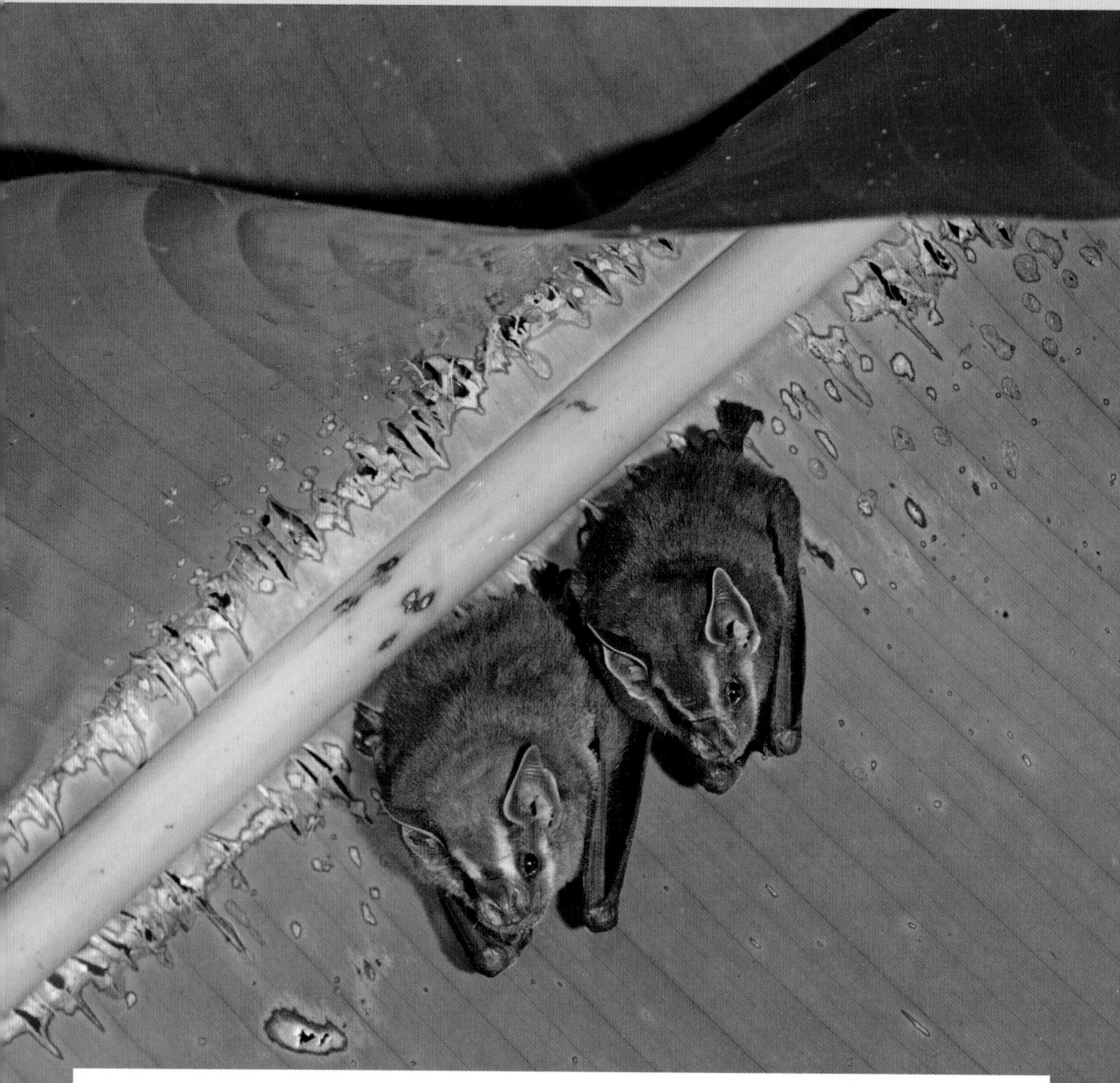

***Artibeus cinereus***

# Gervais's Fruit-eating Bat

**IUCN STATUS** Least Concern
**LENGTH** 2⅛in (5.5cm)
**WEIGHT** ⅜–½oz (10–13g)

**THIS BAT OCCURS** in the north and northeast of South America. It is a small *Artibeus*, with a long and slightly forward-curving nose-leaf, pointed ears with a yellow edge and prominent white stripes above and below each eye. It is found in mature forests, but also occurs in secondary growth and sometimes plantations. It roosts on the undersides of leaves, and will construct several different kinds of leaf-tent roosts, which it may "time-share" with another phyllostomid bat, *Mesophylla macconnelli*. Groups usually consist of a male and several females, but pictured here is a mother and her pup. It is primarily a fruit-eater, but almost certainly consumes some insects. Although its preferred habitat is threatened in many areas, the species can tolerate some degree of habitat disturbance.

***Artibeus hirsutus***

# Hairy Fruit-eating Bat

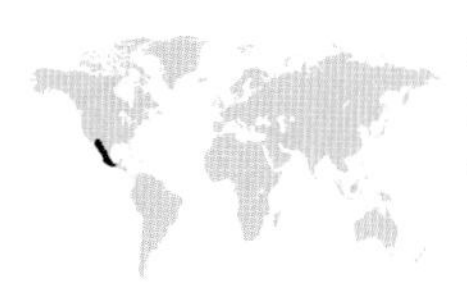

**IUCN STATUS** Least Concern
**LENGTH** 2¾–3⅛in (7–8cm)
**WEIGHT** 1⅛–1⅝oz (32–47g)

**THIS SPECIES IS ENDEMIC** to Mexico, occurring on the Pacific seaboard. It is stout with dense, warm-brown fur and grayish wing membranes; it has no tail. There are two pale, vertical stripes from the eyes up to the base of the ears. It has a flattish muzzle with a small nose-leaf, and small ears. It occurs in uplands and lowlands, and both arid and well-vegetated habitats. It forages around ponds and streams, and in orchards. Day roosts have been discovered in abandoned mines, caves, inside buildings, and among boulders. By night it may rest on the undersides of large leaves. It is primarily a fruit-eater (seen here carrying a fig) but also preys on some insects. This bat was previously classified as Vulnerable as it has lost up to 13 percent of its habitat over the last decade.

***Artibeus jamaicensis***

# Jamaican Fruit-eating Bat

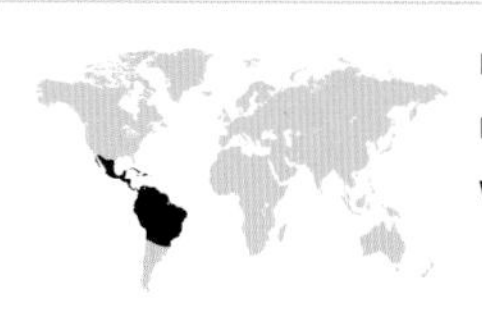

**IUCN STATUS** Least Concern
**LENGTH** 3⅛–3½in (7.8–8.9cm)
**WEIGHT** 1⅜–2⅛oz (40–60g)

THIS BAT, THE BEST-KNOWN *Artibeus*, occurs from south Mexico through central America into north Colombia and Venezuela, and also parts of the Caribbean. It is a typical *Artibeus* bat with faint, pale facial stripes. It occurs in warm, humid forests, up to 7,020ft (2,140m) in elevation. Recorded day roosts include caves and buildings; night roosts are often within foliage. It feeds on mangos, figs, and other non-citrus fruits, such as this sapodilla fruit, as well as pollen, nectar, and insects—it can carry fruits more than half its own weight. Males are territorial at their roosts, the best territories attracting more females. Subordinate males (often sons of the territory holder) are tolerated and occasionally mate with harem females. Two pups are born each year, during the dry season.

***Artibeus lituratus***

# Great Fruit-eating Bat

**IUCN STATUS** Least Concern
**LENGTH** 3⅜–4in (8.7–10.1cm)
**WEIGHT** 1⅞–2⅝oz (53–73g)

**THIS BAT'S EXTENSIVE** range stretches from central Mexico south through Central America and most of the northern half of South America; it reaches Trinidad and Tobago as well as some other Caribbean islands. It has white stripes on its forehead and cheeks; it has a wide mouth, and the lower lip has prominent warts. The ears are short and wide with a distinct yellow tragus. This species is a forest-dweller and feeds mainly on fruit, both wild and cultivated. Strong for its size, it has large, sharp teeth that allow it to carry sizeable fruits. This bat is an important seed disperser for wild plants such as the tree *Manilkara bidentata* in Trinidad. Females produce a single pup after just over two months' gestation, and breed twice a year. This species is common over most of its range, though threatened locally by deforestation.

***Artibeus watsoni***

# Thomas's Fruit-eating Bat

**IUCN STATUS** Least Concern
**LENGTH** 2–2¼in (5–5.8cm)
**WEIGHT** ⅜–½oz (9–15g)

ANOTHER SMALL ARTIBEUS, this bat is distributed from the far south of Mexico through Central America (primarily the Atlantic side) and possibly down to northwest Colombia and extreme northwest Ecuador. It has rather long, fluffy, light brown or gray-brown fur, prominent white facial stripes, and largish ears that sometimes show a pale creamy edge to the bases. It prefers tall, humid forest, but also uses secondary growth, plantations, and larger gardens. It constructs leaf-tent roosts in various styles (depending on leaf type), and has not been found roosting anywhere but in such tents. The tents are easily found in suitable habitat and may be located just a few feet above ground level. Like other *Artibeus* bats it mainly feeds on fruits, including *Cecropia*, a pioneer plant of lowland forest.

***Artibeus toltecus***

# Toltec Fruit-eating Bat

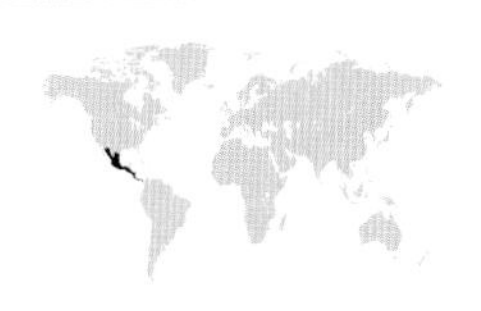

**IUCN STATUS** Least Concern
**LENGTH** 2⅜–2½in (5.9–6.5cm)
**WEIGHT** ½–¾oz (14–20g)

**THIS BAT IS FOUND** in west and east Mexico and through parts of Central America to Panama. It occurs in evergreen forests at mid altitudes, mainly 2,000–5,000ft (600–1,500m), though up to 6,500ft (2,000m) in places, and also visits fruit plantations and woodland edges. It roosts in caves, tree-trunk crevices, and sometimes in buildings; it also constructs leaf tents from banana leaves. The diet includes some insects and pollen, but is mainly of fruit, especially figs; at least fourteen other fruit types are eaten, including the important reforesting pioneer plant *Solanum umbellatum* (seen below). In Costa Rica, and perhaps elsewhere, this bat appears to have two breeding seasons a year, with most young born in either late spring or early fall. These are times of peak production of different fruits.

***Artibeus phaeotis***

# Pygmy Fruit-eating Bat

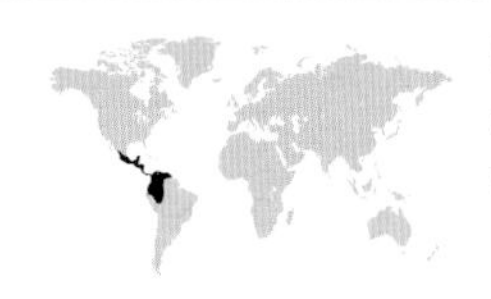

**IUCN STATUS** Least Concern
**LENGTH** 2–2½in (5.1–6cm)
**WEIGHT** ¼–½oz (8–15.6g)

**THIS BAT IS FOUND** from west and south Mexico through Central America into northwestern South America, reaching Ecuador in the west and Suriname in the east. It has mid-gray-brown fur, blackish membranes, and prominent white facial stripes. The ears have a white edge, and the nose-leaf is relatively broad. This is a lowland forest species, occurring only rarely in the uplands, but occasionally up to 4,000ft (1,200m). It forages within the mid-story, flying with considerable agility between plants. It mainly feeds on fruits, but also some insects, and flowers for nectar and pollen. This is a tent-making species that usually roosts in small, clustered groups; it has also been observed roosting within caves. Little is known of its reproductive cycle, but pregnant females have been noted through late summer into early winter. Seen here are a mother and her pup.

***Chiroderma villosum***

## Hairy Big-eyed Bat

**THIS BAT OCCURS FROM** Mexico through most of Central America and South America, south to Paraguay, and also on Trinidad and Tobago. It has drab gray fur, small, yellowish funnel-shaped ears, and a wide, short snout. The nose-leaf has a distinctive notch at the tip unlike similar genera and there are vague pale stripes above the large brown eyes. It inhabits lowland, dense, evergreen rainforest, foraging around the canopy and roosting in tree hollows. The diet is primarily fruit, especially figs. Unlike most phyllostomids, *Chiroderma* bats chew and digest the seeds of the fruit, deriving extra nutrition from them. This "cheating" behavior relies on other bat species that excrete the seeds whole and serve as effective seed dispersers. Without them, there would be strong selective pressure on the plants to evolve defenses against the "cheats."

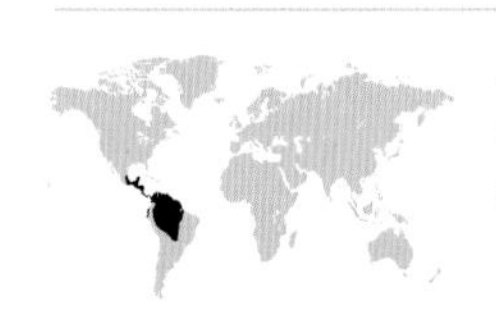

**IUCN STATUS** Least Concern
**LENGTH** 2½–2⅝in (6.5–6.8cm)
**WEIGHT** ¾oz (21–22.9g)

***Chiroderma salvini***

## Salvin's Big-eyed Bat

**THIS SPECIES OCCURS FROM** west Mexico patchily through Central America into north and northwestern South America, reaching north Venezuela and south to Bolivia; it is almost absent from Brazil. It is a medium-sized phyllostomid with rather gray-brown fur. The eyes are large, the snout short, the ears rounded, and the nose-leaf rather tall. Like *C. villosum* there is a slight notch in the nose-leaf but it is easily recognizable by its white facial stripes. This bat is associated with several habitat types, occurring mainly in forest with still or flowing water, but also plantations, more open savanna, and semi-desert with streams. It is likely to roost in tree foliage. A fruit-eater, it appears to forage mainly in the canopy and subcanopy. Although often scarce the species has a wide distribution, and is adaptable enough not yet to make habitat loss a significant threat.

**IUCN STATUS** Least Concern
**LENGTH** 2⅝–3½in (6.6–8.8cm)
**WEIGHT** 1–1¼oz (30–36g)

***Platyrrhinus fusciventris***

## Brown-bellied Broad-nosed Bat

**THIS BAT, ONE OF SEVERAL** forms recently split from *P. helleri* (a species complex), is found in northeastern South America, including Trinidad and Tobago. It is pale, with a ginger tint to its fur, and broad but diffuse white stripes above and below the eye. The head is typical of the fruit-eating *Platyrrhinus* genus, with a short, strong and broad muzzle, rather large eyes, and smallish, round-topped ears. The wide, frilly-based nose-leaf, and the ears, are yellowish around the edges. It favors thick, humid, lowland forest, and roosts in caves and crevices, tree holes, tunnels, hollow logs, and buildings. The species probably mostly feeds on figs, like its close relatives. It seems to be common in many areas and is present at a number of protected areas. Its taxonomy probably requires some further investigation.

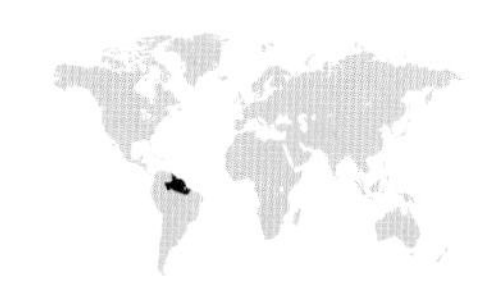

**IUCN STATUS** Least Concern
**LENGTH** unknown
**WEIGHT** unknown

***Platyrrhinus lineatus***

## White-lined Bat

**THIS SOUTH AMERICAN BAT'S** range extends from far east Brazil south to Uruguay, west across to Bolivia, then north in a narrow band through Peru, Ecuador, and Colombia to far west Venezuela. It is a boldly patterned bat with white facial lines above the eyes, pale edges to the ears, and a wavering white line running from the crown down the spine; the fur is otherwise dark gray-brown. It has a blunt snout with a broad nose-leaf. Found in forest habitats, it roosts within foliage and forages for fruit; it also feeds on some nectar, pollen, and insects. Its harem roosts hold seven to fifteen females and a single male. There are two birth peaks per year in some parts of the range, just one in others. It has a stable population.

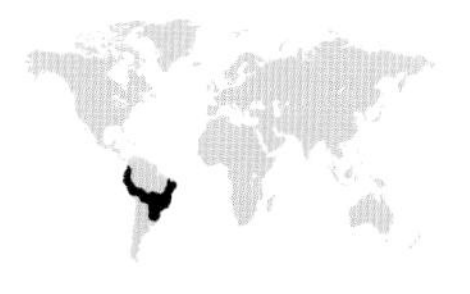

**IUCN STATUS** Least Concern
**LENGTH** 2¾in (7cm)
**WEIGHT** 1oz (26.7g)

***Platyrrhinus helleri***

# Heller's Broad-nosed Bat

**IUCN STATUS** Least Concern
**LENGTH** 2⅛in (5.8cm)
**WEIGHT** ½oz (15.6g)

THIS BAT OCCURS FROM south Mexico through most of Central America, then broadly across South America. Its taxonomy has undergone considerable revision, with several populations formerly assigned to *P. helleri* split as new species. It has light gray to reddish brown fur, with obvious white facial stripes above and below each eye. The wide nose-leaf is yellow at its base, as are the smallish, round-topped ears. The species is found in moist, tropical forest and is also a regular visitor to fruit plantations. It roosts among foliage as well as in tree hollows, possibly making leaf tents. It is frugivorous but occasionally feeds on insects. Females produce one pup at the start of the rainy season. Adults live together in small harems or bachelor roosts.

***Uroderma bilobatum***

# Tent-making Bat

**IUCN STATUS** Least Concern
**LENGTH** 2⅜–2¾in (5.9–6.9cm)
**WEIGHT** ½–¾oz (13–20g)

THIS BAT OCCURS IN Central America and the northern two-thirds of South America, though it may be a species complex. It is an attractive bat with a well-marked face. It has white stripes above and below each eye and white edges on its smallish, pointed ears create a bold pattern. It occurs in all kinds of lowland forest, and can also be found in second growth woodland, plantations, and clearings.

The best-known tent-making species, it constructs shelters in various styles from palm and other leaves, and is not known to use any other roost. Its pale facial stripes help provide camouflage as it roosts, in groups of usually fewer than ten, but occasionally more than fifty. The diet is mainly fruit, with some insects and nectar. It is an important seed disperser in many habitat types.

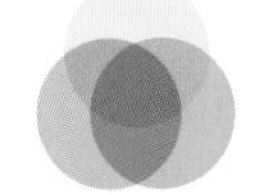

## Vampyressa & Vampyriscus

There are five species of bats in the genus *Vampyressa*, and another three in the closely related and similar *Vampyriscus*. Despite their names, they are not allied to vampire bats, but are similar to *Artibeus* and related genera in appearance; they are known collectively as the "yellow-eared bats." They are generally rather poorly known, and their taxonomy requires further revision. The species are *V. elisabethae* (a recent split from *V. melissa*, not assessed but present in west Panama), *V. melissa* (declining, found from central Colombia to central Peru), *V. sinchi* (another recent split from *V. melissa*, present in the Colombian Andes), *V. pusilla* (Argentina, Brazil and Paraguay), and *V. thyone* (Mexico to the north of South America). Those in *Vampyriscus* are *V. bidens* (widespread in northeast and central South America), *V. brocki* (from the Guianas west through north Brazil to north Peru), and *V. nymphaea* (from Nicaragua to west Ecuador).

***Vampyressa thyone***

## Vampyrodes

The genus *Vampyrodes* was formerly monotypic, but has recently been split, with *V. caraccioli* (see page 150) in the north of South America, and *V. major* in Mexico, extending through Central America to northwest Colombia. They are *Artibeus*-like frugivorous bats with pale facial stripes. These bats occur in dense, complex forest where they roost in small harem groups (usually one male with about three females) among foliage. They are also able to thrive in plantations and gardens.

## Sturnira

*Sturnira*, the "yellow-shouldered bats," or "American epauletted bats," is a large genus of twenty-three species, of which two are examined in more detail: *Sturnira lilium* (see page 152) and *Sturnira ludovici* (page 153). As a group, they range from Mexico across Central America and through most of the northern two-thirds of South America. The name "yellow-shouldered" relates to staining often observed around the shoulders of adult males, from secretions produced by glands in this area. The secretions produce a scent that attracts females. When not meeting to mate, these bats tend to be solitary. They are fruit-eaters, with small ears and wide nose-leaves that sit against their flat, broad faces.

**Sturnira**

The species *S. burtonlimi* (from Costa Rica and Panama, newly described), *S. koopmanhilli* (from west Ecuador and Colombia, newly described), *S. mistratensis* (from Andean Colombia, newly described), *S. perla* (from Ecuador, newly described), and *S. sorianoi* (from Venezuela and Bolivia, newly described) are all little known. New species are *S. adrianae* (a new species from north and west Venezuela and north Colombia), *S. angeli* (a split from *S. lilium*, found in the Windward Isles), *S. bakeri* (from southwest Ecuador, newly described), *S. parvidens* (another split from *S. lilium*, found in Mexico and Central America), and *S. paulsoni* (also split from *S. lilium*, found on St. Vincent and the Grenadines).

Two species in the genus are at risk of extinction: *S. mordax* and *S. nana*. The former species occurs in Costa Rica and Panama; its forest habitat is under threat. The latter occurs patchily in Ecuador and Peru, with a very small known area of occupancy. The forests in which it occurs have no protection, and are constantly being encroached upon by human activity.

The remaining species in the genus are: *S. aratathomasi* (southwest Colombia), *S. bidens* (from west Venezuela, reaching south through west Colombia and Ecuador to Peru), *S. bogotensis* (a similar range to *S. bidens*, but coastal only in Peru), *S. erythromos* (from northeast Venezuela running southwest through west Colombia, Ecuador, Peru, and into northwest Chile and Argentina), *S. hondurensis* (Mexico, Guatemala, Honduras, and Nicaragua), *S. luisi* (Costa Rica through Panama, then south along the South American coast to northwest Peru), *S. magna* (south Colombia, Ecuador, Peru, and northwest Bolivia), *S. oporaphilum* (Peru, then through Bolivia to north Argentina), and *S. tildae* (widespread over the north of South America, including Trinidad).

***Sturnira tildae***

***Vampyrodes caraccioli***

# Great Stripe-faced Bat

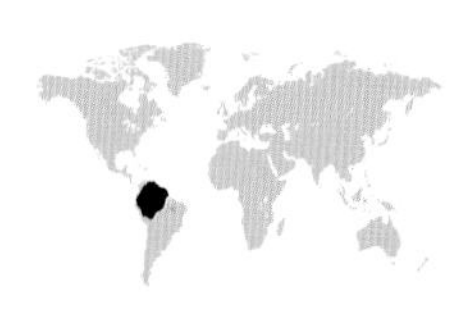

**IUCN STATUS** Least Concern
**LENGTH** 3–3½in (7.7–8.9cm)
**WEIGHT** 1–1⅝oz (30–47g)

**THIS SPECIES IS FOUND** in northern South America—from east Colombia, Venezuela, the Guianas, and northwest and central Brazil to east Peru and north Bolivia. The only species in its genus, it resembles and is related to several other genera, such as *Artibeus*, but is larger than most similar species. It has mid-gray-brown fur with dark wing membranes, and contrasting pale arms and fingers. The face bears broad, pale stripes above the eyes and narrower ones below; it has ears with yellow edging and a short, wide nose-leaf. This bat occurs in tropical, evergreen forests, also plantations and larger gardens. It roosts within tree foliage under the canopy, usually in small harems of one male and also in three or four females. It mainly forages at about 10ft (3m) above ground, searching for figs and other fruit, which it plucks and consumes elsewhere. It also takes nectar, pollen, and insects. It most likely has two breeding peaks each year, as do most tropical-forest, fruit-eating phyllostomids. The species is not significantly threatened, although habitat loss from deforestation is a problem in some areas.

*Sturnira lilium*

# Little Yellow-shouldered Bat

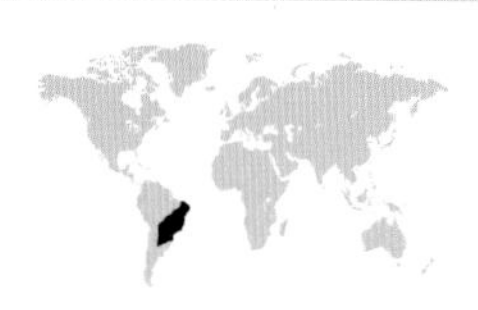

**IUCN STATUS** Least Concern
**LENGTH** 2½in (6.2–6.5cm)
**WEIGHT** ½–⅝oz (13–18g)

THIS BAT'S RANGE EXTENDS from the far east of Brazil to Uruguay and north Argentina, and to east Bolivia. Some subspecies have been split as separate species, so its range is smaller than formerly thought. Males have scent glands on the shoulders; secretions can leave yellowish staining on their brown fur. This bat occurs in humid, lowland forest and more sparsely in uplands. It roosts in tree hollows, often alone, but also in groups of up to ten. The diet is mainly fruit, especially fruits from plants of the families Solanaceae, Piperaceae, and Cecropiaceae, and it is an important, efficient seed disperser. It also feeds on some nectar. Females give birth once a year, either in spring or fall, and carry their young on foraging expeditions, at least for the first few weeks

***Sturnira ludovici***

# Highland Yellow-shouldered Bat

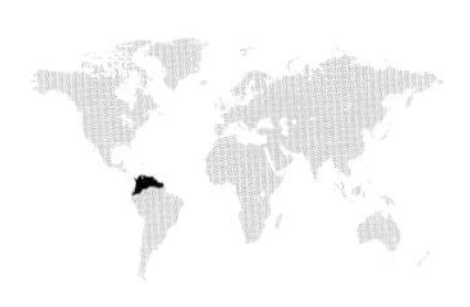

**IUCN STATUS** Least Concern
**LENGTH** 2⅝–2¾in (6.6–7cm)
**WEIGHT** ⅝–¾oz (17–23g)

**THIS SPECIES IS FOUND** in the far northwest of South America, from Venezuela to Colombia and Ecuador. It is a typical *Sturnira* bat with rather plain brown fur, slightly pale on the underside with a short, broad snout and rather small, broad-based nose-leaf. The shoulders of males are often stained yellow or orange from glandular secretions. It occurs in humid, evergreen forest, usually at above 1,600ft (500m), and in Andean oak forest. It is frugivorous, and disperses seed of pioneer plants such as *Solanum umbellatum* into clearings, allowing forest regeneration to begin. It is likely also to feed on pollen and nectar, and possibly insects. It roosts in tree hollows and lives in harems; the use of scent signals by males is likely to allow individual recognition and contribute to group stability. Loss of upland Andean forest is a potential threat to this rather poorly known bat.

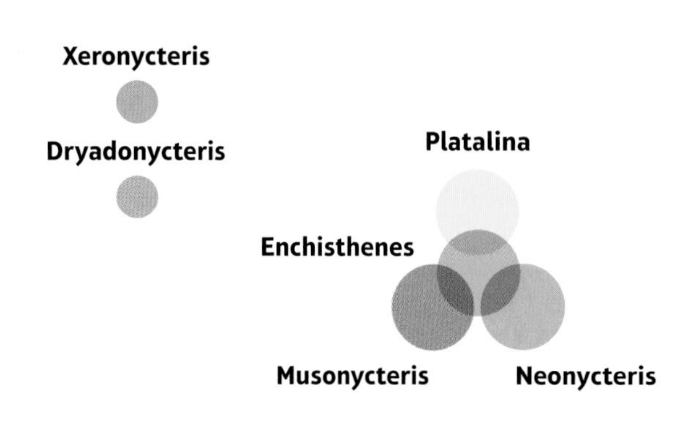

## Monotypic genera

Several monotypic genera are recognized within Phyllostomidae—among them some of the most distinctive and well-known bats in the family, and indeed the world. Nineteen of these species are covered in detail. They are: *Desmodus rotundus* (see page 156), *Diaemus youngi* (page 157), *Diphylla ecaudata* (page 157), *Choeronycteris mexicana* (page 158), *Scleronycteris ega* (page 159), *Hylonycteris underwoodi* (page 160), *Trinycteris nicefori* (page 161), *Lionycteris spurrelli* (page 159), *Lampronycteris brachyotis* (page 162), *Macrophyllum macrophyllum* (page 164), *Trachops cirrhosus* (page 163), *Phylloderma stenops* (page 164), *Chrotopterus auritus* (page 165), *Vampyrum spectrum* (page 165), *Ectophylla alba* (page 167), *Ametrida centurio* (page 168), *Centurio senex* (page 168), *Sphaeronycteris toxophyllum* (page 169), and *Mesophylla macconnelli* (page 169). The following eleven species are also classed within monotypic genera.

## Dryadonycteris

*Dryadonycteris capixaba* is a nectar-feeder; it was described in 2012 from specimens taken in two nature reserves (Reserva Natural Vale and the Floresta Nacional de Goytacazes) in lowland Atlantic forest in southeast Brazil. It resembles bats of genera such as *Glossophaga*, but its anatomy is distinct from all described genera.

## Musonycteris

*Musonycteris harrisoni* is a vulnerable species, found in thorny and deciduous forest in Mexico and under threat because of habitat loss. It has an extraordinarily elongated, slender snout, giving the impression of an anteater's head attached to a bat's body. The ears and nose-leaf are both rather small. This bat is adapted to take nectar from flowers with long corollas, and is an important pollinator of bananas and several other plants.

## Platalina & Xeronycteris

*Platalina genovensium* is another unusually long-nosed nectivorous species; it is found in west Peru and north Chile. Climate change and hunting for folk medicine are among the threats it faces. *Xeronycteris vieirai* is also nectivorous, and likely to be an important pollinator in the Brazilian *caatinga* habitat where it occurs. It is currently poorly known.

## Neonycteris

*Neonycteris pusilla* is a little-known species, found in northwest Brazil and east Colombia. Its large ears, short face, and tall nose-leaf indicate that it is an insectivore.

## Enchisthenes

*Enchisthenes hartii* is an *Artibeus*-like frugivore, with very dark fur and prominent white facial stripes. It was formerly classed as a member of *Artibeus*, but split on the basis of genetic

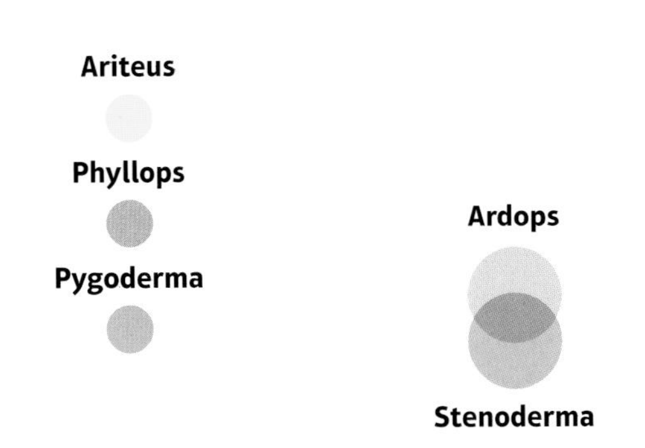

differences; its parotid (salivary) glands are also structurally different to those of any other phyllostomids. This bat's distribution extends from Mexico through much of Central America into the north and west of South America, close to the coast.

## Ardops & Ariteus

*Ardops nichollsi* is an unusual-looking, fruit-eating bat. It has an almost teddy-bear-like appearance due to its thick, gray-brown fur and very wide, round head. The snout is short with a wide nose-leaf, the ears small with pointed tips, and the smallish eyes are light brown. It has white shoulder spots. This species is believed to roost hanging from branches, rather than in shelters of any kind, but it is not well known. It occurs in the Lesser Antilles. *Ariteus flavescens* is rather similar in appearance. It has pale yellowish fur and apparently similar roosting habits. This bat is found on Jamaica.

## Phyllops

Another short-faced, broad-headed fruit-eater, *Phyllops falcatus*, has a particular fondness for figs. It is rather rare and occurs on the Cayman Islands, Cuba, the Dominican Republic, and Haiti.

## Pygoderma

*Pygoderma bilabiatum* is a species that occurs in Bolivia, southeast Brazil, Paraguay, and north Argentina. It has a flat face with large, round ears and a long nose-leaf; the eyes are pale, as is the fur, and the bare parts have a strong yellowish tint. Its eyelids have a curious swollen appearance. It is a fruit-eater and occurs around habitations as well as in forests.

## Stenoderma

*Stenoderma rufum* is another fruit-eating bat; it suffers from decline and habitat loss on Puerto Rico, where it occurs. It is a pale reddish bat with white markings around its shoulders and a short face. The nose-leaf is long and has warts at its base.

***Ariteus flavescens***

***Desmodus rotundus***

# Common Vampire Bat

**IUCN STATUS** Least Concern
**LENGTH** 2¾–3½in (7–9cm)
**WEIGHT** ½–1¾oz (15–50g)

THIS SPECIES OCCURS FROM north Mexico through Central and South America to north Argentina and Chile. With long legs and elongated thumbs, it can run and jump easily on the ground when approaching prey—and make a quick, vertical take-off. It has large, sharp incisors and canines. It roosts mainly in small groups in hollow trees and caves in a wide variety of tropical and subtropical habitats. It depends on large mammals, particularly cattle or livestock, to obtain blood—the saliva contains anticoagulants, so the blood flows freely. At the roosts, territorial males defend groups of females. The females have close social ties, maintaining friendships with mutual grooming and reciprocal sharing of blood meals through regurgitation. This bat is overpopulated, has become a costly pest, and unfortunately requires control in ranching areas. Direct threats to humans, however, are minuscule.

***Diaemus youngi***

## White-winged Vampire Bat

**THIS SPECIES OCCURS** patchily in east and south Mexico, extending south through Central and South America, and reaching north Argentina. It is fairly large with silky-looking, light-brown fur and similar-colored bare parts; the wing's leading edge is white. The snout is short and nearly flat, with no obvious nose-leaf; the lower jaw projects a little further than the upper. It has widely spaced, triangular ears; its front incisors are long and sharp. It occurs in forests, forest edges, and plantations, roosting colonially in caves and tree holes. It feeds on blood, mainly from large birds; as with other vampire bats, the saliva contains anticlotting agents. Its large brain is probably an adaptation to the challenges of attacking its hosts undetected. It is sometimes persecuted, but poses no danger to people.

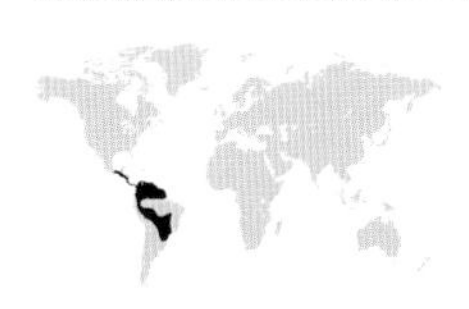

**IUCN STATUS** Least Concern
**LENGTH** 3¼in (8.3cm)
**WEIGHT** 1⅛oz (31–32g)

***Diphylla ecaudata***

## Hairy-legged Vampire Bat

**THIS VAMPIRE BAT IS PRESENT** in a wide variety of forested habitats from northern Mexico south through Central and South America to southeast Brazil. It has gray fur and pronounced skin folds around the nose and beneath the large black eyes. The flat, wide nose is sensitive to temperature, allowing the bat to find the best spot to bite its prey: where there is a rich blood supply close to the skin's surface. This widespread, but rather scarce species, hunts in forest and more open habitats. It roosts in groups of about ten in caves, derelict buildings, and tree holes. Roosting groups show a consistent social hierarchy. They usually take blood from large birds roosting in trees. The bat settles on the same branch's underside and feeds from the bird's foot for about a half hour.

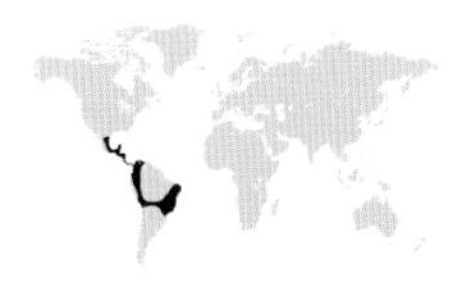

**IUCN STATUS** Least Concern
**LENGTH** 2⅞–3⅛in (7.3–7.9cm)
**WEIGHT** ⅞oz (25–26g)

*Choeronycteris mexicana*

# Mexican Long-tongued Bat

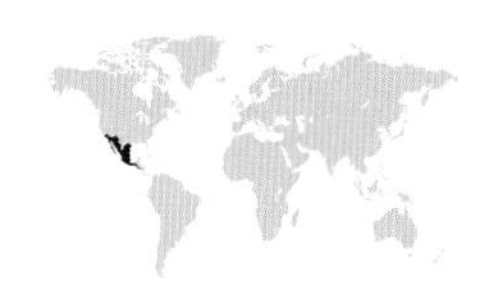

**IUCN STATUS** Near Threatened
**LENGTH** 3⅜in (8.5cm)
**WEIGHT** ¾–⅞oz (20–25g)

THIS BAT RANGES FROM the southern United States to Honduras and El Salvador. The muzzle is about 50 percent of the head's length and square-tipped. It has large, broad wings, so can hover when feeding from flowers. It roosts in underground caves in dry forest and thorn scrub, and feeds on nectar and pollen; it also takes fruits. With a tongue about one-third as long as its total body length, it can access nectar in flowers with corollas. It is an important pollinator for agaves (seen here) and some cacti; it also visits hummingbird feeders in gardens. It usually breeds once a year, sometimes twice in Central America. Populations in the north migrate south, directed by food availability. Declining and dependent on a fragile habitat, it is also threatened by disturbance at its roosts.

***Scleronycteris ega***

## Ega Long-tongued Bat

**THIS BAT IS KNOWN** from eastern Venezuela, the Guianas, and northwest Brazil. It is related to the genus *Glossophaga* and other nectar-feeding phyllostomids, but is distinctly smaller than most other nectar-feeders. It is mid-gray to dark-brown in color, with a rotund body and head. The ears are small and simple, the eyes smallish, and the snout very long and slender, with a small, pointed nose-leaf at the tip. The habitat is evergreen tropical forest with plenty of flowering plants. Its diet is assumed to be nectar and pollen, probably with some insects. This bat is little known; it has only been recorded with certainty from four locations (other records have proved erroneous, and some specimens are still unverified). It is probably declining, but at present no meaningful conservation actions can be taken beyond protecting suitable forests.

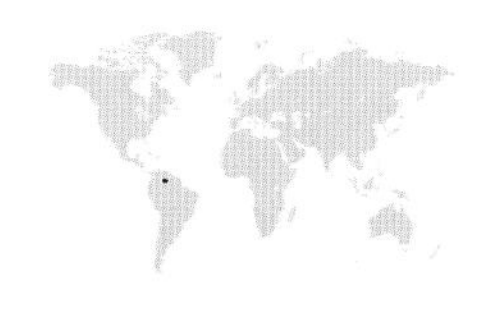

**IUCN STATUS** Data Deficient
**LENGTH** unknown
**WEIGHT** unknown

***Lionycteris spurrelli***

## Chestnut Long-tongued Bat

**THIS WIDESPREAD, LITTLE-KNOWN** species is the only member of the genus *Lionycteris*; it is probably a close relative of nectar bats in the genus *Lonchophylla*. It ranges from east Panama through a large swathe of north South America, reaching the north of Bolivia. It is rather a dark bat with reddish or brown fur, gray-black face, ears, and membranes with contrasting pink limbs; its snout has a small, forward-pointing, triangular nose-leaf. A habitat generalist with a varied diet of fruit, nectar, pollen, and insects, this bat may be encountered in forests, plantations, savanna, and gardens; it normally roosts in caves and crevices within rocky outcrops. It is common in at least some parts of its extensive range and is not considered threatened, partly due to its ability to thrive in varied habitats.

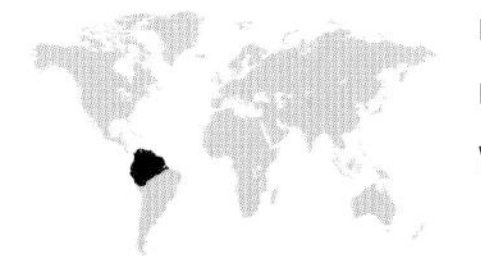

**IUCN STATUS** Least Concern
**LENGTH** unknown
**WEIGHT** unknown

***Hylonycteris underwoodi***

# Underwood's Long-tongued Bat

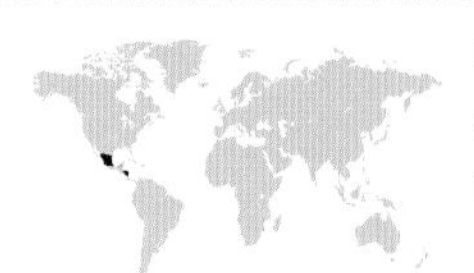

**IUCN STATUS** Least Concern
**LENGTH** 2⅛in (5.4cm)
**WEIGHT** ¼–⅜oz (6–9g)

THIS SMALL BAT OCCURS from the far south of Mexico through Guatemala to northwest Honduras, and separately in Costa Rica and Panama. It has dense, dark red-brown fur, small ears, and a long, slim snout, which gives its face a mole-like appearance. It mainly occurs in evergreen forests on steep ground, up to 8,661ft (2,640m), occasionally also deciduous forest. This bat roosts in small caves, rock crevices and tree hollows. It feeds mainly on pollen and nectar which it takes in flight, using its long tongue to probe hanging clusters of flowers with long corollas, such as this *Matisia ochrocalyx*; however, it also eats fruit, and is a significant seed disperser of certain forest plants. It also preys on some insects. The species may breed twice a year in the south, in both the dry and rainy seasons.

*Trinycteris nicefori*

# Niceforo's Big-eared Bat

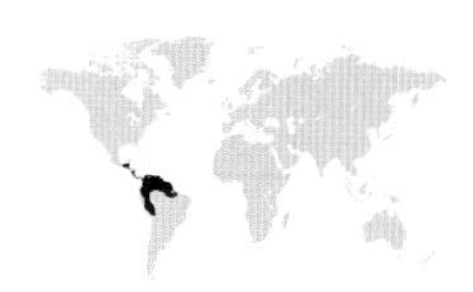

**IUCN STATUS** Least Concern
**LENGTH** 2–2¼in (5.1–5.8cm)
**WEIGHT** ¼–⅜oz (7–11g)

THIS BAT IS THE ONLY species in its genus, having been separated from the closely related genus *Micronycteris*. It occurs in Belize and through Central America to much of South America. It is rather small, with rufous fur and dark brown bare parts; there is the hint of a darkish eye-mask. It has a large, domed head, modest-length snout, and a short, upright nose-leaf. The ears are large, wide, and funnel shaped. It occurs in lowland forest and roosts in hollow trees or buildings. It emerges soon after sunset and is most active for the first hour of the night; activity peaks again just before dawn. The diet includes fruit, insects, pollen, and nectar. It appears to be rare, especially in Central America, but is also a species of conservation concern in Bolivia.

*Lampronycteris brachyotis*

# Yellow-throated Big-eared Bat

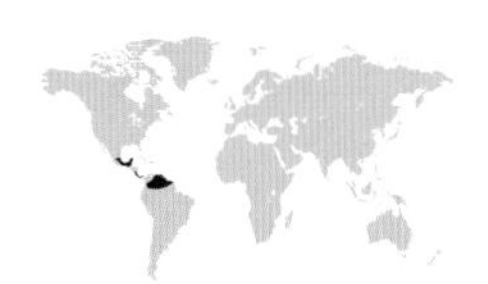

**IUCN STATUS** Least Concern
**LENGTH** 1⅞–2½in (4.8–6.2cm)
**WEIGHT** ⅜–½oz (12–14g)

**THIS BAT APPEARS** to have a rather patchy distribution in the Neotropics, being present in south Mexico and north and south Central America, and separately in northern South America from Venezuela to north Brazil. It is a small bat with a short but robust, squarish snout that projects almost at right angles from its steep forehead. The fur is light brown with a strikingly vivid, orange-yellow tint on the throat and chest. It has small eyes, large pointed ears, and a tall nose-leaf. This species is found in damp lowland forest (evergreen and deciduous) and roosts in caves, tree hollows, and similar shelters—often in harems comprising a male with up to nine females. Breeding mostly occurs at the start of the rainy season. It is rather rare, but not known to be threatened.

***Trachops cirrhosus***

# Fringe-lipped Bat

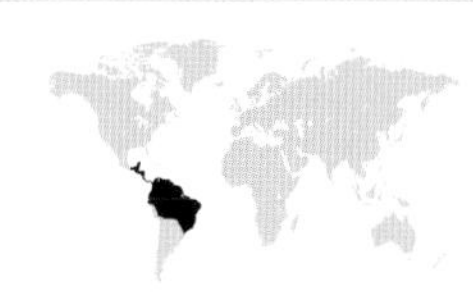

**IUCN STATUS** Least Concern
**LENGTH** 3–3½in (7.6–8.8cm)
**WEIGHT** 1⅛oz (32.3g)

THIS BAT, THE ONLY *Trachops* species, ranges from south Mexico through most of Central America, and across much of north, central, and western South America. It has broad wings and very large, broad-based ears; the snout is longish with a small, medium-sized nose-leaf used to direct its echolocation signals. Its upper lip has prominent warty bumps and bristles. A forest species, this bat prefers tropical, evergreen forest, but sometimes occurs in dry, deciduous forest near water. It roosts in small groups of about ten in caves, tree holes, and abandoned buildings. Sometimes larger maternity roosts form in deep caves. A versatile gleaning predator, it is also known as the Frog-eating or Lizard-eating Bat—individual bats can become specialist frog-hunters, even learning to differentiate between poisonous and edible frogs by their calls. It is fairly common in most of its range.

***Macrophyllum macrophyllum***

## Long-legged Bat

**THIS UNUSUAL *PHYLLOSTOMID*** has an extremely broad, continuous distribution from south Mexico throughout Central America and the northern half of South America, just reaching north Argentina. It has a long tail within its large uropatagium, and long legs. The fur is dark brown, the snout short, and the nose-leaf broad; the lower lip is studded with warts in a V-pattern. This species requires standing water for hunting, and is found in forests and forest clearings. Roosts are often in irrigation tunnels, as well as caves, ruined buildings, and tree hollows. The largest roosts hold up to seventy individuals. This bat appears to hunt for emerging insects, and possibly tadpoles, over water with a slow, fluttering flight, using the legs and uropatagium to sweep surface-dwelling insects toward its mouth; it also hawks for flying insects.

**IUCN STATUS** Least Concern
**LENGTH** 3⅛–4in (8–10cm)
**WEIGHT** ¼–⅜oz (7.5–9g)

***Phylloderma stenops***

## Pale-faced Bat

**THIS BAT, CLOSELY RELATED** to *Phyllostomus*, is sometimes placed in that genus. It occurs in south Mexico and patchily through Central America, then across north, central, and eastern South America. It is sturdy, with light gray-brown fur and light gray-pink face, ears, and limbs; the membranes are darker gray. It has triangular ears, a large, wide snout with a deep chin, and a broad-based, forward-angled nose-leaf. The male has a glandular throat-sac. The species occurs in dense and complex damp, tropical forest; it has also been observed in forest clearings and other moderately disturbed habitats. It often hunts around swamps and marshes. The diet includes both fruit and insects, and even small vertebrate; also nectar and pollen. It will tackle active wasps' nests. Its roosting behavior is not documented; its breeding biology is also unstudied.

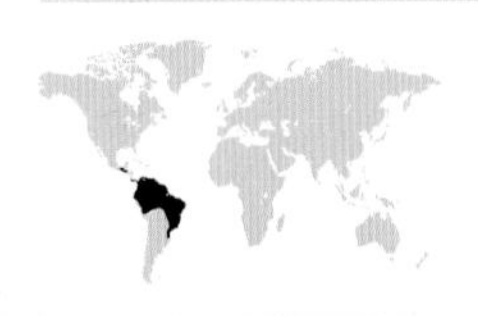

**IUCN STATUS** Least Concern
**LENGTH** 3⅜–4½in (8.7–11.5cm)
**WEIGHT** 1¾–2¼oz (51–64g)

***Chrotopterus auritus***

## Woolly False Vampire Bat

**THIS LARGE, DISTINCTIVE BAT** ranges from Mexico south through Central America to South America, reaching the far north of Argentina. It has dense, light gray-brown fur; the face is almost hairless and the round-tipped ears are very large. It has a short snout, projecting and large lower jaw, and a small nose-leaf. It occurs in various kinds of forests and roosts in tree hollows and caves, and also ruined buildings. The diet includes more insects in the wet season, and larger vertebrate prey, such as lizards and small birds, in the dry season; it also feeds on some fruit. It mainly hunts down low, and will seize prey from the ground. Food items are carried back to a night roost to be eaten. Breeding occurs once a year, with a single large pup (almost one-third of the mother's weight) born in midsummer after a gestation of about 220 days.

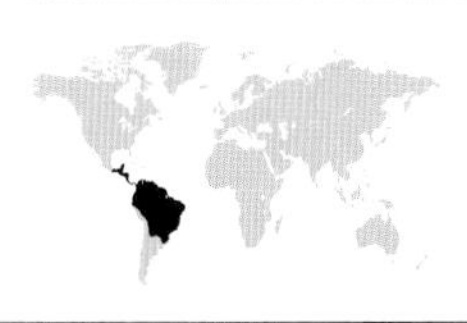

**IUCN STATUS** Least Concern
**LENGTH** 4–4⅜in (10.1–11cm)
**WEIGHT** 2¾–3⅜oz (79–96.3g)

***Vampyrum spectrum***

## Spectral Bat

**THIS LARGE, HIGHLY PREDATORY BAT** occurs in south Central America and across north, west, and central South America. It has rather short, gray-brown fur and large, broad wings that provide exceptional lift and agility, but no tail. The ears are large, broad, and rounded, the snout long, sturdy, and hammer-shaped, bare and pink at its tip with a small nose-leaf. It prefers evergreen forest in the lowlands, but also occurs in cloud-forest, deciduous forest, or more open swamplands. The species roosts in hollow trees, often by water, in groups comprising a bonded adult breeding pair with a couple of offspring. Both parents care for the pups, taking turns to hunt birds, rodents, and small opossums. It stalks prey, attacking with a rapid dash or pounce, then brings kill back to the day roost to share with the family. It relies heavily on primary forest, which is a depleted habitat.

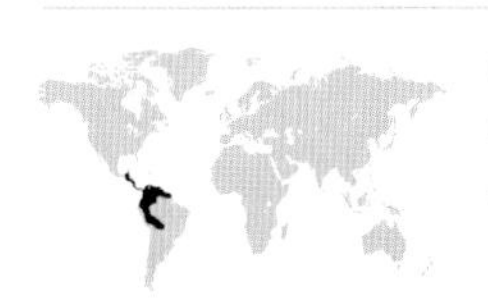

**IUCN STATUS** Near Threatened
**LENGTH** 5–5⅜in (12.5–13.5cm)
**WEIGHT** 6–6⅜oz (170–180g)

*Ectophylla alba*

# Honduran White Bat

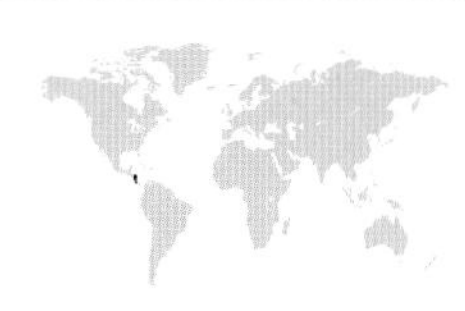

**IUCN STATUS** Near Threatened
**LENGTH** 1½–1⅞in (3.7–4.7cm)
**WEIGHT** ¼oz (5.7g)

THIS SPECIES OCCURS IN Central America, on the Caribbean side from east Honduras through to northwest Panama. It is a small distinctive species with white fur (sometimes light gray) and yellow arms, fingers, and ears. The wing membranes are contrastingly dark. It has a short face with a large, yellowish-pink nose-leaf and large, black eyes. It occurs in damp evergreen primary forest and secondary growth. This species roosts clustered into the tops of leaf tents of the apical style, which it constructs from *Heliconia* leaves. Their white fur reflects the green color of their surroundings, providing camouflage. Roosting groups are mixed-sex, except when females are nursing pups, when the males depart for a time. Females in a maternity colony will sometimes take care of one another's young when not out foraging. They breed in the rainy season, producing one pup after a short gestation period. These little bats are frugivores—especially fond of figs, which they pluck and carry to a night roost to eat. Their Near Threatened status is due to a sustained, significant rate of habitat loss in their range.

***Ametrida centurio***

## Little White-shouldered Bat

**THIS BAT, THE ONLY SPECIES** in its genus, is found from Panama into north and northeastern South America, including Trinidad and Tobago. Its range extends south as far as central Brazil. Females are much larger than males—2⁄5oz (12g), compared to 3⁄10oz (8g)—and until the 1960s were thought to belong to two separate species. It has gray-brown fur, and white patches where the wings meet the shoulders. The ears and face are yellowish, the eyes dark, yellowish brown and rather large. The species occurs in wet tropical forest with a dense understory and mid-story, in lowlands and uplands, especially close to streams. This bat's jaw anatomy suggests it is mainly a fruit-eater, and it may roost inside large, rolled-up leaves. Deforestation and degradation of forest habitats may be affecting this species.

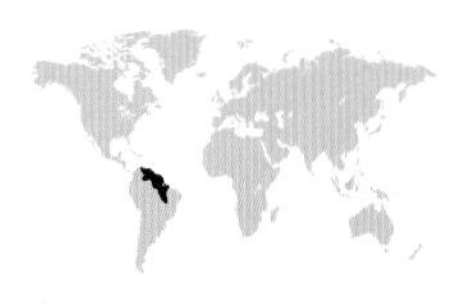

**IUCN STATUS** Least Concern
**LENGTH** 1⅜–1⅞ (3.5–4.7cm)
**WEIGHT** ¼–½oz (7.8–12.6g)

***Centurio senex***

## Wrinkle-faced Bat

**THIS BIZARRE-LOOKING** phyllostomid, the only species in its genus, is present from east and west Mexico through Central America to north Colombia and Venezuela. It has pale gray-brown fur, with white spots around its shoulders. The face bears many prominent folds and wrinkles, inspiring the species name (*senex*, meaning "old man"). The wrinkles are more prominent in males, who also have a flap of fur-covered skin to cover their faces. It occurs in tropical forest with streams, in both primary and secondary growth, and also in plantations. It usually roosts amid tangles of vine stems, females in large groups but males alone, or in small groups. The diet is mainly fruit. This bat appears to suck juices from fruits rather than taking bites (its facial folds may help channel fruit juice into the mouth). Although rather rare, it is not thought to be under threat at present.

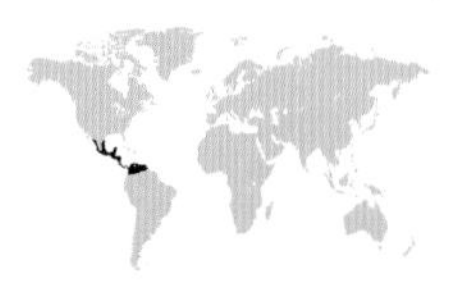

**IUCN STATUS** Least Concern
**LENGTH** 2⅛–2¾in (5.5–7cm)
**WEIGHT** ½–⅞oz (13–26g)

***Sphaeronycteris toxophyllum***

## Visored Bat

**THIS BAT OCCURS WIDELY** in tropical South America. It has light creamy-gray fur and no tail. The almost bare, pinkish-yellow face looks somewhat monkey-like, with a deep upper lip and a large space between the nostrils and mouth: there is a large fold of skin under the chin. It has a horn-shaped, fleshy projection from the forehead (giving rise to its English name), much bigger in males. The species occurs in primary, tropical, evergreen forest, sometimes in secondary forest and some other adjacent habitats, but is extremely rare and little known. It appears to be a fruit-eater, but its social system is unknown, as are the functions of its unusual facial ornamentations. This bat appears to be rare and may well be declining. Surveys to establish exact distribution and research into its ecology are required.

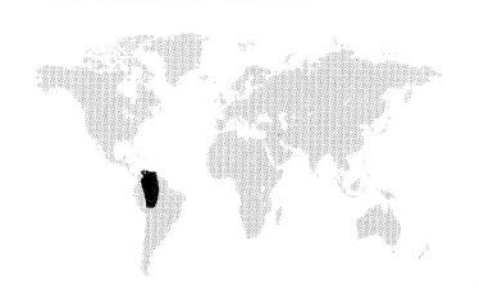

**IUCN STATUS** Data Deficient
**LENGTH** 2¼in (5.6–5.8cm)
**WEIGHT** ⅝oz (17g)

***Mesophylla macconnelli***

## Macconnell's Bat

**THIS BAT, THOUGH FORMERLY** considered an *Ectophylla*, is now the only member of its genus. It occurs from Nicaragua south through Central America to Amazonian Brazil and Bolivia. It has pale, creamy-gray, woolly fur, and yellowish ears, and is one of the smallest fruit-eating phyllostomids. The species occurs in mature, humid, tropical forest, and also in secondary growth; it roosts in leaf tents that it constructs from palm and arum leaves, as well as in hollow trees. The leaf-tent roosts, made in several different styles, hold up to eight individuals. They may be used for five or six months before replacements are needed, though females nursing young may change roosts more frequently. The diet is primarily fruit, perhaps also some insects. This is a fairly common bat over most of its range.

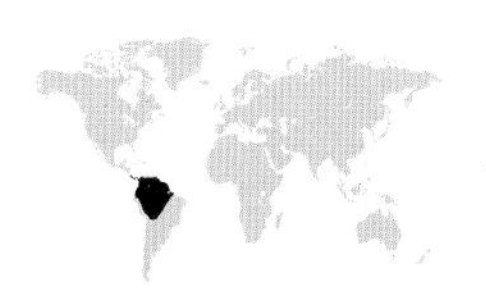

**IUCN STATUS** Least Concern
**LENGTH** 1⅝–2in (4–5cm)
**WEIGHT** ⅛–⅜oz (5–9g)

**Thyroptera**

## THYROPTERIDAE

The disk-winged bats, family Thyropteridae, form a small, distinct group containing just one genus. They are present in the Neotropics, and because of their roosting habits, require forested habitats. Their most notable trait is the presence of skin disks or suction cups at the ankles and wrists. The disks have a convex shape and secrete a fluid from glands around their edges, which allow them to adhere to smooth vertical surfaces when pressed against them. The suction is strong enough to support the bat by just one of the four disks. These bats use the disks to attach themselves head-up inside large, partly unfurled plant leaves, thus providing themselves with safe and readily available places to roost. Though because the leaves open rapidly, these bats must find new homes often. Thyropterids are probably related to the Old World sucker-winged bats (Myzopodidae). However, their wing disks are believed to have evolved independently as they are significantly different—in particular, myzopodid wing disks work only through wet adhesion rather than suction.

Thyropterids are small bats with relatively large wings and tail membranes. They are brown with pale undersides, which may help to provide camouflage by making their shape less apparent when sunlight shines through their roost leaf. The face is compressed and simple, with a short, pointed snout, tiny eyes, and small, triangular ears that angle forward. The forehead is steep and the fur thick, giving a very rounded appearance to the head and body. These bats are quick and highly maneuverable fliers, able to catch insects on the wing.

This book recognizes five species in the genus *Thyroptera*. *T. tricolor* is described opposite. Of the other four, three are very poorly known. *T. devivoi* was described in 2006 from a few specimens taken in Brazil and Guyana; its habits in the wild are still unknown. *T. lavali* appears to be present over an extensive area of north South America, but although it has been caught in ones or twos from many locations, the total number of specimens and observations is still very low. As a forest species, this is likely to be affected by habitat loss. The third species is *T. wynneae*, described only in 2014. It is known from northeast Peru and northeast Brazil, though its true distribution is probably more extensive. The final species, *T. discifera*, is present sparsely across northwestern South America.

The discovery of two new *Thyroptera* bats since the turn of the century indicates what a difficult group this is to study. It hints tantalizingly that more of these remarkable bats may exist in the forests of the Neotropics, awaiting discovery.

***Thyroptera tricolor***

# Spix's Disk-winged Bat

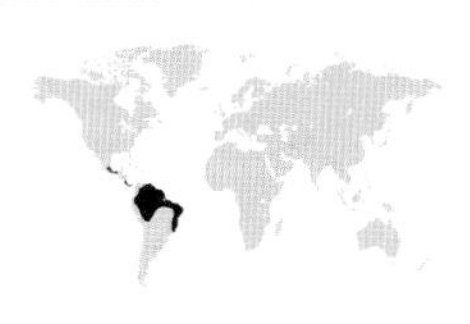

**IUCN STATUS** Least Concern
**LENGTH** 1⅛–1½in (2.7–3.8cm)
**WEIGHT** ⅛oz (4–5g)

**THIS BAT RANGES FROM** south Mexico through parts of Central America to much of northern South America. It has mid-brown upper parts and a clearly demarcated white underside. Small, convex skin disks at the ankles and wrists enable the bat to attach itself though suction to the vertical surface of a *Heliconia* or other leaf. It is found in humid forest and roosts in small groups within partly rolled-up leaves, usually changing to a new leaf every night as the previous one will have opened up. Females carry their young to new roosts as necessary. Breeding occurs twice a year, with a single pup born. It hunts flying insects, pursuing them with rapid twists and turns. It is a difficult species to trap and study, although it is common in at least part of its range.

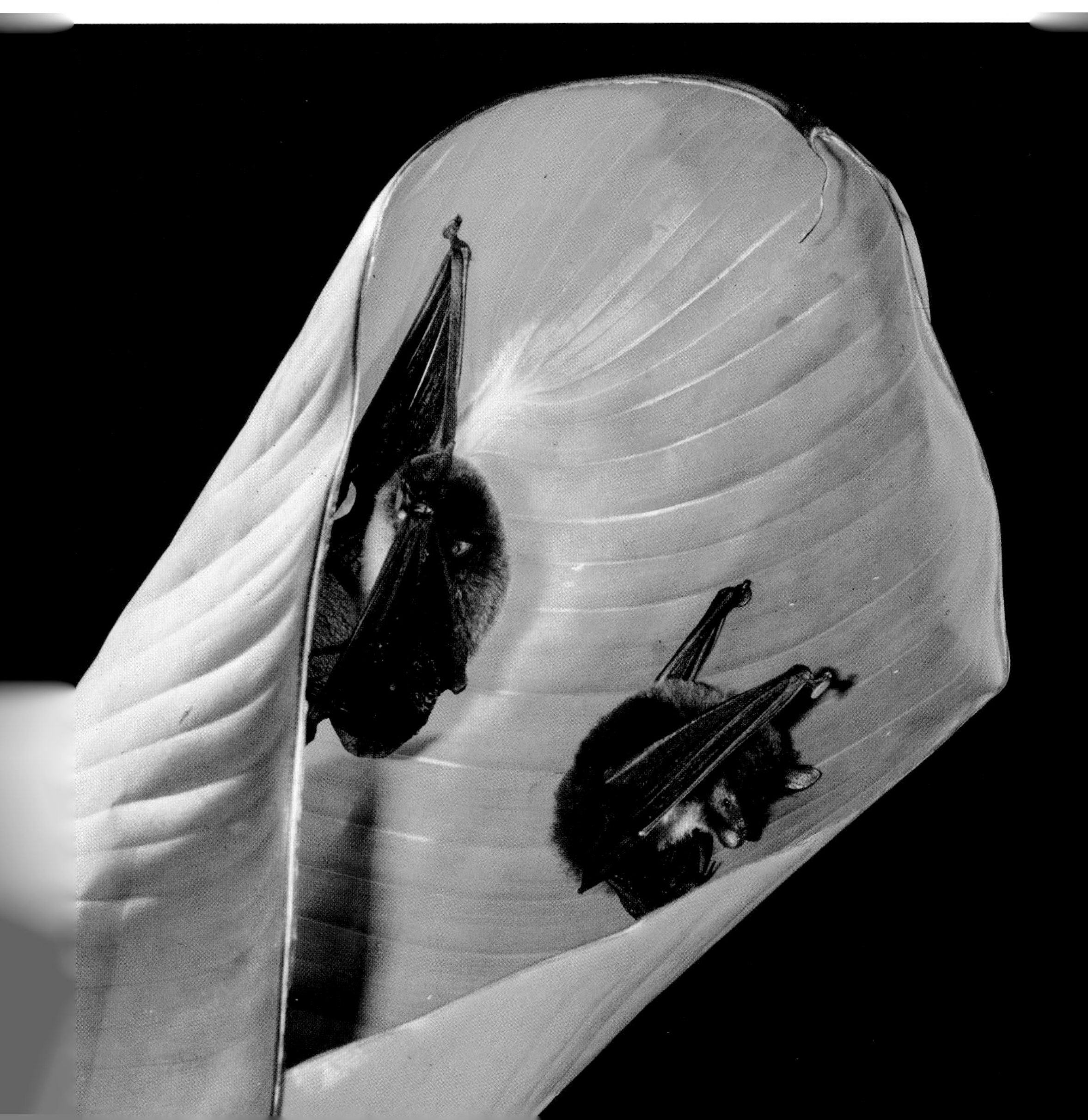

Cistugo

## CISTUGIDAE

Cistugidae is a small family of African bats with just one genus, holding two species. Bats of the genus *Cistugo* are known as "hairy bats," or "wing-gland bats" in recognition of their most distinctive anatomical traits. Their taxonomy has been disputed, with some authorities placing them within the family Vespertilionidae, and considering them closely related to the large Vespertilionidae genus *Myotis* ("mouse-eared bats"). Some earlier authors place the two *Cistugo* species actually within *Myotis*.

Today the group is usually separated as a family in its own right, thanks to a mitochondrial DNA sequencing study of 2010. This analysis showed that Cistugidae diverged from all vespertilionids some thirty-five million years ago. This evidence regarding its evolutionary history, along with a strong genetic distinctiveness today, is deemed sufficient to warrant family status.

*Cistugo* bats certainly do bear a marked—though superficial—resemblance to *Myotis*, and to other vespertilionid genera such as *Eptesicus*. They are small bats with short, pointed, simple snouts, lacking fleshy ornamentation. They have quite deep and broad lower lips, very small eyes, and smallish triangular ears with small, pointed tragi. The fur is rather long, soft, and fluffy, with individual hairs that tend to stand away from the body. The coat is on the pale side with light gingery or yellowish tones on the upperside; it is gray-white or creamy white on the belly. The wings and tail membranes are broad. They possess a small, pimple-like wing-gland within the upper inner part of the wing membrane; these are not evident in museum specimens and are rather inconspicuous in the living bat. They are presumably involved in the bat's olfactory communications, but their function has not been studied.

These bats are elusive and difficult to observe. Only a few museum specimens of either species exist, and their biology is not well studied. They are nocturnal insectivores, which hawk and glean prey within edge-type habitats; they will also hunt over water. The diet consists mainly of flies and true bugs (Hemiptera); they also take some beetles and caddisflies. They emerge shortly after sunset to begin foraging. *Cistugo* bats roost in relatively small groups (around thirty, or fewer), and the females give birth to singleton pups, in the middle of the austral summer. Researchers who have handled these bats report that they are quieter and more docile in the hand than most small bats.

The two species of *Cistugo* both occur in the south of Africa and both are rare. *C. lesueuri*, the type specimen of which was killed by a pet cat (and is named for J. S. le Sueur, the cat's

**Miniopterus**

owner, who recovered the specimen, which is now in the Transvaal Museum), occurs widely but sparsely in South Africa and Lesotho. It occurs in open, rocky habitats around rivers, marshland, and other areas of fresh water. It usually roosts in rock crevices in cliff faces, and sometimes in buildings. It is declining, with habitat loss a major factor. There also appear to be high numbers of wind farm casualties in some areas. However, this bat does occur in several protected areas. Studies are needed to assess its rate of decline and the impact of wind farms on its population.

*C. seabrae*, the smaller species in the genus, is found from southwest Angola south through west Namibia and into South Africa (Northern Cape). It occurs in open, dry habitats, but close to freshwater sources. Its behavior is little known, but it appears to glean prey from tree foliage. It frequently roosts inside buildings, and also in caves. Mining operations may damage its roosts, but overall this species is not thought to be threatened.

## MINIOPTERIDAE

The family Miniopteridae—the "long-fingered" or "bent-winged" bats—contains a single, large genus, *Miniopterus*, which holds thirty-four species. These are small bats which, by their genetics, are closely allied to Vespertilionidae, and Miniopteridae is sometimes still considered a subfamily of that grouping rather than a family in its own right. Compared to typical vesper bats, however, the miniopterids show a range of distinctive anatomical traits.

The wing structure is particularly striking. Miniopterids have exceptionally long third fingers, making for an elongated wingtip. The long wings of miniopterids power their fast, direct flight, as they hunt flying insects in the open air. Some miniopterids are also long-distance migrants. The third finger can fold back on itself when the wing is drawn in; hence the name "bent-winged bats." Other traits include a very steeply domed skull, a pair of small, vestigial, premolar teeth in the upper jaw, the absence of a tendon-locking mechanism in the feet, and an unusually small genome (half the size of the average mammalian genome).

These bats are found in Europe, Asia, Africa, and Australasia. A number of new species have been described since the turn of the twentieth century, some of which have yet to be assessed by the IUCN. Seven species of *Miniopterus* are described in detail. These are *M. australis* (see page 176), *M. gleni* (page 176), *M. majori* (page 177), *M. manavi* (page 177), *M. medius* (page 175), *M. natalensis* (page 178), and *M. schreibersii* (page 178).

The rest are *M. aelleni* (northwest Madagascar), *M. africanus* (west Kenya), *M. ambohitrensis* (from north Madagascar, newly described), *M. brachytragos* (north and west Madagascar, newly described), *M. egeri* (east Madagascar), *M. fraterculus* (south and southeast Africa), *M. fuliginosus* (widespread in south, east, and southeast Asia), *M. inflatus* (extensive though patchy range across sub-Saharan Africa), *M. magnater* (southeast Asia across to New Guinea), *M. mahafaliensis* (from south Madagascar, newly described), *M. orianae* (Australia and southeast Asia), *M. pusillus* (found in insular southeast Asia, New Guinea and northeast Australia), *M. sororculus* (widespread in the highlands of Madagascar), and *M. tristis* (Indonesia, Papua New Guinea, Philippines, Solomon Islands, and Vanuatu).

All of the known cave roosts of *M. maghrebensis* (Morocco and Tunisia), are endangered by various human activities. This species may also be affected by heavy pesticide use in its scrubland habitat. Two more endangered miniopterids are *M. fuscus* (Japan, where several of its colonies have shrunk dramatically or disappeared over the last few decades), and *M. robustior* (found on two of the Loyalty Islands, New Caledonia, where its habitat is being degraded). There are several miniopterids that are little known. They are *M. griffithsi* (from south Madagascar, newly described), *M. griveaudi* (Grand Comore Island, Comoros), *M. macrocneme* (Indonesia, New Caledonia, Papua New Guinea, Solomon Islands, and Vanuatu), *M. minor* (recorded sparsely in east Africa), *M. newtoni* (São Tomé island), *M. paululus* (Philippines and Lesser Sundas), *M. petersoni* (southeast Madagascar), and *M. shortridgei* (various Indonesian islands). This book also recognizes two species yet to be assessed by the IUCN—*M. mossambicus* (from southeast Africa, newly discovered) and *M. pallidus* (Turkey, recently split from *M. schreibersii*).

***Miniopterus pusillus***

***Miniopterus medius***

# Medium Bent-winged Bat

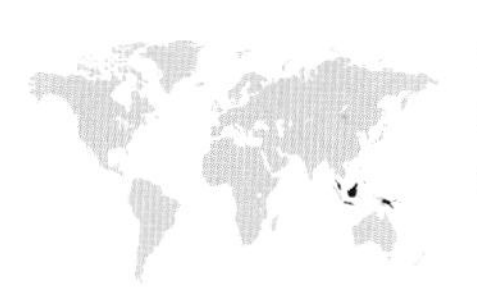

**IUCN STATUS** Least Concern
**LENGTH** unknown
**WEIGHT** unknown

THIS BAT OCCURS IN PARTS of southeast Asia, including various islands. It is small with blackish-brown fur (paler on the belly), a short, compressed snout, small eyes, and low-set, rounded ears with whitish, rounded tragi. It occurs in lowland and upland tropical forest, where it roosts in closely packed groups in various caves, crevices, and hollows—sometimes alongside other bats in mixed roosts, which can hold more than 10,000 individuals. It catches insects in flight, using echolocation. Almost nothing is known of its social behavior and breeding biology. One out of five females examined at a roost in January was found to be in early pregnancy, suggesting that pups are born in spring. This bat may face disturbance at its roosts due to limestone extraction, and habitat loss from deforestation.

*Miniopterus australis*

## Little Long-fingered Bat

**THIS BAT IS FOUND** extensively in southeast Asia and Australasia, including Malaysia, Indonesia, Borneo, the Philippines, New Guinea, Vanuatu, the Solomon Islands, and coastal east Australia. It has dense gray-brown fur, a domed head, and short, compressed snout, with small eyes and low-set, smallish, rounded ears. It occurs in coastal areas, foraging for insects over rainforest canopy and forest edges, and roosting in large caves alongside the larger *M. schreibersii*. It may rely on its sister species to raise the cave temperature to its requirements. Colonies can hold thousands of individuals. Females give birth annually, early in the austral summer, to a single pup; in some areas adults enter torpor in the austral winter. In the Philippines, its roost caves are affected by disturbance, but it is not generally threatened.

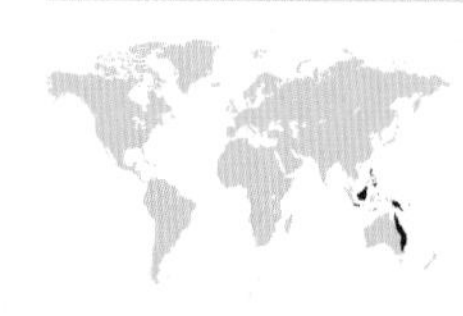

**IUCN STATUS** Least Concern
**LENGTH** 3⅜–3¾in (8.6–9.6cm)
**WEIGHT** ¼oz (7–8g)

*Miniopterus gleni*

## Glen's Long-fingered Bat

**THIS RELATIVELY LARGE** *Miniopterus* species occurs only on Madagascar, and has been recorded from most parts of the island. The lack of records from the southeast may reflect lower recording effort rather than genuine absence. It has the typical *Miniopterus* wing anatomy, with an elongated third finger and a compressed face with a short, pointed snout, low-set triangular ears and small eyes. This bat is dark-colored with blackish membranes. It roosts in caves close to water, usually in small groups (the largest roost recorded held some ninety individuals), and forages over diverse vegetation, from forest to arid scrubland. In some areas this bat is hunted for bushmeat, and some of its roost caves are also subjected to disturbance, but it is generally thought to be faring well.

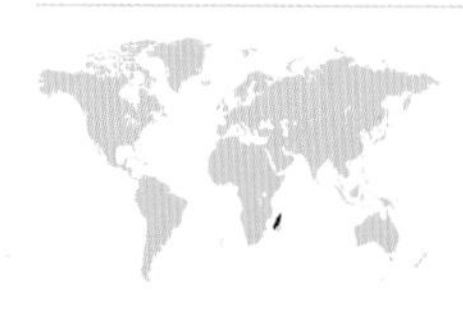

**IUCN STATUS** Least Concern
**LENGTH** unknown
**WEIGHT** 5–5¾oz (12.5–14.5g)

***Miniopterus majori***

## Major's Long-fingered Bat

**THIS BAT OCCURS** only on Madagascar, and is known from upland regions in the northeast, east, and south; its distribution is extensive but patchy. Records of it from the Comoros Islands are now thought to be erroneous. It is a dark *Miniopterus* bat, very like *M. schreibersii*, with dense, velvety, brown fur. It has a strongly domed head, a short and compressed face with small eyes, and rather narrow, low-set ears with whitish edges and rounded tragi. A colonial cave-rooster, it has been found sharing caves within deep humid forests with several other Madagascan *Miniopterus* species. It forages over rainforest, scrub, and most other habitat types, and is a fast-flying, aerial hunter. Otherwise, very little is known of its biology and behavior, and its taxonomy also requires investigation.

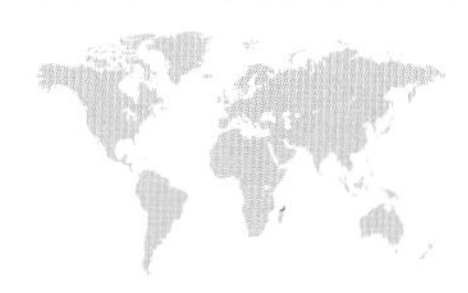

**IUCN STATUS** Least Concern
**LENGTH** unknown
**WEIGHT** ⅜oz (11.5g)

***Miniopterus manavi***

## Manavil Long-fingered Bat

**THIS BAT IS WIDESPREAD** across Madagascar, only becoming scarcer on the eastern and southern coasts. However, it may actually be a species complex, as there are differences in the forms found over the island. It is also present on the Comoros Islands. It has a typical *Miniopterus* appearance with a highly domed head, small compressed face, and small eyes and ears. The fur is mid-brown. This bat is adaptable and may be seen foraging over various well-vegetated habitats, from rainforest to plantations and large gardens, as well as farmland and dry deciduous forest. It roosts in caves or under rocky overhangs, in groups of up to 4,000, sometimes with other *Miniopterus* bats. It is sometimes hunted for food, but is not under particular threat at present.

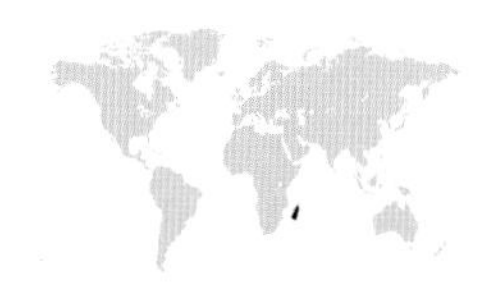

**IUCN STATUS** Least Concern
**LENGTH** unknown
**WEIGHT** unknown

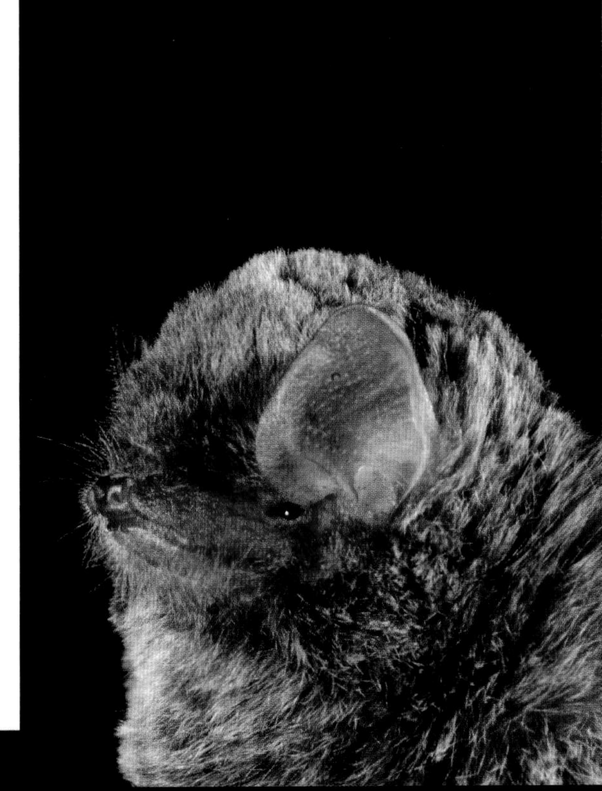

*Miniopterus natalensis*

## Natal Long-fingered Bat

**THIS BAT HAS A PATCHY** distribution in the south and east of Africa, as well as the far southwest Arabian peninsula. Its fur is quite dark gray-brown, with a blackish undercoat and paler tips. The head is rounded, with a steep forehead, short pointed snout, and small, low-set ears. It mainly occurs in savanna, scrub, and semi-desert, and roosts in caves and old mines—those used for hibernation being cooler than maternity roost caves. Females have one pup a year, in fall or winter (depending on latitude). It preys on insects, which it hawks in flight. It is quite common, with roost site availability the main factor determining its presence and abundance. Good caves can hold more than 2,500 bats, but few sites are protected, and the colonies are easily disturbed.

**IUCN STATUS** Least Concern
**LENGTH** 3⅝–5 (9.1–12.3cm)
**WEIGHT** ¼–⅜oz (7.5–11.3g)

*Miniopterus schreibersii*

## Schreiber's Bent-winged Bat

**THIS BAT HAS AN EXTENSIVE** distribution across the south of Europe, from Spain through to Turkey and into the north Middle East. It also occurs in northwest Africa, and patchily in west Africa. The species formerly included several forms found further south and east, now split into separate species. This is a rather grayish *Miniopterus* in new molt with the typical wing and facial features of its genus. It is a fast-flying aerial hunter that forages over a wide range of habitat types; it roosts in caves, sometimes in very large colonies (exceeding 100,000 individuals). Its Near Threatened status reflects the vulnerability of these large cave roosts, and declines due to heavy pesticide use. The species has also suffered several recent mass mortality events of unknown cause—possibly severe winter weather, or a viral disease.

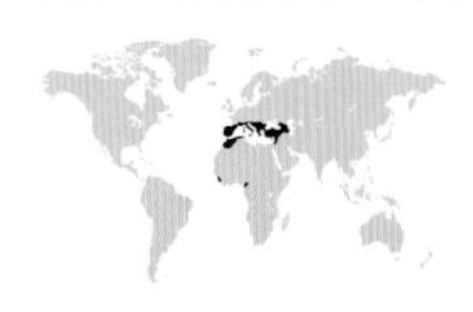

**IUCN STATUS** Near Threatened
**LENGTH** 2–2½in (5.2–6.3cm)
**WEIGHT** ¼–⅜oz (8–11g)

## MOLOSSIDAE

The family Molossidae contains 122 species in seventeen genera, and is one of the largest bat families. Members of the group are known collectively as the "free-tailed" bats because most show a length of thin, bare tail extending beyond the uropatagium. They are long-winged, fast-flying bats that hawk for their prey in the high open air, using sophisticated echolocation. The typical molossid face features a short, broad muzzle with a deep, wrinkled top lip, reminiscent of a mastiff or other short-faced dog. The folds in the upper lip enable the mouth to expand, in order to manipulate large prey. Molossids typically have very broad, rounded ears that are pushed forward, sometimes extending even beyond the snout tip in profile. The molossid ear has a small tragus but a large antitragus (the fold opposite the tragus), which forms a brow-like ridge over the medium-sized eye.

Some species are rather small, but many are large. Researchers have found that each species has a strong and distinct—but often pleasant—odor. For example, *Tadarida brasiliensis* (see page 206) smells like cornflour; it has been found to emit the aromatic chemical compound 2-aminoacetophenone, which also occurs in cornflour. Most molossids have rather plain, short, velvety dark fur. They are typically gray in new molt, gradually bleaching to rusty or reddish brown after molting. The group also includes the almost furless naked bats (genus *Cheiromeles*) and *Chaerephon chapini* (see page 182), in which the male sports a remarkable, erectile head crest.

Molossid bats occur in both the Old and New Worlds, including most of North America and all of South America, the south of Europe and Asia, Australasia, and central, south, and east Africa. The genus *Tadarida* originated in the Old World, but includes a very widespread New World species, *T. brasiliensis*—indicating that these strong-flying bats have high potential to spread and colonize new areas. They also include some of the most highly migratory bat species. Some species are more or less solitary, and their high-flying habits and high-level roosts can make them very difficult to observe and study. Some others form extremely large roosting colonies and perform spectacular sky-shows at dusk as they leave en masse to begin foraging. These large colonies consume enormous quantities of flying insects (including crop pests) around the roosts, making them a huge asset to farmers.

This bat family includes some exceptionally fast fliers, with the well-studied *T. brasiliensis* clocked at 100mph (160kmph) in level flight, undoubtedly aided by tail winds, but their speed comes at the expense of agility. Their roosts are therefore typically well above ground level, so the bats can fall into the open air and thus pick up enough speed to generate lift for powered flight. They are comfortable moving around in the roost; they have short but strong legs and are good, quick climbers. These bats are also often observed to crawl backward, apparently using their extended tail-tip as a "feeler."

Molossids have echolocation calls and ear anatomy adapted to their open-air hunting habits. The ears are angled to pick up sounds coming primarily from in front—the echoes of their calls as they fly, in a generally straight path. The calls they make are usually at relatively low frequency, in some cases audible to human ears, and many of the well-studied species call within the same rather narrow frequency band (20 to 35 kHz), despite considerable variation in body size. However, many molossids also demonstrate an ability to modify their search call quite considerably,

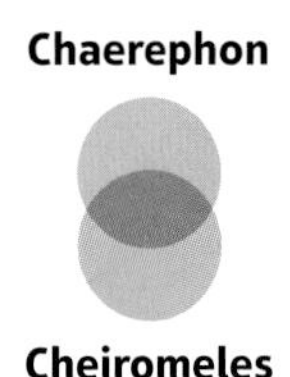

employing frequency shifts, alternating their peak frequencies, and alternating upward- and downward-modulated calls within call sequences. Using these tricks helps these bats to find a wide variety of flying prey types, which are thinly and unevenly distributed across otherwise empty space. Molossids are also very vocal in and near their roosts, giving a variety of other calls that have a social function.

## Chaerephon

The genus *Chaerephon* is the largest in the family, with twenty-one species distributed across Africa, Asia, and Australasia. They are known variously as the "wrinkle-lipped" or "free-tailed" bats. One of the most notable traits of the genus is the head crest possessed by males of most species. The crest reaches its most extreme development in *C. chapini*, but is quite pronounced in several other species too, such as *C. gallagheri.* It is believed to have a role in scent dispersal.

Already described are *C. bivittatus* (see page 182), *C. leucogaster* (page 183), *C. plicatus* (page 184), and *C. pumilus* (page 183). Of the remainder, *C. bregullae* occurs only on Fiji and Vanuatu, with just one known roost on Fiji and two in Vanuatu, collectively holding about 5,000 bats. None of the three sites are protected, so the bats are highly vulnerable to disturbance and exploitation for food. The species *C. johorensis*, of Malaysia and Sumatra, is known from only one regular roost, though it has been trapped at night in a variety of locations and probably has other, hitherto undiscovered roosts. It is very rare and there is a high rate of deforestation taking place within its known range. *C. tomensis* of São Tomé island is known only from three specimens, all taken in the 1990s. Two species, *C. gallagheri* (known only from one forest site in Democratic Republic of Congo), and *C. russatus* (recorded from five widely separated sites in sub-Saharan Africa) are little known, while *C. shortridgei* and *C. pusillus* have no IUCN status (the IUCN considers them subspecies of *C. chapini* and *C. pumilus* respectively).

The remaining Chaerephon species are: *C. aloysiisabaudiae* (west and central Africa), *C. ansorgei* (central and southeast Africa), *C. atsinanana* (eastern Madagascar), *C. bemmeleni* (patchily in western and parts of southeastern Africa), *C. jobensis* (northern Australia and parts of New Guinea), *C. jobimena* (western Madagascar), *C. major* (western and eastern Africa), *C. nigeriae* (western and south-central Africa), and *C. solomonis* (Solomon Islands).

## Cheiromeles

The two large bat species that form the genus *Cheiromeles* are known as the "naked" or "hairless" bats. They are not in fact completely hairless, but possess some sparse hair on the throat and feet, and on the wing membranes.

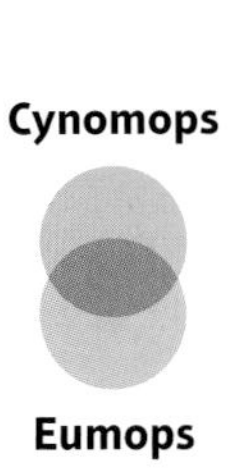

The genus is confined to southeast Asia. *C. torquatus* is described on page 186. Its smaller but very similar relative, *C. parvidens*, occurs in the Philippines, on Sulawesi, and a few other nearby islands. It is a little-known species, with a wide range, but in all likelihood declining because of forest loss, which deprives it of both roosting and hunting habitat.

## Cynomops

The genus *Cynomops*, or "dog-faced bats," are a group of mostly small molossids that range from Mexico through Central and South America. They have particularly sturdy, boxy snouts and large eyes. The group is closely allied to the genus *Molossus*. This book recognizes seven species, treating *C. mastivus* as a full species. *C. abrasus* and *C. paranus* are little known, while the remaining four (*C. greenhalli*, *C. mexicanus*, *C. milleri*, and *C. planirostris*) have wide geographical ranges, although none of the four are very well studied. A revisionary study of the genus in 2018 proposed that two new species—*C. freemani* from Panama and *C. tonkigui* from the eastern Andes—should also be recognized.

## Eumops

The genus *Eumops* holds sixteen species. They are known variously as the "mastiff bats" or "bonneted bats," and include some well-known species, including *E. floridanus* (see page 189)—one of the most threatened bat species of North America, and the subject of a variety of conservation initiatives. Four more well-studied *Eumops* bats are described in detail; *E. auripendulus* (page 188), *E. glaucinus* (page 187), *E. perotis* (page 191), and *E. underwoodi* (page 190). Although most free-tailed bats are gentle animals, the larger species can exhibit quick and intense tempers and bites if roughly handled. *E. ferox*, found in Central America and parts of the Caribbean, has the English name of Fierce Bonneted Bat. The naturalist Philip Henry Gosse, writing in 1851, observed that "When handled, its impatience of confinement is manifested by a continuous screeching, not very loud, but excessively harsh and shrill . . . The mouth also is then opened widely and threateningly, and a sufficiently grim armature of teeth developed."

As a whole, *Eumops* ranges through North and South America. The IUCN has yet to assess the recently described *E. chiribaya*, discovered in southwest Peru in 2010, and formally described in 2014. Two others, *E. maurus* and *E. wilsoni*, are poorly known. The remaining species are: *E. bonariensis* (Argentina), *E. dabbenei* (Venezuela, Colombia, Bolivia, Paraguay and Argentina), *E. delticus* (northern Brazil and nearby regions), *E. hansae* (patchily in Central America, and in much of northwest South America), *E. nanus* (Central America and South America), *E. patagonicus* (Bolivia, Paraguay and northern Argentina), and *E. trumbulli* (northeastern South America, to eastern Peru and northern Bolivia).

***Chaerephon bivittatus***

## Spotted Wrinkle-lipped Bat

**THIS BAT OCCURS** in east Africa, from Eritrea and Ethiopia south to Zimbabwe, in a relatively narrow band mainly away from the coast. It is a dark-furred, slim-winged molossid, with a long, free tail, and scattered, rather inconspicuous pale creamy spots on its upperside and underside. The undersides of the arms and legs are also pale. The face is robust and bulldog-like, with a deep and wrinkled upper lip. The species is found in savanna habitats, both moist and dry, and roosts in crevices within inselbergs (isolated rocky hillocks); it is probably dependent on these for roosting, but has occasionally been found roosting in old mining works. Its ecology is barely known and its distribution not fully studied, but it appears to be quite common with a stable population.

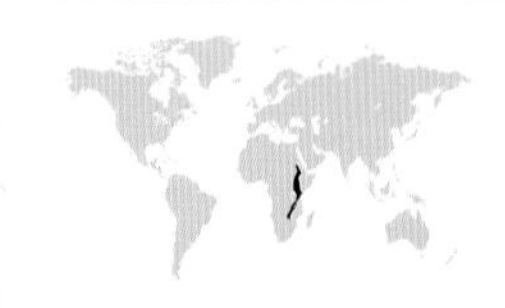

**IUCN STATUS** Least Concern
**LENGTH** unknown
**WEIGHT** ½–1⅛oz (15–32g)

***Chaerephon chapini***

## Long-crested Free-tailed Bat

**THIS BAT APPEARS TO** be widespread in Africa south of the Sahara, but exact distribution and abundance are not yet known, as it is elusive and poorly studied. It is small with short, light gray fur, a pink face with a short, turned-up snout and corrugated lips, large eyes, very wide, forward-tilted ears, and a long, free tail. During courtship the male can extrude and fan out a crest of long, white, brown-based hairs, which disperses scent from a gland at its base. The female has no obvious crest. This species has been observed in savanna, woodland, and riverine habitats, where it roosts in small groups. It is an insectivore, locating its prey by echolocation. This bat occurs in many protected areas, although is probably affected by deforestation in some areas.

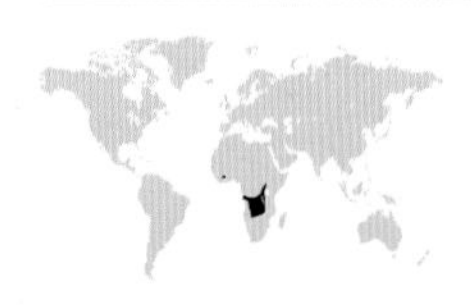

**IUCN STATUS** Least Concern
**LENGTH** unknown
**WEIGHT** unknown

***Chaerephon leucogaster***

## Grandidier's Free-tailed Bat

**THIS BAT IS CONSIDERED** to be a subspecies of *C. pumilus* by some authorities, including the IUCN. It has different echolocation calls to *C. pumilus*, as well as consistent differences in measurements. It is endemic to Madagascar and the nearby islands of Mayotte and Pemba. It is the smallest species of *Chaerephon*, with a dusky gray-brown upperside and crisply demarcated white chin and belly. The wings are whitish gray. It has broad-based, triangular, round-tipped ears, with a prominent central ridge. The species occurs in tropical dry forest and savanna, and roosts in tree holes. Its habits are not well studied, but it is probably a highly aerial insectivore. Its abundance and interactions with other Malagasy bats are in need of further study, particularly as so much of its habitat is under threat.

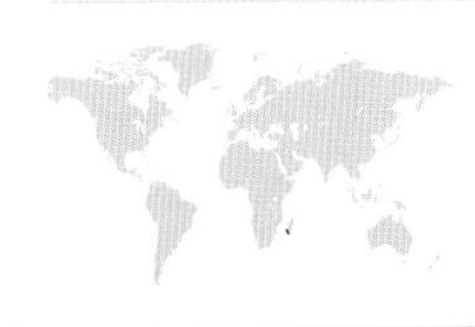

**IUCN STATUS** Not assessed
**LENGTH** 2¾–3⅝in (7.1–9.1cm)
**WEIGHT** ¼–⅜oz (7–11g)

***Chaerephon pumilus***

## Little Free-tailed Bat

**THIS BAT HAS A PATCHY** range across sub-Saharan Africa; it is also present in west and northeast Madagascar and other islands, and the far southwest Arabian Peninsula. It is a small, long-tailed molossid with gray fur, becoming white on the flanks and center of the belly. The ears are broad and flattened-looking; the face is more pointed than other *Chaerophon* bats. It occurs in various habitats including forest and savanna, and roosts in tree holes and buildings, or in the open, within dense clusters of palm leaves. The species hunts by night and is itself hunted by predators such as the Bat Hawk (*Macheiramphus alcinus*), a specialized hunter of bats. Within colonies, males guard harems of three to twelve females, which give birth up to three times a year (usually to one pup, but occasionally to twins).

**IUCN STATUS** Least Concern
**LENGTH** 2⅛–4in (5.4–10.2cm)
**WEIGHT** ⅜oz (11g)

***Chaerephon plicatus***

# Wrinkle-lipped Free-tailed Bat

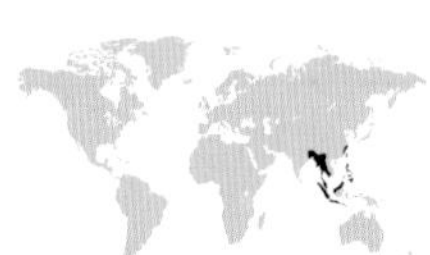

**IUCN STATUS** Least Concern
**LENGTH** unknown
**WEIGHT** unknown

**THIS IS A WIDESPREAD**, though localized, bat, with several subspecies. It occurs across large areas of south and southeast Asia, including India, Sri Lanka, Thailand, Myanmar, Vietnam, Malaysia, Indonesia, and parts of the Philippines. A dusky gray-brown molossid, it has a long, free tail and square snout, with prominent folds on the sides of the upper lip and outer edge of the ears. It roosts in large colonies inside caves, sometimes hundreds of thousands strong. The bats emerge in a spectacular stream at dusk, dispersing to hunt over forest and open countryside; it uses echolocation to target winged insects in a fast, high-level, and powerful flight. It is estimated that a colony of some 275,000 bats will consume nearly 1,100 tons (1,000 metric tons) of insects per year, making them key components of the local ecosystem as well as invaluable pest controllers. The roosts attract guano-harvesters, and can also be tourist attractions, but repeated disturbance will cause the bats to abandon the site. Some caves are protected by local people, but entrance routes must be kept clear of encroaching vegetation.

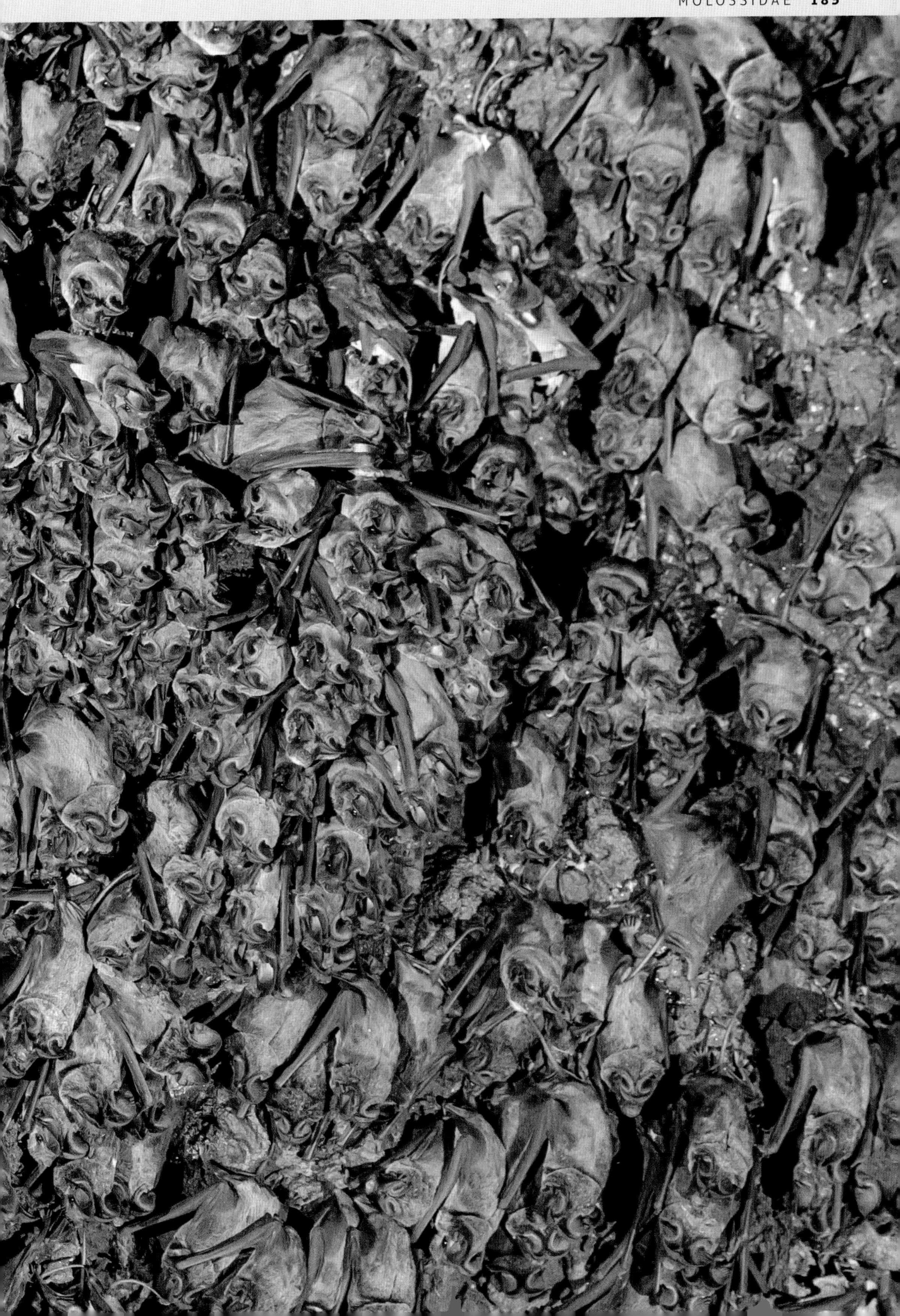

***Cheiromeles torquatus***

# Greater Naked Bat

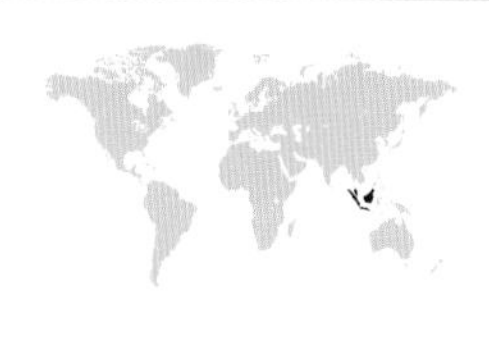

**IUCN STATUS** Least Concern
**LENGTH** 4½–5¾in (11.5–14.5cm)
**WEIGHT** 5⅞–6⅞oz (167–196g)

**THIS BAT IS FOUND IN** peninsular Malaysia, south Thailand, Sumatra, Java, Borneo, and the Philippines. It is among the world's largest microbats and has mostly bare, dark gray skin, with patches of hair around glands on the throat to help disperse scent. The muzzle is long, heavy, and wedge-shaped. It has pouches on the flanks into which it can tuck its wings, allowing it to climb more freely. It forms day roosts (which may hold 1,000 individuals) in various crevices, and forages in and over woodland and fields, giving audible echolocation clicks as it hunts. Females produce two young twice a year. It shares some roosts with swiftlets, whose nests are harvested for food, which causes disturbance. It is also hunted for food in Malaysia, and erroneously thought to damage rice crops.

***Eumops glaucinus***

# Wagner's Bonneted Bat

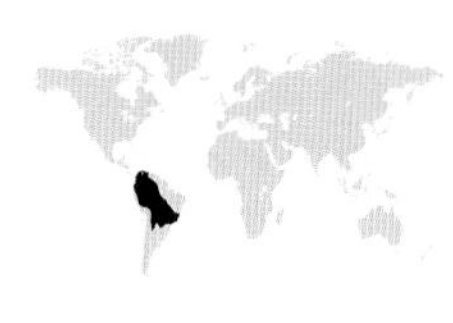

**IUCN STATUS** Least Concern
**LENGTH** 4⅝–6½in (11.7–16.5cm)
**WEIGHT** ⅞–2oz (25–55g)

**THIS BAT IS FOUND FROM** Colombia and Venezuela down to Paraguay, and to southeast Brazil and north Argentina. Prominent folds on the lower edges of the ears meet on the forehead to form its "bonnet." Males possess a gland on the throat that produces a musky-smelling secretion. Mainly found in tropical and subtropical forest, the species also occurs in urban and suburban habitats, and more arid, open places. It roosts in tree holes and in the canopy, sometimes in buildings, often alongside other species. It hunts insects high in the air, using echolocation and a fast, direct searching flight. This bat also produces an audible screeching call, most likely to communicate with others, as this is also heard at roosts. Currently it appears to face no serious threats.

***Eumops auripendulus***

# Shaw's Mastiff Bat

**IUCN STATUS** Least Concern
**LENGTH** 3–3⅝in (7.5–9.2cm)
**WEIGHT** ⅞–1¼oz (25–35g)

THIS BAT IS FOUND FROM Mexico south through Central America to South America, reaching northern Argentina in the east and northern Peru in the west. It is also present on Trinidad and Tobago, and Jamaica. It is a dark molossid, with glossy, blackish-brown fur. The short, robust face is almost hairless and has a broad, mastiff-like jaw. Primarily a forest-dweller, the species is also regularly encountered in open areas, from lowlands up to nearly 6,500ft (2,000m). Roosts are usually small and can be in all kinds of cracks and crevices, including in buildings. They are sometimes too small for the bat to fly in and hang head-down, making it crawl in and out on its feet and thumbs. This bat's long, narrow wings make it a strong, fast flier, suited to catching flying insects in open air.

*Eumops floridanus*

# Florida Bonneted Bat

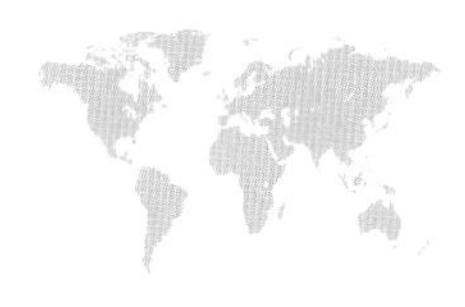

**IUCN STATUS** Vulnerable
**LENGTH** unknown
**WEIGHT** unknown

**THIS SPECIES, NOW KNOWN** only from a few locations in south Florida, is the rarest bat in the United States. Formerly more widespread, it occurs in varied habitats including forest, but also around towns, roosting in both natural cavities and crevices in buildings; it also uses bat boxes. This bat may roost alone or in groups, with a single, dominant male defending a harem of up to twenty females. Females produce one pup a year, typically in summer. Habitat loss is one of the reasons for its decline: large, decaying trees with suitable roosting cavities are often removed or blown down in hurricanes. Excessive spraying for mosquitoes also affects this bat. With more than 3,000 to 5,000 individuals left, all existing roosts must be discovered and protected as soon as possible.

***Eumops underwoodi***

# Underwood's Bonneted Bat

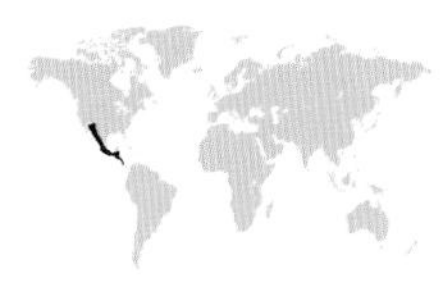

**IUCN STATUS** Least Concern
**LENGTH** 6⅜–6½in (16–16.5cm)
**WEIGHT** 1⅜–2¼oz (40–65g)

**THIS BAT'S RANGE JUST** reaches the southwest of the United States, extending south along the Pacific seaboard of Mexico, through to Costa Rica and Nicaragua. It is the second-largest bat found in the United States after the Western Mastiff Bat, its close relative. Its long rump hairs are diagnostic of the species. It has long, narrow wings that enable fast, powerful flight—at least 27mph (45km/h). The species prefers drier habitats, including woodland (with hollow trees for roosting), but also desert. It needs large pools from which to drink, as it lacks the maneuverability to dip down to smaller ponds. It catches large (up to 2½in (6cm) and hard-bodied beetles and grasshoppers in flight, as well as moths. Females produce one pup a year, in June or July. Though rare, it occurs in several protected areas.

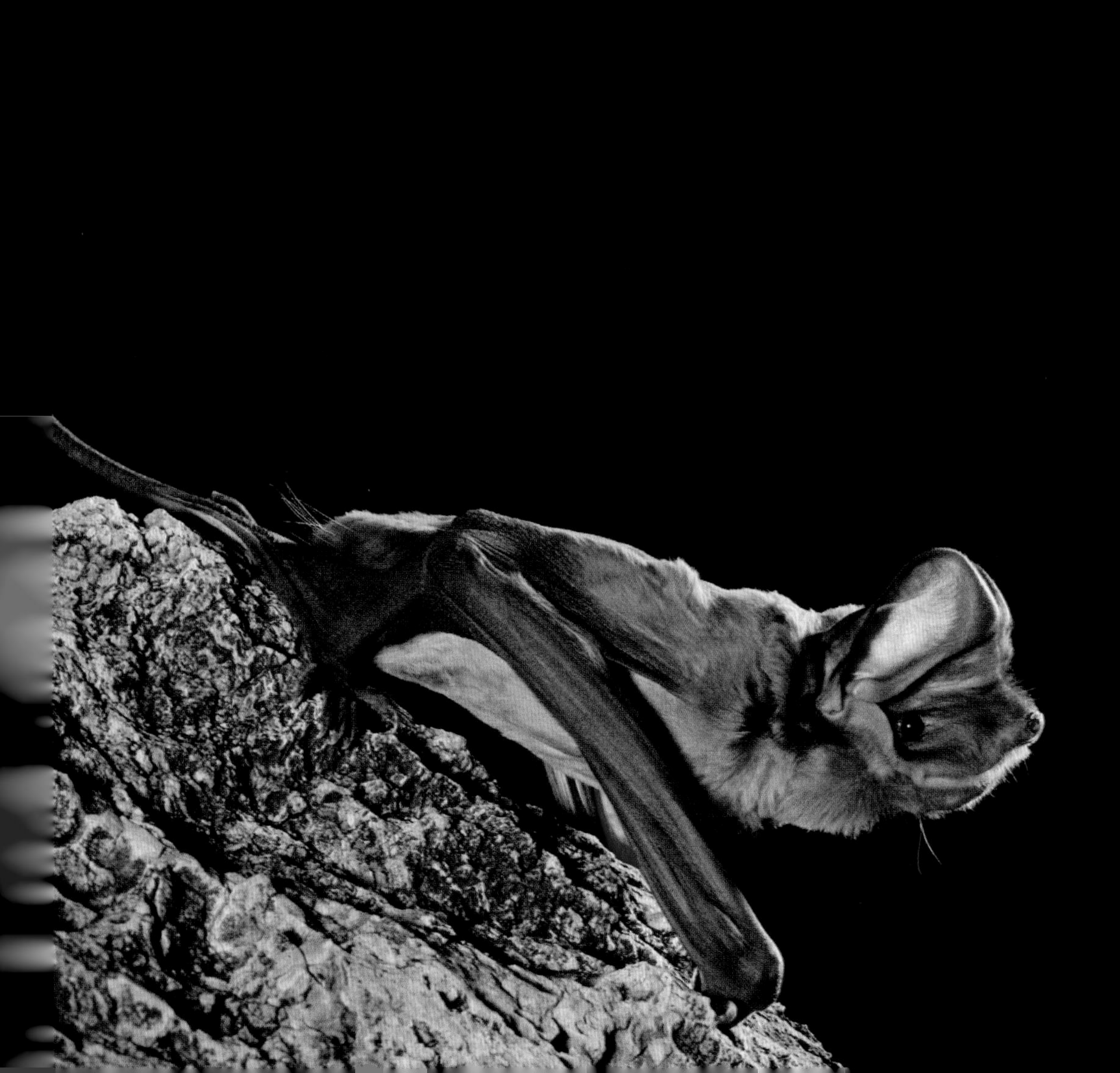

***Eumops perotis***

# Western Mastiff Bat

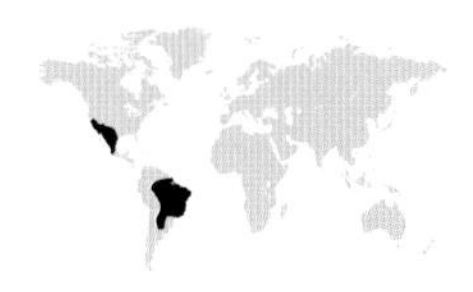

**IUCN STATUS** Vulnerable
**LENGTH** 6¼–7⅜in (15.9–18.7cm)
**WEIGHT** 2–2⅛oz (55–60g)

THIS VERY LARGE BAT occurs in the southwest of the United States and Mexico, possibly Cuba, and much of northern South America (though this population may represent a separate species). It mainly hunts in open habitats, and roosts in crevices on steep cliff faces, rocky outcrops, and occasionally buildings; it also requires large open water bodies, 100ft (30m) in length, to drink. It hunts insects in high flight in open sky, using very loud but low-frequency echolocation (less than 10kHz), and may also be guided by the prey's sounds, particularly in the hunt's last phase. Prey may also be gleaned from tree trunks and canyon walls. Roosts hold up to a hundred individuals of both sexes, all year round. Mating occurs in early spring, and singleton pups are born in summer.

***Molossops temminckii***

## Dwarf Dog-faced Bat

**THIS BAT HAS AN** extensive range in South America. It is very small, with light gingery-brown fur, fading to whitish on the belly. The ears are small and rather narrow, the face broad with a pointed nose and deep chin. It is broader-winged, and so more maneuverable on the wing than most molossids, and its echolocation calls are also adapted to suit hunting in a more cluttered environment. It occurs in rainforest and other woodland habitats, and may visit agricultural areas and villages. Groups often hunt together, targeting swarms of termites, and hawking around artificial lights for moths and beetles. Roosts are under tree bark and often inside buildings. Females form maternity roosts and give birth to one pup in late fall or winter, depending on region.

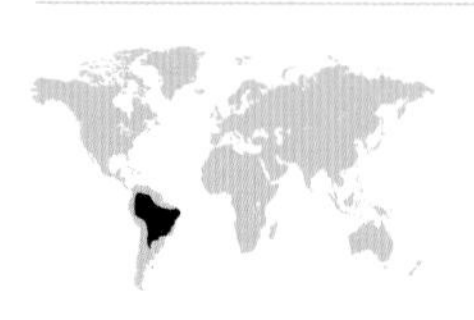

**IUCN STATUS** Least Concern
**LENGTH** 1¼in (3.3cm)
**WEIGHT** ⅛– ¼oz (4–7g)

***Molossus molossus***

## Pallas's Mastiff Bat

**THIS SPECIES RANGES** from the Florida Keys and northeast Mexico through Central America and the Caribbean (where it is common) to as far south as north Argentina. A dark brown molossid, it has a dense, velvety coat and a short, free tail. The face is broad and wedge-shaped, the rounded ears are folded forward and crumpled-looking. It occurs in all kinds of habitats and can be common in urban areas, often establishing large roosts inside buildings. It will also roost in palm trees. It hawks flying insects in the open air, mainly hunting just after dusk and immediately before dawn, with a four-hour period of torpor around midnight. This bat may be persecuted when it roosts in buildings, but overall it faces few serious threats at present.

**IUCN STATUS** Least Concern
**LENGTH** 3½in (9cm)
**WEIGHT** ⅜–½oz (11–14g)

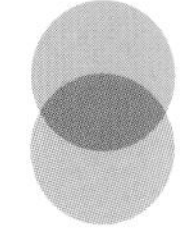

Molossus

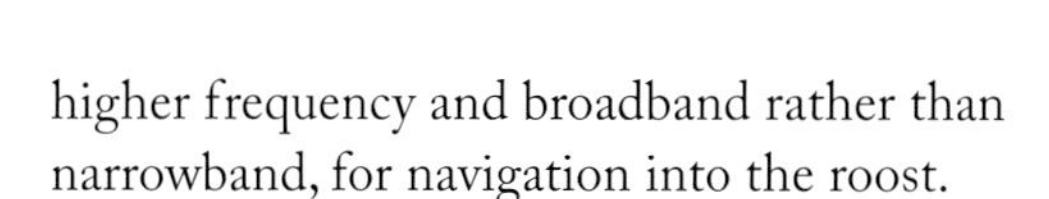

## Molossops

Another Neotropical genus is *Molossops*. These are rather small molossids, known (as are some other genera) as "dog-faced bats." This book recognizes three species—*M. aequatorianus*, *M. neglectus*, and the widespread *M. temminckii*, described opposite. *M. aequatorianus* has an apparently tiny range within west-central Ecuador. It occurs in dry tropical forest, which is rapidly being destroyed, with the land then converted to agricultural use. *M. neglectus* has been recorded from several countries in the north of South America, but its precise distribution remains unknown and it is barely studied.

## Molossus

The eight bats in the genus *Molossus* are found from Mexico through to the north of Argentina, with some species in the Caribbean. Most are known as "mastiff bats," and in appearance are typical molossids, with blunt, robust snouts and rounded, forward-angled ears. They are open-air hunters that, like others in the group, have a versatile echolocation system. That of *M. molossus* is well studied: in searching flight it produces relatively long, narrowband calls, adapted to pick up moving prey a long distance away in otherwise empty air. These calls are given in pairs of pulses that alternate in frequency: the first signal is given at 34.5 kHz, the second at 39.6 kHz. When this bat nears its roost, the echolocation calls become of significantly higher frequency and broadband rather than narrowband, for navigation into the roost.

*M. molossus* and *M. rufus* are described in detail opposite and on page 196, respectively. *M. alvarezi* is a little-known species, only recently recognized (2011), and so far lacking a regularly used English name. It occurs in south Mexico, and possibly more widely. The remaining species are: *M. aztecus* (southern Mexico and Central America), *M. coibensis* (Pacific Central America through to northern South America), *M. currentium* (a narrow strip through South America from Brazil south to Uruguay and Argentina), *M. pretiosus* (southern Mexico and Central America and northwestern South America), and *M. sinaloae* (western Mexico through Central America to northern Venezuela and Colombia).

***Molossus currentium***

## Mops

*Mops*—a curious name deriving from a Malayan word simply meaning "bat"—is a large genus of sixteen species, which occur in Africa and Asia. It is possible that the genus should be further subdivided when more has been learned about the group's genetics. These bats show quite a range in body size. Males typically possess a forehead crest of stiff hairs, like *Chaerephon* bats, and both sexes often have a pale marking along the flank sides where the wing membranes join the body. Their coats are otherwise dark gray or brown, and usually short and velvety. The ear folds are joined above the bridge of the nose, and the upper lip is prominently wrinkled.

These bats are mainly found roosting in forest habitats. Some of them are very poorly known, as their roosts are small, located high in tree holes, and their high-altitude, open-air hunting habits means they are rarely netted by researchers carrying out general bat surveys. The species *M. congicus*, for example, was discovered by chance in 1910. Naturalists discovered it roosting within a large, hollow tree that was blown down in a storm near their forest camp in Central Africa.

This book features *M. midas* and *M. mops* on page 197, and *M. condylurus* on page 196. The little-known species *M. bakarii*, relatively new to science and described in 2008, occurs on Pemba island, Tanzania; here it seems to be uncommon, but its habit of roosting in buildings may help to guard against the impact of deforestation. *M. leucostigma*, which has a white belly, is found in Madagascar. Several widespread African species exist, these are: *M. brachypterus*, *M. demonstrator*, *M. nanulus*, *M. niveiventer*, *M. spurrelli*, and *M. thersites*. Of the remaining four species, *M. niangarae* is known from only one specimen from the Democratic Republic of Congo. It was formerly classed by the IUCN as Critically Endangered, but that status has been revised to Data Deficient to reflect its uncertain taxonomy. It may be conspecific with *M. trevori*, a more widespread but apparently scarce African species which is also little known. Another African *Mops* bat is *M. petersoni*, a species found in west Africa in areas undergoing heavy deforestation. The final species is *M. sarasinorum*, which is known from Sulawesi, and may be conspecific with *M. mops*.

## Mormopterus

This book recognizes eighteen species in the genus *Mormopterus*, with the proviso that recent molecular sequencing studies suggest it may warrant division into four distinct new genera. This is a very widespread group of molossids, with representatives in Africa, across south Asia to Australia, and even a few species in the Neotropics. In general these bats are more delicate-looking than most molossids, with pointed rather than boxy snouts, less

Myopterus

elaborately folded ears, paler fur and skin (making the large eyes more prominent), and little or no obvious lip wrinkling.

*M. lumsdenae* and *M. jugularis* are both described on page 198. Several other *Mormopterus* species are considered to be threatened, these are *M. acetabulosus* of Mauritius, and *M. phrudus*, *M. norfolkensis*, and *M. minutus*. *M. phrudus* has only been recorded from a site at Machu Picchu, Peru, while *M. norfolkensis* is a rare Australian endemic, and *M. minutus* occurs only on Cuba and seems able only to form roosts in one species of palm tree. *M. eleryi*, a recently discovered Australian species, is rare. *M. loriae* of New Guinea, *M. halli* of Australia, and *M. doriae* of north Sumatra are all little known. *M. beccarii*, also poorly known, occurs on the Moluccan islands of Halmahera, Ambon, and Seram (Indonesia), and patchily in parts of mainland New Guinea and nearby islands. It has recently been split from *M. lumsdenae*, widespread in Australia; further splits may be warranted.

The remaining species are *M. coubourgianus*, *M. francoismoutoui*, *M. kalinowskii*, *M. kitcheneri*, *M. petersi*, *M. planiceps*, and *M. ridei*. *M. francoismoutoui* occurs on Réunion island and was formerly thought to be the same *Mormopterus* species present on nearby Mauritius (*M. acetabulosus*), but studies show the two are distinct. Unlike its close relative, *M. francoismoutoui* is abundant and increasing. It is a familiar sight on Réunion and has many known roosts, including one holding more than 65,000 individuals. Most of the others—*M. coubourgianus*, *M. kitcheneri*, *M. petersi*, *M. planiceps, and M. ridei*—are present in Australia, and their relationships may require taxonomic revision. The last, *M. kalinowskii*, occurs in Peru and Chile.

**Myopterus**

*Myopterus* is a small genus of just two species—*M. daubentonii* and *M. whitleyi*. They are small, pale-furred bats with almost translucent wing membranes. They have square-ended muzzles and longer, simpler, and narrower ears than most molossids. *M. daubentonii* is poorly known—it has been recorded from several locations in west and central Africa. *M. whitleyi* is also present in this region of Africa, but is a little better known. A solitary bat of moist, lowland forest, it is probably declining because of habitat loss.

***Molossus rufus***

## Black Mastiff Bat

**THIS IS A WIDESPREAD** bat of South America; its range includes north Argentina, much of Central America, and south and coastal Mexico. It is a rather large molossid with blackish-brown fur, which becomes lighter and redder through gradual ammonia bleaching at the roost. The face and other bare parts are blackish. It has a triangular head in profile with a deep, broad snout coming to a pointed tip, and broad, forward-folded ears. It occurs mainly in forested habitats, from rainforest to woodland edges and scrubland, and is often seen around houses. Roosts are often inside buildings, and can hold hundreds of individuals. It is an aerial insectivore which appears to specialize in flying beetles, but it takes other prey too. Most pups are born in spring or summer.

**IUCN STATUS** Least Concern
**LENGTH** 4½–5¾in (11.3–14.4cm)
**WEIGHT** 1–1¼oz (28–37g)

***Mops condylurus***

## Angolan Mops Bat

**THIS IS A SMALL AFRICAN** molossid, with an extensive distribution south of the Sahara. It is a light brown bat with a paler belly. The face, ears, and membranes are blackish-brown. It has a strong jaw, pointed snout, and smallish, round-tipped, forward-pressed ears. The upper lip has prominent vertical folds. A savanna-dweller, it avoids heavily forested areas though may forage along woodland edges. This bat also regularly visits agricultural fields, eating many kinds of crop pests, and hunts around villages and towns. Roosts are in caves, buildings, rock cracks, and tree holes. Some of its roosts become extremely hot, exceeding temperatures that most bats could not tolerate (104°F/40°C or higher); it is also tolerant of cold conditions. Its adaptability to temperature variation allows it to use roost sites that other bats cannot.

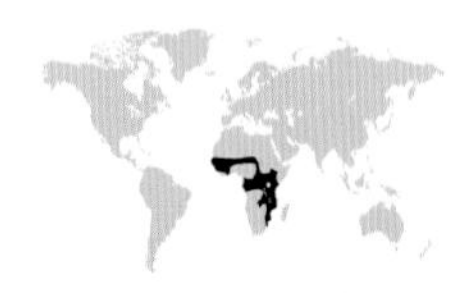

**IUCN STATUS** Least Concern
**LENGTH** unknown
**WEIGHT** unknown

***Mops midas***

## Midas Mops Bat

**THIS RATHER RARE**, large molossid bat of Africa and Madagascar also occurs in the far southwest of Saudi Arabia. In Africa it has a fragmented distribution south of the Sahara, and in Madagascar is mainly confined to the southwest. The Madagascar population is sometimes considered a separate subspecies (*M. m.* subsp. *miarensi*). This is a typical *Mops* bat; the sides of its upper lip are prominently wrinkled. It is found in open and wooded savanna habitats in the lowlands, and roosts in dark crevices within rock faces, damaged tree trunks, and inside buildings. The largest nursery roosts may number one hundred or so females—the largest-known roost held about 600—but groups of around twenty are more usual. It hawks for flying insects high in the air, often over river valleys.

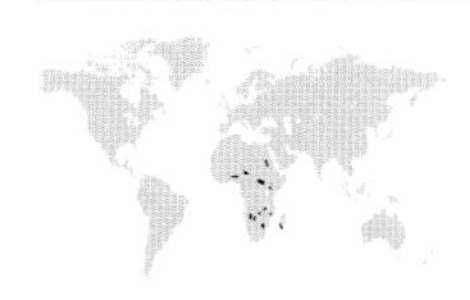

**IUCN STATUS** Least Concern
**LENGTH** 5⅝in (14.2cm)
**WEIGHT** 1⅜–2¼oz (40–61.5g)

***Mops mops***

## Malayan Free-tailed Bat

**THIS BAT RANGES** from south Thailand through the Malay Peninsula to Sumatra and Borneo. It has dark brown or reddish-brown fur and bare parts, and a short, robust snout. The lip edges are somewhat wrinkled—a trait seen in many molossids—which may help it cope with the hard parts of prey such as beetles. It has a long, free tail and large feet. It occurs in low-lying primary forest, with rivers. Roosts are usually inside tree hollows, which it sometimes shares with *Cheiromeles torquatus*. It hunts high over the canopy and above rivers. Although not well studied, this bat is likely to be in decline. Because of its fairly restricted distribution and apparent natural rarity, as well as rapid deforestation taking place in its range, it is classed as Near Threatened.

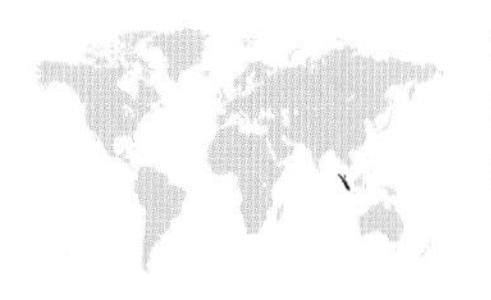

**IUCN STATUS** Near Threatened
**LENGTH** unknown
**WEIGHT** 1–1¼oz (28–35g)

***Mormopterus lumsdenae***

## Northern Free-tailed Bat

**THIS BAT IS A WIDESPREAD** species in Australia, occurring across a broad band along the north. It has rather pale, brownish-gray fur, becoming whitish on the belly, and the bare parts are dusky pinkish-brown. Its ears are rather small, pressed back, simple and rounded; the snout is pointed and wedge-shaped with a rather deep chin. This bat has large feet and can scramble quickly on all fours. It makes use of various habitats, including rainforest, eucalyptus forest, lighter woodland, and savanna. It usually roosts in tree holes, competing with nesting bees and birds for access to them, but also sometimes inside buildings. It may be threatened by loss of tree holes for roosting, due to deforestation, and general habitat degradation.

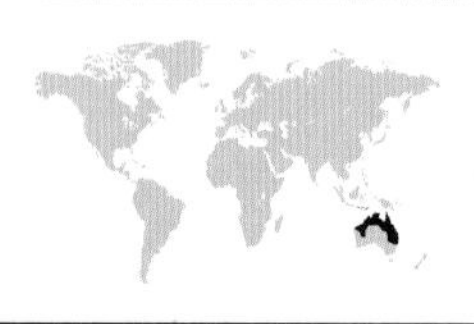

**IUCN STATUS** Least Concern
**LENGTH** unknown
**WEIGHT** unknown

***Mormopterus jugularis***

## Peters's Wrinkle-lipped Bat

**THIS BAT OCCURS FROM** the south of Madagascar to the north tip, on eastern slopes only. It has a soft-textured, mid-gray-brown coat, with a paler belly; males are larger than females. The triangular ears have rounded tips; the face is long and wedge-shaped with a bulging chin, wrinkled lips, and slightly upturned tip to the snout. The species hunts in typical molossid high-level flight over open grassland, farmland, gallery, and spiny forest, as well as rivers and open water. Most roosts are in buildings, but it also roosts inside rock fissures. Roosts are quite often shared with other molossids such as *Mops leucostigma*. Where space allows these can be very large—a maternity roost of some 1,000 females with their pups has been documented. The diet is predominantly beetles and true bugs (Hemiptera).

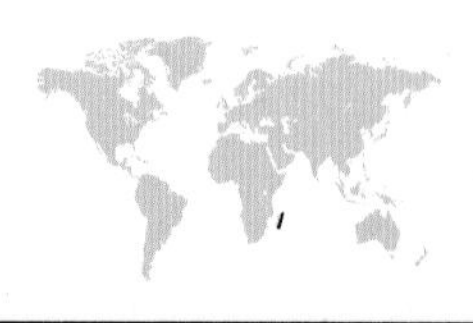

**IUCN STATUS** Least Concern
**LENGTH** 3½–3¾in (9–9.5cm)
**WEIGHT** ¼–½oz (8–14g)

Nyctinomops

Otomops

## Nyctinomops

The widespread American genus *Nyctinomops* holds just four species. These are fairly large, dark-furred, and short-faced molossids. All four species have extensive ranges. *N. femorosaccus* is described on page 200, and *N. macrotis* on page 201. The other two are *N. aurispinosus*, present in Mexico and discontinuously in coastal northwest and eastern South America, and *N. laticaudatus*, which is found in Mexico and through Central America and a large swathe of South America.

## Otomops

Bats of the genus *Otomops* exhibit particularly extreme facial anatomy. The very large, round ears are angled forward so strongly that their tips project well beyond the tip of the sharply upturned snout. The eyes are small and nearly hidden by the ear folds, while the top lip is large; it is wrinkled and pendulous at the sides. The fur is usually dark reddish gray or brown, sometimes with white or blackish markings around the shoulders and flanks.

This genus is present in Africa, Madagascar, and south and southeast Asia. *O. martiensseni* is described on page 203. Of the remaining species that this book recognizes here, five (*O. papuensis* and *O. secundus* of Papua New Guinea, *O. formosus* of Java, *O. johnstonei* of Indonesia, and *O. wroughtoni* of India and Cambodia) are very little known. *O. madagascariensis* is endemic to Madagascar; its distribution appears highly fragmented, but more surveys are needed to clarify this. *O. harrisoni* is a rare bat of east Africa and Arabia, which roosts in largish colonies in montane cave systems. Disturbance at its roosts due to guano-mining is a serious problem.

***Nyctinomops femorosaccus***

# Pocketed Free-tailed Bat

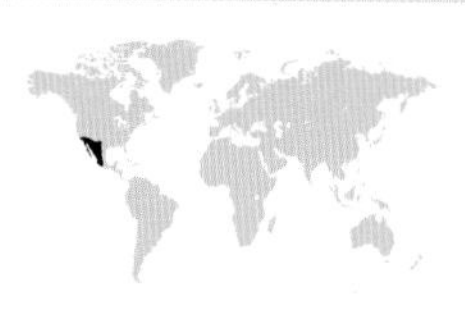

**IUCN STATUS** Least Concern
**LENGTH** 3⅞–4⅝in (9.9–11.8cm)
**WEIGHT** ⅜–½oz (10–14g)

**THIS BAT OCCURS IN** the far southwest of the United States and northwest Mexico, including Baja California. It has short, silky brown fur and a long, free tail. It also has an extra fold of skin between the femur and tibia that forms a pocket, hence the name. The wings are long and narrow, and its flight is fast and powerful. It is found in open, dry country, with rocky crevices and caves for roosting, and some standing water for drinking. This species roosts in small groups, usually fewer than one hundred individuals. In midsummer, females have one pup, which is flying and feeding itself by late August. Emerging well after sunset, the bat calls repeatedly as it leaves the roost, then heads off to spend an hour or so hunting various flying insects. A second round of hunting activity takes place near dawn.

***Nyctinomops macrotis***

# Big Free-tailed Bat

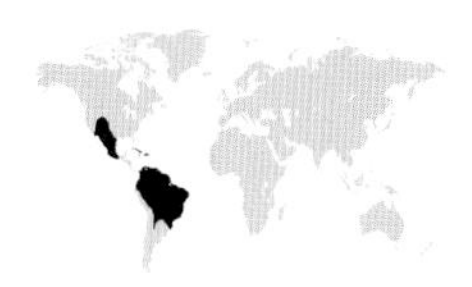

**IUCN STATUS** Least Concern
**LENGTH** 4¾–6⅜in (12–16cm)
**WEIGHT** ¾–1oz (22–30g)

**THIS LARGE BAT IS** present in the south and west of North America, and discontinuously over much of north and central South America; it also occurs on Cuba, Jamaica, and Hispaniola. Those in North America are migratory, wintering in Mexico. It is chestnut-brown with a dark gray face, ears, and membranes, and a long, free tail. Males are markedly larger and heavier. It occurs in rugged, open countryside in North America, more forested areas in South America, and up to 8,500ft (2,600m) altitude. Roosts, holding around one hundred individuals, are in caves and crevices, and quite often in buildings. This bat's roosting and constant hunting calls are audible to humans. A fast, powerful flier, it leaves the roost after dark and mainly hawks large moths.

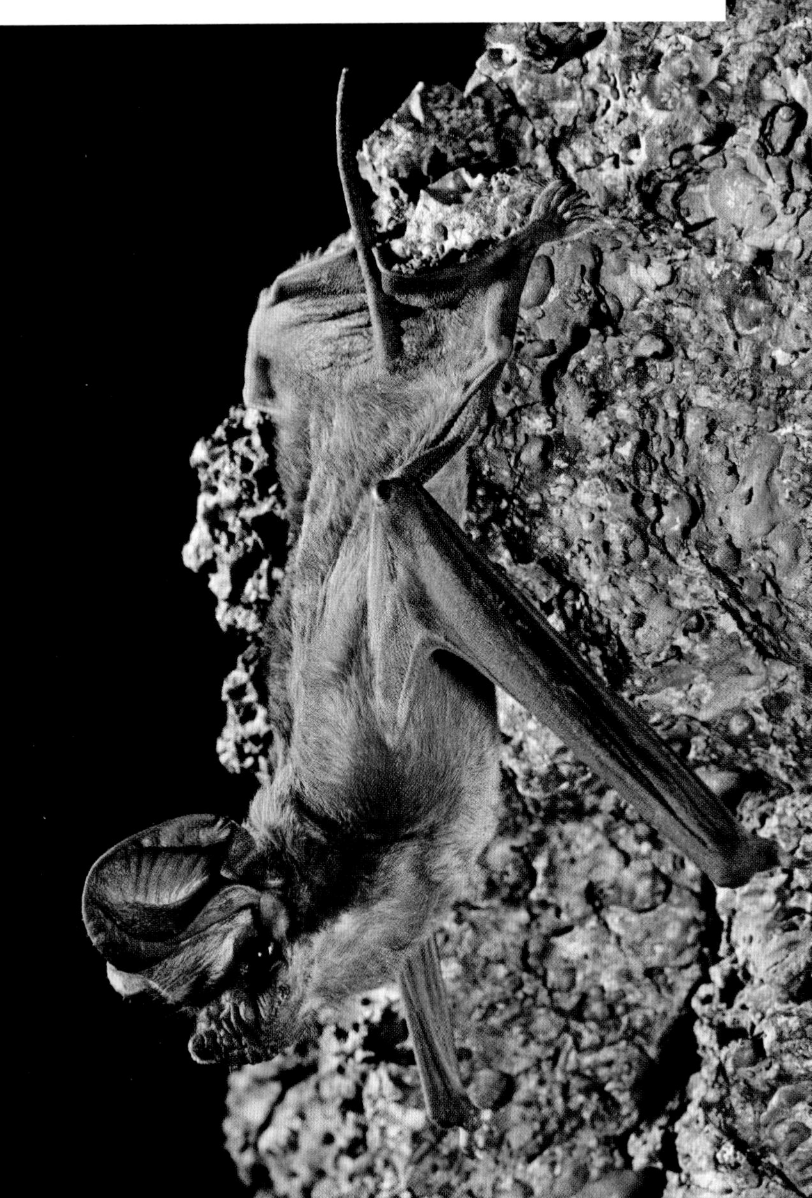

***Otomops martiensseni***

# Large-eared Free-tailed Bat

**IUCN STATUS** Near Threatened
**LENGTH** 4¾–5½in (12–13.8cm)
**WEIGHT** ¾–1¼oz (20–36g)

THIS BAT IS RECORDED sparsely from Ghana across to Uganda and down to South Africa. A large bat, it has big, broad-based ears and long, narrow wings; its tail projects well beyond the uropatagium. It occurs in various habitats, from rainforest in west Africa to savanna further east, and roosts in caves, sometimes in large groups. It also occurs around towns in South Africa, roosting in buildings. It is a fast flyer that catches moths on the wing in open air. Adult males associate with their harem of females, and any dependent young, all year round; its calls when leaving the roost are unique to individuals, and appear to have a social function. The species is believed to have declined substantially in east Africa, but may be increasing in South Africa.

## Promops

*Promops* is a Neotropical genus, represented by three species—*P. centralis, P. davisoni*, and *P. nasutus*. These bats are robust molossids with very short, blunt faces and rather small ears. *P. centralis* occurs from the south of Mexico through parts of Central America to the north of South America. It is generally common, and able to thrive in a variety of habitat types. *P. nasatus*, also a widespread species, is recorded from across much of northern South America, except the Amazon basin. However, it is not well studied and almost nothing is known of its habitat and habits. Even more mysterious is *P. davisoni*, which is known from dry forest areas in coastal Peru and Ecuador.

## Tadarida

The genus *Tadarida* includes one of the best-known bats in the world, and certainly in the Americas: *T. brasiliensis*. This bat, described on page 206, is a celebrated, long-distance migrant—a record-breaker in terms of flight speed and altitude. It is perhaps best known for its enormous summer roosting colonies. The other *Tadarida* species occur in the Old World only, and are much less well studied. However, among them is Europe's only molossid, *T. teniotis*. Its range includes much of Mediterranean Europe and north Africa, as well as parts of central and south Asia. It is a typical *Tadarida* in appearance—small and rather drab gray-brown with wrinkled lips, broad ears with bases that curve around the eyes on the outer edges, not quite meeting across the forehead on the inner edges, and a short, stout, and turned-up snout. Females form small maternity roosts, but it is otherwise solitary. This species is difficult to observe because, like most other molossids, it hunts high up in the open air.

This book recognizes another eight *Tadarida* species. *T. aegyptiaca* is found in the south of Africa, but also Arabia, west India, and patchily elsewhere in Africa. It is a very adaptable species, as likely to be encountered in the open desert or in towns as in more lushly vegetated habitats, and roosts in buildings as well as natural crevices. *T. fulminans* is found in Madagascar and also east and southeast Africa. With its long wings and narrow wingtips, it is a fast flier and also agile on the wing. *T. lobata* is another east African species with a more restricted range. It can be told from the other African *Tadarida* bats by a diamond of white fur on its back, between the shoulders. A fourth African species, *T. ventralis*, is known from Ethiopia south to Zimbabwe, but has rarely been observed alive.

The IUCN places two *Tadarida* species in a different genus—*Austranomus*. By this book's taxonomy they remain in *Tadarida*, but are distinct from the other species. *T. australis* is endemic to Australia. It is a widespread bat, notable for the bold white stripes on its flanks

that are revealed in flight. The similar *T. kuboriensis* is found in New Guinea.

The remaining two species are found in east Asia. They are rather poorly studied bats. *T. insignis* occurs in east China including Taiwan, and North and South Korea. *T. latouchei* is present in China, Thailand, and Laos.

### Neoplatymops

The family Molossidae also includes four monotypic genera. *Neoplatymops mattogrossensis* is closely related to *Molossops*. It is a cream-colored bat with much darker membranes, ears, and face, and a rather long, slim snout for a molossid; it also has rather short and broad wings for a molossid and forages more around vegetation than is typical for the family. This bat is widespread in South America.

### Platymops & Sauromys

*Platymops setiger* and *Sauromys petrophilus* are both rather similar to *Neoplatymops mattogrossensis* in appearance, but occur in east Africa and southern Africa respectively.

### Tomopeas

The final species, *Tomopeas ravus*, is a small, short-faced molossid with pale fur and dark bare parts. It has a short, slim snout and rather simple, large triangular ears. It occurs in Peru in coastal desert areas where it roosts in caves and rocky crevices. Declining because of habitat loss, including fumigation of its roosting caves in a bid to control rabies, this unusual molossid bat is endangered.

*Tadarida brasiliensis*

# Brazilian Free-tailed Bat

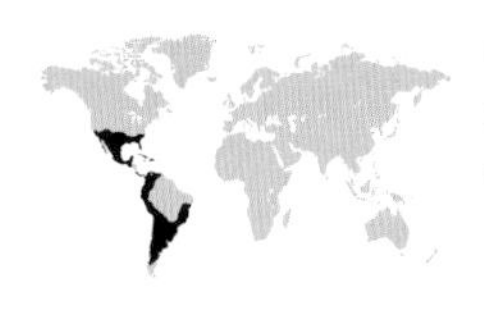

**IUCN STATUS** Least Concern
**LENGTH** 3⅛–3⅞in (7.9–9.8cm)
**WEIGHT** ¼–⅜oz (7–12g)

THIS WELL-KNOWN BAT occurs in the southern half of North America, through Central America and the Caribbean, on to much of South America. Many of the populations in the north migrate south for winter. A dusky, gray- to reddish-brown molossid, it has long, narrow wings that power the fastest flight recorded in a bat. It may be seen in almost any habitat, and famously forms huge roosts—most notably the 1.5 million that roost under the Congress Avenue Bridge in Austin, Texas, in the summer months (seen here). The sight of these bats leaving their roost at sunset attracts 100,000 tourists each year. Cave roosts can be even larger: the largest-known, in Bracken Cave, Texas, holds some ten to twenty million. Roosts contain a high level of ammonia gas from the bats' waste. Their metabolism is adapted to deal with this, and their flight patterns entering and leaving the cave encourage better air circulation. This is a highly aerial insectivore, which on migration exploits high-flying migrating moths; it may hunt as high as 10,800ft (3,300m). It breeds once a year, with each female bearing a singleton pup.

## VESPERTILIONIDAE

The family Vespertilionidae, or vesper bats, is the largest of all bat families and the second largest mammalian families in the world, surpassed only by Muridae (the Old World rats and mice). It holds 485 species in fifty-one genera. The family is represented worldwide, being established on every continent except Australia, and with a presence on many oceanic islands. The closest-related family to Vespertilionidae is Molossidae, with Miniopteridae and Cistugidae also closely allied— together these four families form the superfamily Vespertilionoidea. The species *Tomopeas ravus* is considered the most basal molossid bat, providing the closest living link between Vespertilionidae and Molossidae. English names for the group include "common bats," "plain-faced bats," and "evening bats." Some of the smaller species are known as "pipistrelles." This name applies most often to the genus *Pipistrellus*, but also to some other genera. The name "serotine" is also used for a variety of species from different genera.

With so many species and genera within the family, there is great diversity in vespertilionid appearance, anatomy, and ecology. The majority of species are small, and almost all are insectivorous, though a few regularly take larger prey. They lack the complex nose-leaves and other facial ornamentation seen in families such as Rhinolophidae and Phyllostomidae. Instead these bats have short, simple snouts, with just a few bumps or warts at most, and their echolocation calls are "shouted" through the open mouth, rather than emitted through the nose. The eyes are usually small and the ears often are relatively small too, though there are some notable exceptions. They have long tails, fully enclosed within the uropatagium. The fur color is usually rather plain brown or gray-brown, often with the face, ears, and wings just a shade darker. However, the family also includes some of the most distinctively marked and highly colored bat species in the world.

Some genera of vesper bats have very large ears. Some use their acute hearing to detect their arthropod prey on the ground or on foliage, picking up the tiny sounds that these creatures make as they move around. The bats can use a hovering flight as they pick prey from surfaces; they will even raid spiders' webs. These species often take quite large prey (in some cases even vertebrates), which they carry back to a night-feeding roost to consume. For those that capture insects in flight, the large uropatagium is often used as a sort of scoop, gathering in prey before it can be seized in the mouth. The vespertilionids include some large and powerful species that are adept hunters of birds. Among them is the only bat species known actively to chase and catch birds in flight (*Nyctalus lasiopterus*); these redoubtable ariel hunters, unlike the

gleaning species with their slow, fluttering flight, have impressive speed in the air, thanks to their long, narrow wings.

A large proportion of vespertilionids live in temperate climates, and their usual strategy for dealing with winter is to hibernate, rather than migrate. However, some do undertake lengthy migrations. The details of their migrations are mostly unknown, but can be inferred by the bats' disappearance from breeding areas in winter. The group includes many species that roost in buildings, especially over winter in the case of the hibernators; summer maternity roosts are more likely to be within tree hollows. A few tropical species habitually roost in the open, among clumps of dead leaves or other shelters. Others have modified thumbs and feet with pads of skin, enabling them to attach to a leaf or other smooth surface.

The majority of familiar bat species in temperate Europe and North America are vespertilionids. In Britain, for example, fifteen of the seventeen breeding bat species are members of this family. Through their sheer abundance, they are of great ecological importance. These bats are also of economic importance. They consume vast amounts of insects, and so serve as natural pest controllers; they control populations of insects that can spread diseases to people. The group also includes those bats worst affected by white-nose syndrome—a lethal fungal infection that spreads with devastating ease through bat hibernacula. Many species of vesper bats habitually roost inside buildings, a habit that makes their roosts very vulnerable to disturbance or even complete destruction. They do benefit from strict wholesale protection in many countries in Europe, and in some others elsewhere.

The social systems of vesper bats are quite varied. Often the sexes live apart until the fall, when males hold a territory and advertise it with a kind of "song," as well as through emitting scents from body glands. The most successful males attract several females while the rest attract none, but the harems tend to be short-lived. Bucking the general bat trend for bearing singleton young, some vespertilionids produce twins as standard; a few may have three or even four young.

**Kerivoula**

The genus *Kerivoula* contains twenty-four species. These are sometimes known as the "woolly bats," because of their dense, fluffy coats. The fur is quite long even on the face, where most other bats become quite sparsely haired, if not completely bare. This genus is distributed across Asia (mainly southeast Asia), with a few species in Africa and some more in New Guinea. They are forest species and tend to roost in small groups in hanging dead foliage.

There are full accounts for *K. furva* (opposite), *K. hardwickii* (page 212), *K. lanosa* (page 213), *K. minuta* (page 216), *K. pellucida* (page 216), and *K. picta* (page 214). Aside from these well-studied species, the genus is not particularly well known. Five species are little known, they are: *K. agnella* (islands off mainland Papua New Guinea), *K. cuprosa* (central and west Africa), *K. eriophora* (Ethiopia), *K. krauensis* (peninsular Malaysia, newly discovered), and *K. myrella* (Manus and other islands off Papua New Guinea); *K. eriophora*, is only known from its type specimen, taken in the nineteenth century. Yet another species, *K. depressa* of Myanmar, Cambodia and possibly elsewhere in mainland southeast Asia, is not yet assessed; it has only recently (2017) been found to warrant full species status, having formerly been classed as a subspecies of *K. hardwickii*.

The following three species are all threatened to a greater or lesser extent. *K. africana* is confined to a small area of Tanzania, east Africa. *K. flora* is present in Borneo, the Lesser Sundas, Bali, Sumbawa, and Sumba in Indonesia, and also in parts of Vietnam, though this population may represent a separate species. This bat is threatened by forest loss, due to logging and clearance to make space for farmland. Finally, *K. intermedia* is a primary forest-dependent species, found in Borneo and the Malay Peninsula.

The remaining species are: *K. argentata* (south and east Africa), *K. kachinensis* (Myanmar, Thailand, Laos, Cambodia, and Vietnam), *K. lenis* (India, peninsular Malaysia, and Borneo), *K. muscina* (New Guinea), *K. papillosa* (Thailand, Cambodia, Vietnam, peninsular Malaysia, Sumatra, Java, Sulawesi, and north Borneo), *K. phalaena* (west and central Africa), *K. smithii* (Africa, extending from Nigeria and Cameroon across to Uganda and Kenya), *K. titania* (Myanmar, Thailand, Cambodia, Laos, and Vietnam), and *K. whiteheadi* (Thailand, Borneo, and the Philippines).

***Kerivoula furva***

# Dark Woolly Bat

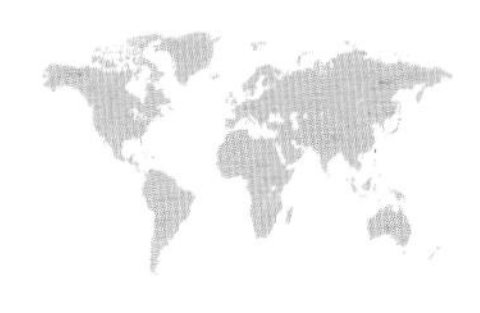

**IUCN STATUS** Not assessed
**LENGTH** unknown
**WEIGHT** ⅛–¼oz (4.3–7.3g)

**THIS SPECIES WAS** described in 2017 as a new species—it had been considered to be a form of either *K. hardwickii* or *K. titania*, but mitochrondrial and nuclear DNA differences indicated that it was distinct from both. It is found on the island of Taiwan (China), where it is the only *Kerivoula* species, and possibly occurs elsewhere in southeast Asia. Like other *Kerivoula* bats it has long, dense fur, darker brownish gray than other Asian *Kerivoula* bats, a small pinkish face, and large, wide-based, triangular ears. Females are consistently a little heavier than males. Very little is known of its ecology at this time, but it is likely to be highly dependent on forest habitats, to roost inside caves and buildings, and to forage with a slow, fluttering flight within forest foliage.

***Kerivoula hardwickii***

# Common Woolly Bat

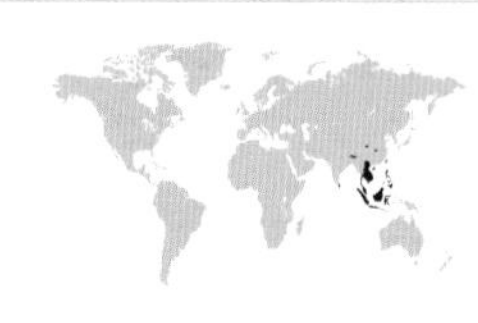

**IUCN STATUS** Least Concern
**LENGTH** 3⅛–3½in (7.8–8.9cm)
**WEIGHT** ⅛oz (4g)

**THIS TINY BAT IS COMMON** and widespread in south and southeast Asia. It occurs in both swamp and dry forests, and also farmland and around habitation. It has dense, very soft, reddish-brown fur, tiny eyes and a short pink snout. It hunts by night, catching flying insects, and typically roosts in caves and buildings. However, in Borneo, the species has formed a fascinating symbiotic association with the carnivorous pitcher plant *Nepenthes rafflesiana* var. *elongata*. Radio-tagged bats were found to roost every day inside the plant's prey-trapping pitchers, one individual (or a mother and baby) inside each. The bats' feces supply the plant with a valuable extra source of nitrogen; about 33 percent of the plant's total nitrogen is derived from this. This adaptable bat species is not thought to be in decline.

***Kerivoula lanosa***

# Lesser Woolly Bat

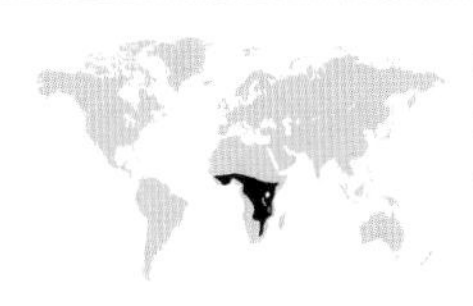

**IUCN STATUS** Least Concern
**LENGTH** 3⅛in (8cm)
**WEIGHT** ¼oz (6–8g)

**THIS WIDESPREAD SPECIES** is known from west and most of southeast Africa, down to the tip of South Africa. Its taxonomy is not well understood, and it may comprise a complex of several species. Its long guard hairs are white-tipped, giving a frosted appearance. The long fur, short snout, and steep forehead combine to produce an almost spherical head shape in profile. Many records of the Lesser Woolly Bat are from riverine and other wetland habitats; it has also been observed in dry woodland and savanna. It sometimes roosts in small groups inside old weaver and sunbird nests. It has a slow, fluttering flight, and takes insects on the wing as well as from foliage. Although widespread, this bat is uncommon and some populations may be threatened.

***Kerivoula picta***

# Painted Bat

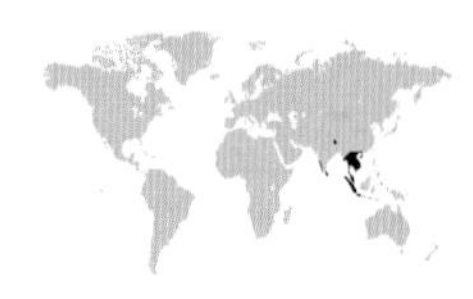

**IUCN STATUS** Least Concern
**LENGTH** 1¼–2¼in (3.1–5.7cm)
**WEIGHT** ⅛oz (4.5g)

**THIS SMALL, ATTRACTIVE** bat occurs across much of southeast Asia, from Myanmar and south China south through to Malaysia and western Indonesia. It is also known from parts of India, Sri Lanka, Nepal, and Bangladesh. It is a striking and colorful species, particularly in flight; the arms, fingers, and adjacent parts of the membranes are bright orange-red, standing out against the blackish central parts of the wing membranes. The function of its unusual coloration is unknown. The body and tail membranes are also orange, becoming white on the belly.

This is a species of dry forest and also visits plantations. It emerges after dark to undertake hunting flights that last an hour or two. It takes prey from on and around foliage near the ground, such as grasses and bushes, using a slow, fluttering flight. This bat has been observed roosting in male–female pairs (the female often accompanied by her single offspring) within folds of dying banana leaves. Other day roosts include old birds' nests. This species is affected by loss of dry forest and changes to plantation management, but remains fairly common.

*Kerivoula minuta*

## Least Woolly Bat

**THIS RARE, POORLY KNOWN** bat occurs in peninsular Malaysia, Borneo, and Thailand. A tiny species, its size range overlaps that of Kitti's Hog-nosed Bat, so it has a claim to be the world's smallest mammal. However, too few specimens have been measured so far to generate reliable average figures. In structure, it is a typical "woolly bat," with very small eyes and a short, pointed snout—almost lost within the dense, light-brown fur on its round head. This is a species of undisturbed primary forest. Here it hunts in the understory, flying slowly and with great agility as it takes insect prey from leaves. Its habitat is under huge pressure, as forest is cleared for logging and to make room for agriculture. This bat is almost certainly declining as a result.

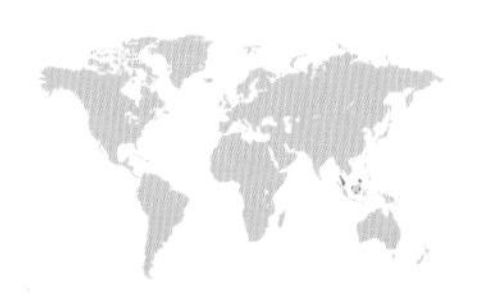

**IUCN STATUS** Near Threatened
**LENGTH** unknown
**WEIGHT** unknown

*Kerivoula pellucida*

## Clear-winged Woolly Bat

**THIS VERY SMALL BAT** occurs widely in southeast Asia, including southern Thailand, the Malaysian peninsula, Java, Sumatra, Borneo, and the Philippines. It is a rather pale, grayish-yellow species with markedly translucent wing membranes and ears. The snout is delicate and pointed, the eyes very small. It occurs in primary forest, sometimes also secondary, and roosts among dead foliage as well as other sheltered spots. It is a slow but agile flier that hunts for insects in the understory, using echolocation calls that are louder but lower-pitched than those of other *Kerivoula* species. Females give birth to a single, well-developed pup each year, which they initially carry with them on foraging flights. Extensive losses of primary forest throughout its range have caused a population decline of roughly 30 percent, hence its Near Threatened status.

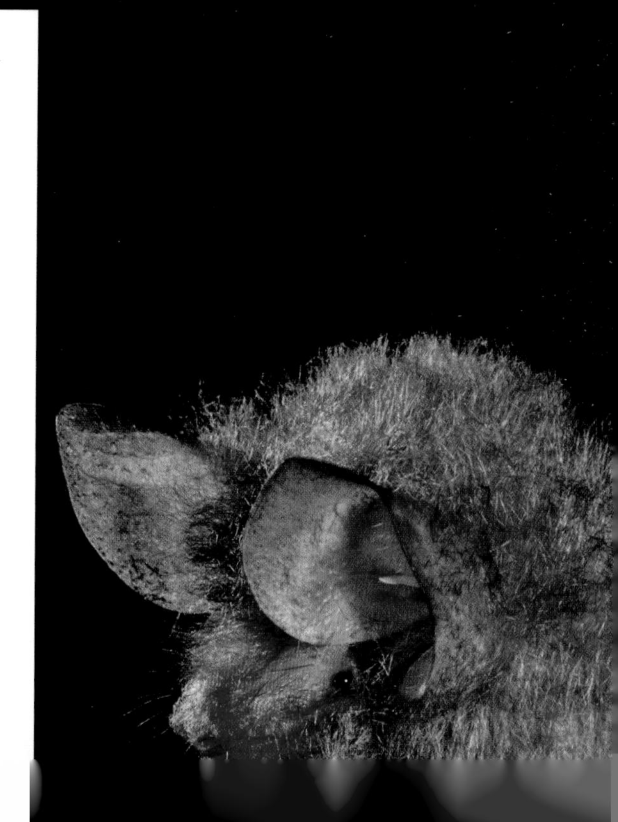

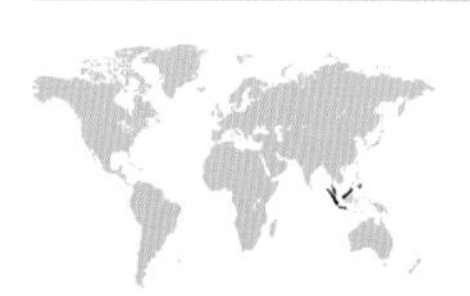

**IUCN STATUS** Near Threatened
**LENGTH** 2⅛–3½in (5.5–9cm)
**WEIGHT** ⅛oz (4.5g)

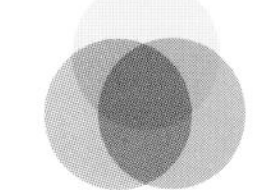


## Phoniscus

*Phoniscus* is a small genus of just four species. They are sometimes known as the "trumpet-eared bats" and have rather funnel-shaped ears with slim, pale-tipped tragi. Pale golden tips to the brown body hairs give a sparkling, gilded appearance in a couple of species. Two of the four, *P. jagorii* (southeast Asia) and *P. papuensis* (New Guinea and east Australia), are widespread. *P. atrox* is threatened because of its strict dependence on primary forest, which is dwindling within its range across the south of Thailand, the Malay peninsula, Sumatra, and parts of Borneo. The final species, *P. aerosa*, is known from only two specimens, which may have been taken in south Africa or in Asia. Unsurprisingly, it is little known; its English name of Dubious Trumpet-eared Bat reflecting the uncertainty of its origin, as well as doubt over its continued existence.

## Harpiocephalus

The Asian genus *Harpiocephalus*, the "hairy-winged bats," is often considered to be monotypic, holding just *H. harpia*—occurring in south and southeast Asia. A second, very similar species, *H. mordax*, is considered a subspecies of *H. harpia* by many authorities, but is recognized as a full species in this book's taxonomy. It has quite fluffy fur, a distinctly wide-ended nose, small eyes, and small ears; it also has very prominent upper canines. This bat occurs patchily in south Asia, mainland southeast Asia, and on Borneo. Its fragmented distribution and its reliance on primary forest in a region where rapid deforestation is ongoing puts it in danger.

## Harpiola

*Harpiola* is another small genus, closely related to *Murina*. It holds just two range-restricted species: *H. grisea* and *H. isodon*. These are delicate, small bats with fluffy fur, the hairs tipped yellowish to give a gilded appearance. They have small ears and short, dainty snouts with slightly tubular nostrils. Both species are little known. *H. grisea* occurs in a small area of north India, and *H. isodon* is found only on Taiwan (China).

***Harpiocephalus harpia***

## Murina

The genus *Murina* is large, with thirty-eight species (a significant increase on the number of species recognized by the IUCN, as several new *Murina* species have been discovered or split since 2009). This genus has an Asian distribution, with many species present in southeast Asia and several being island endemics. The group are known collectively as the "tube-nosed bats" because the nostrils are slightly fleshy and protuberant. They have very small eyes, and delicate pointed snouts, and tend to have smallish, rounded ears. They usually roost in the open, among dead foliage, in small groups. Four species—*M. cyclotis*, *M. rozendaali*, *M. suilla*, and *M. walstoni*—are described in detail on pages 220–21.

Fifteen *Murina* species are newly described species and very little is known of them so far. They are as follows: *M. annamitica* (from Laos), *M. beelzebub* (from Vietnam), *M. chrysochaetes* (from south China), *M. eleryi* (from Vietnam), *M. fanjingshanensis* (from China), *M. feae* (from Myanmar), *M. fionae* (from Laos), *M. guilleni* (from Thailand), *M. harpioloides* (from Vietnam), *M. hkakaboraziensis* (from Myanmar), *M. kontumensis* (from Vietnam), *M. lorelieae* (from China), *M. peninsularis* (Malaysia; split from *M. cyclotis*), *M. shuipuensis* (from China), and *M. silvatica* (Japan; split from *M. ussuriensis*).

Another three are poorly known. They are: *M. fusca* (China), *M. jaintiana* (northeast India and northwest Myanmar), and *M. pluvialis* (known only from its type specimen, taken in northeast India).

The following five *Murina* bats are considered threatened. *M. balaensis* is known only from a single forest in south Thailand; its entire geographic range is no more than 1½ square miles (4 km$^2$), making it very vulnerable to extinction. *M. tenebrosa* is known only from its type specimen, taken on Tsushima island, Japan, in 1962. *M. ryukyuana*, discovered in 1996, exists only on Okinawa, Tokunoshima, and Amami Ōshima islands in Japan; it is reliant on mature forest, which is scarce and dwindling on the islands. *M. aenea* is a rare and declining species found in Borneo, the Malay peninsula, and Thailand. Finally, *M. puta* of the upland forests of Taiwan (China) is considered threatened because of its very small geographic range.

The remaining species of *Murina* bats are: *M. aurata* (the north of south Asia and the east mainland of southeast Asia), *M. bicolor* (Taiwan (China)), *M. florium* (Indonesia, New Guinea, and Australia), *M. gracilis* (Taiwan (China)), *M. harrisoni* (Myanmar, China, Laos, Vietnam, Cambodia, and Thailand), *M. hilgendorfi* (China, Japan, Korea, and Russia), *M. huttoni* (the north

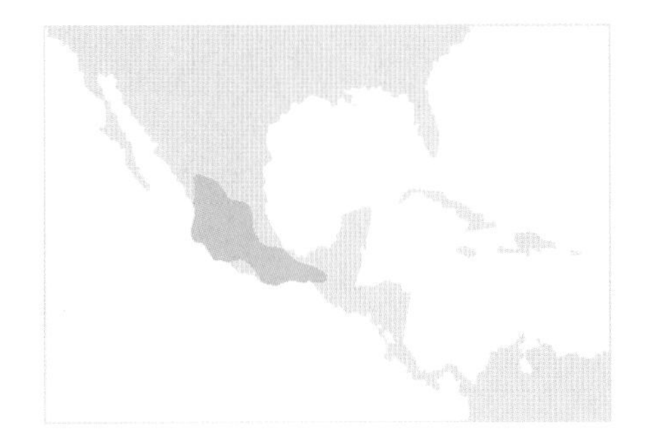

**Baeodon**

of south Asia and southeast Asia), *M. leucogaster* (India, China, and Vietnam); *M. recondita* (Taiwan (China)), *M. tubinaris* (the north of south Asia and mainland of southeast Asia), and *M. ussuriensis* (east Russia, Korea, Japan, and possibly China).

## Baeodon

The two bats in the genus *Baeodon* are closely allied to the genus *Rhogeessa* (the "yellow bats"), and are still placed within that group by the IUCN and other authorities. *B. alleni* is a rare species found only in Mexico; its population is quite fragmented, and some specimens show evidence of congenital problems related to inbreeding. *B. gracilis* is also a rare Mexico endemic.

*Murina cyclotis*

## Round-eared Tube-nosed Bat

**THIS SPECIES OCCURS** patchily in India, Sri Lanka, and Nepal, and extensively through mainland southeast Asia and parts of Borneo, the Philippines, and Indonesia. At least three subspecies are recognized. It has light brown fur with a whitish belly, round-tipped ears with slim, spear-shaped tragi, small eyes, and tubular nostrils. The membranes are blackish. It is found in primary and secondary forest and plantations, tending to roost in clumps of foliage in small groups (up to five); it may hibernate in parts of its range. This bat has a slow and highly maneuverable flight; it tends to hunt within the understory at no more than 6ft (2m) above ground level. It will sometimes eat large prey while hanging by feet and thumbs, leaving uneaten parts on its uropatagium as it feeds.

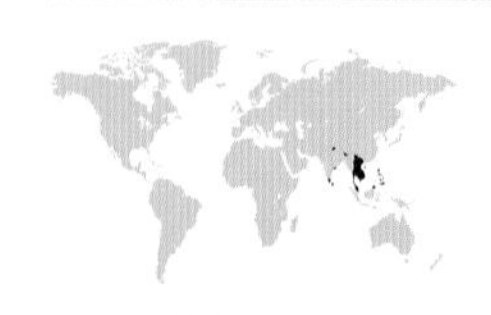

**IUCN STATUS** Least Concern
**LENGTH** unknown
**WEIGHT** unknown

*Murina rozendaali*

## Gilded Tube-nosed Bat

**THIS DISTINCTIVE SMALL** bat occurs in southeast Asia—on the Malay peninsula and northern Borneo (Sabah). It has light brown fur on the upperside, the hairs tipped light golden. This gives an almost sparkling appearance to the coat; its underside is paler. It has medium-sized ears with prominent tragi, small eyes, and a short, pointed muzzle tipped with short, tubular nostrils. As with other *Murina* bats, the tail is long, but fully enclosed within the large uropatagium. It occurs in lowland primary forest with streams, over which it will hunt, and roosts within foliage or in old birds' nests, most likely in small groups. It is a rare species, seriously threatened by habitat loss from logging and conversion of forest to farmland throughout its rather small range.

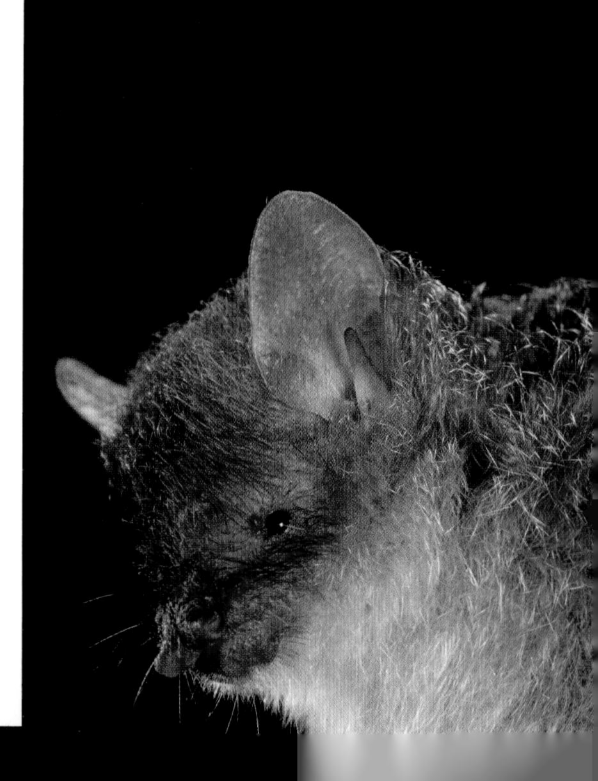

**IUCN STATUS** Vulnerable
**LENGTH** unknown
**WEIGHT** ⅛oz (3.8–4.5g)

***Murina suilla***

## Brown Tube-nosed Bat

**THIS SPECIES IS FOUND** in south Thailand, on the Malay peninsula, Borneo, Java, and Sumatra; also on Nias Island. Some biologists split it into two distinct species. It is a mid-brown *Murina* species with a paler belly, and has large broad wings and uropatagium. The eyes are very small, the ears rounded, and its nostrils enclosed within rather long, fleshy tubes. It occurs in dipterocarp forest from the lowlands up to about 5,000ft (1,500m), and may also forage in plantations and other well-vegetated habitats. It hunts within the foliage, flying slowly and using echolocation to pinpoint insect prey on, or flying around, leaves. It usually has a single pup, but is known to give birth to twins at least occasionally; its roosts are formed within tree foliage. It is an uncommon species.

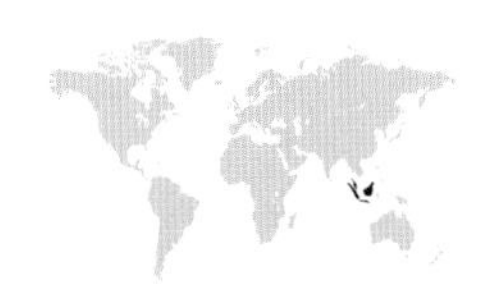

**IUCN STATUS** Least Concern
**LENGTH** 1⅜–2⅜in (3.5–6cm)
**WEIGHT** ⅛oz (3–5g)

***Murina walstoni***

## Walston's Tube-nosed Bat

**THIS BAT WAS DISCOVERED** in 2011, along with two other new *Murina* species and more than one hundred other animal species, in the Greater Mekong region of Vietnam and Cambodia. The exact limits of its distribution are not yet known. It is a very small bat with pale gingery-brown fur, fading to whitish on the belly. It has rather small, round-tipped ears, small eyes, and prominent nostril tubes. Very little is known of its ecology so far, but it is probably similar in most respects to other *Murina* bats. The species occurs in tropical semi-evergreen forest. It is not known whether it can occur in more marginal habitats, but it is likely to be suffering significant population decline as deforestation continues at a rapid rate throughout its known range.

**IUCN STATUS** Not assessed
**LENGTH** unknown
**WEIGHT** ⅛–¼oz (4.5–5.5g)

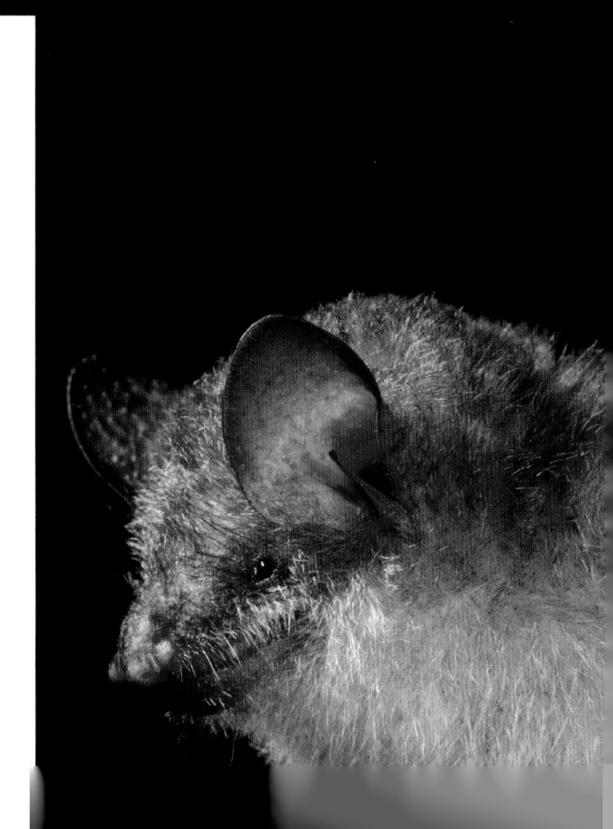

Myotis

## Myotis

The largest genus within Vespertilionidae, and indeed the largest of all bat genera, is *Myotis*. This book recognizes 125 species. *Myotis* bats are also known as "mouse-eared bats." They vary greatly in size, but typically have rather large, round-tipped ears, with long, slim tragi. Nearly all are insect-hunters, but the genus does include a few fishing specialists, with large feet adapted to seize prey while trawling the water's surface.

The *Myotis* species covered in detail are as follows: *M. adversus* (see page 226), *M. albescens* (page 226), *M. austroriparius* (page 225), *M. bechsteinii* (page 228), *M. blythii* (page 227), *M. bocagii* (page 227), *M. californicus* (page 229), *M. capaccinii* (page 235), *M. chinensis* (page 235), *M. ciliolabrum* (page 229), *M. evotis* (page 230), *M. formosus* (page 232), *M. grisescens* (page 234), *M. horsfieldii* (page 234), *M. keenii* (page 236), *M. leibii* (page 236), *M. lucifugus* (page 237), *M. myotis* (page 238), *M. nattereri* (page 242), *M. nigricans* (page 242), *M. occultus* (page 239), *M. pilosus* (page 231), *M. ridleyi* (page 243), *M. riparius* (page 243, *M. secundus* (page 244), *M. septentrionalis* (page 244), *M. sodalis* (page 240), *M. thysanodes* (page 245), *M. tricolor* (page 248), *M. velifer* (page 246), *M. vivesi* (page 248), *M. volans* (page 245), and *M. yumanensis* (page 249).

A number of *Myotis* species are recently discovered or recently split from other species. They are: *M. annatessae* (Vietnam), *M. attenboroughi* (Trinidad and Tobago), *M. badius* (China), *M. bartelsi* (Java and Bali), *M. clydejonesi* (Suriname), *M. gracilis* (Japan), *M. handleyi* (northern South America), *M. hyrcanicus* (Iran), *M. indochinensis* (Vietnam), *M. longicaudatus* (central Asia, Siberia, and the Far East), *M. midastactus* (Bolivia), *M. phanluongi* (Vietnam), *M. pilosatibialis* (Trinidad and Tobago), *M. rufoniger* (east Asia), and *M. weberi* (Sulawesi).

A further group are little known—in some cases because they are recently discovered and yet to be fully surveyed, and in others because only one or a few specimens have ever been obtained. These species are *M. aelleni* (Argentina), *M. alcathoe* (southern Europe), *M. anjouanensis* (Anjouan island, Comoros), *M. annamiticus* (Laos and Vietnam), *M. australis* (uncertain origin), *M. borneoensis* (Borneo), *M. bucharensis* (central Asia), *M. cobanensis* (Guatemala), *M. csorbai* (Nepal), *M. dieteri* (Congo), *M. diminutus* (Ecuador and Colombia), *M. federatus* (peninsular Malaysia), *M. frater* (Siberia, China, Korea), *M. hermani* (Thailand, Malaysia, Indonesia), *M. insularum* (Samoa, though dubious—only one specimen exists), *M. izecksohni* (Brazil), *M. longipes* (south Asia), *M. morrisi* (recorded in Ethiopia and Nigeria), *M. oreias* (Singapore), *M. petax* (eastern Russia, Korea, northeastern China and north Japan) *M. peytoni* (India), *M. punicus* (northwest Africa, Corsica, and Sardinia), *M. rufopictus* (Philippines), *M. schaubi* (west central Asia), *M. simus* (north Argentina, Paraguay, north Brazil, Bolivia, Peru, Ecuador, Colombia), *M. soror* (Taiwan (China)), and *M. stalkeri* (Indonesia).

The following *Myotis* bats are considered threatened: *M. bombinus* (Siberia, Korea, northeast China, Japan), *M. dasycneme* (northeast Europe into Asia), *M. dominicensis* (Dominica and Guadalupe), *M. macrotarsus* (Borneo, Philippines), *M. martiniquensis* (Lesser Antilles), *M. nyctor* (Lesser Antilles), *M. ruber* (Brazil, Paraguay, Argentina), *M. scotti* (Ethiopia), and *M. sicarius* (south Asia).

Six *Myotis* species are considered endangered—namely *M. atacamensis* (south Peru and north Chile), *M. findleyi* (Tres Marias islands, Mexico), *M. planiceps* (Mexico), and *M. pruinosus* (Japan). Also *M. yanbarensis* (confined to forest fragments on three islands on the Ryukyu archipelago, Japan), and *M. hajastanicus* (a barely known species, which occurs only around the Sevan Lake in Armenia).

The remaining *Myotis* species not described already are all classed as Least Concern by the IUCN. They are as follows: *M. altarium* (China and Thailand), *M. annectans* (south and mainland southeast Asia), *M. ater* (Vietnam, Thailand, Malaysia, Indonesia), *M. auriculus* (Arizona and New Mexico to Guatemala), *M. brandtii* (Europe and Asia through to China), *M. chiloensis* (Chile and southwest Argentina), *M. daubentonii* (the

**Myotis auriculus**

**Myotis goudoti**

north of Europe east to China and Japan), *M. davidii* (China), *M. dinellii* (Argentina, south Bolivia, south Brazil), *M. elegans* (Mexico to Costa Rica), *M. emarginatus* (the south of Europe, north Africa, west Asia), *M. escalerai* (southwest Europe), *M. fimbriatus* (China), *M. fortidens* (Mexico to Guatemala), *M. gomantongensis* (Borneo), *M. goudoti* (Madagascar), *M. hasseltii* (south and southeast Asia), *M. ikonnikovi* (central and east Asia), *M. keaysi* (Mexico, Central America, north South America), *M. laniger* (India, China, Vietnam), *M. lavali* (Brazil and Paraguay), *M. levis* (Argentina, south Brazil, Uruguay, Paraguay), *M. muricola* (south and southeast Asia), *M. mystacinus* (Europe, west Asia), *M. moluccarum* (Indonesia, New Guinea, Solomon Islands), *M. montivagus* (south, east, and southeast Asia), *M. macrodactylus* (east Asia), *M. macropus* (Australia), and *M. melanorhinus* (British Columbia south to Mexico), *M. nesopolus* (Netherlands Antilles), *M. oxyotus* (Costa Rica through to the north of South America), *M. pequinius* (China), *M. rosseti* (Cambodia, Thailand, possibly Vietnam), *M. siligorensis* (south and southeast Asia), and *M. welwitschii* (east and south Africa).

***Myotis austroriparius***

# Southeastern Myotis

**IUCN STATUS** Least Concern
**LENGTH** 3–3⅞in (7.7–9.7cm)
**WEIGHT** ¼oz (5.9–6.9g)

THIS BAT'S RANGE encompasses parts of the southeastern United States, reaching west as far as southeast Oklahoma and east Texas. It is non-migratory. In appearance a fairly typical, small *Myotis*, it has a blunt face, smallish narrow and pointed ears, and a thick coat of light, usually rather drab gray-brown fur; some individuals have bright reddish-orange fur. Its bare parts are pale pinkish brown. It hunts in wet swampy forest or drier woodlands close to lakes or large rivers, as it prefers to hunt over open water. It roosts inside warm, humid caves, and sometimes tree hollows or buildings. Over wide areas, colonies seldom include over a hundred individuals. But in northern Florida, nursery colonies include up to 100,000. An insectivore, this bat catches much of its prey low over water; it makes several foraging flights each night, resting in the roost in between. Although classed as Least Concern, with a total population in the hundreds of thousands, the species has undergone a marked population decline; surveys in Florida show a 50 percent decline at some maternity roosts in the later twentieth century.

***Myotis adversus***

## Large-footed Myotis

**THIS BAT SEEMS TO** have a patchy distribution, with records from peninsular Malaysia, Singapore, Borneo, Java, Sulawesi, and some other smaller islands in the region. It has dark membranes and thick gray-brown fur; this is sparse on the face, revealing some pink skin around the small eyes. It has relatively small ears, and large feet with long claws. It is a cave-roosting bat that tends to occur in low-lying forest close to still and flowing water, over which it hunts. Its echolocation allows it to sense water ripples that reveal the presence of aquatic insects or small fish, which the bat then captures by raking the surface with its feet. Its ecology is not well known and its population trend unclear—it may be affected by habitat loss due to deforestation.

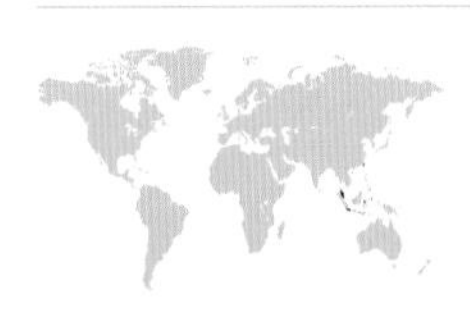

**IUCN STATUS** Least Concern
**LENGTH** 2–2¼in (5.2–5.6cm)
**WEIGHT** ¼–⅜oz (7–12g)

***Myotis albescens***

## Silver-tipped Myotis

**THIS BAT IS FOUND FROM** the south of Mexico through most of Central America, then extensively in tropical South America, reaching to Uruguay and northeast Argentina. It is a distinctive *Myotis*, with its white hair tips giving the otherwise dark fur a beautiful frosted appearance. It has fairly large, triangular ears with round tips. The face and limbs have sparse fur, showing pink skin below. It is most often encountered in forest areas, where it roosts in caves, buildings, and hollow trees. There are also records from plantations, parks, and large gardens. It mainly forages over lakes or streams, preying on flying insects, but stomach-content analysis shows that it also feeds on some fish. Though rarely very abundant, this bat is widespread and adaptable; it is believed to have a stable population.

**IUCN STATUS** Least Concern
**LENGTH** 3⅛–3¾in (7.9–9.6cm)
**WEIGHT** ⅛–¼oz (5–8g)

***Myotis blythii***

## Lesser Mouse-eared Myotis

**THIS BAT'S RANGE** extends from Spain across the south of Europe, on through the north Middle East and Afghanistan to north India; outposts also occur in south and east China. It has pale sandy-gray fur, whiter on the belly, and pinkish-brown membranes. The ears are long and rather narrow with slim tragi, the snout is short and quite "Roman-nosed." It has relatively large eyes. It forages in open habitats such as grassland, and mostly roosts and hibernates in caves and tunnels, or sometimes old buildings. This species has been proved to hybridize occasionally with its close relative *M. myotis*. It is susceptible to disturbance at its roosts, which—together with increased use of agricultural chemicals—is causing a general decline in parts of its range. Protection of key roosting caves is needed.

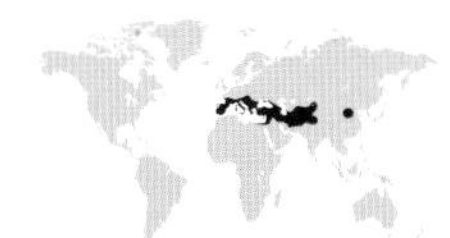

**IUCN STATUS** Least Concern
**LENGTH** unknown
**WEIGHT** unknown

***Myotis bocagii***

## Bocage's Mouse-eared Bat

**THIS AFRICAN *MYOTIS*** has a broad but apparently patchy range south of the Sahara, from Liberia across to Ethiopia (and also Yemen), and south to Swaziland and northeastern South Africa. It is absent from southwest Africa. It is pale sandy-grayish with a distinct rufous tint on the crown and upperside; the wing membranes are dark. The ears are long but narrow, with slender small tragi; the snout is very short. This species occurs in various habitats including tropical forest, wet savanna, plantations, and around wetlands, hunting low over the water. It appears to roost in foliage, either alone or in small groups. Its apparent scarcity is likely to reflect the difficulty surveyors have in finding and catching it; there are not thought to be any serious threats to its long-term survival at present.

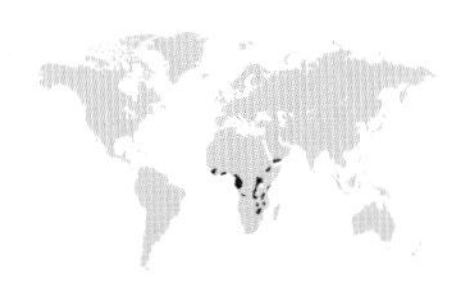

**IUCN STATUS** Least Concern
**LENGTH** 3⅜–4⅛in (8.7–10.5cm)
**WEIGHT** ¼oz (8g)

***Myotis bechsteinii***

# Bechstein's Myotis

**IUCN STATUS** Near Threatened
**LENGTH** 1¾–2in (4.3–5cm)
**WEIGHT** ¼–½oz (7–13g)

THIS SPECIES IS PATCHILY distributed in Europe. It is a rather pale gray-brown *Myotis*, with very large, broadly round-tipped ears. The face is pink and sparsely furred, the eyes small but prominent. The species is closely tied to deciduous woodland habitats, especially beechwoods. It mainly forages within the wood, roosting and hibernating in tree hollows and caves. Pregnant females and those with young frequently switch roost, with those at the same stage of their reproductive cycle tending to roost together, but it usually hibernates alone. It hunts above and within the tree canopy, hawking and gleaning for varied insect prey. Loss of mature woodland has led to declines across Europe. It is protected in most countries, and internationally through the Bonn Convention and Bern Convention. Management methods include protecting old trees with hollows.

***Myotis californicus***

## California Myotis

**THIS SMALL MYOTIS** bat occurs in the west of North America, from British Columbia to the far south of Mexico. It has fluffy, light tawny-brown fur (juveniles are slightly grayer), with contrasting blackish membranes, ears, and face. It occurs in all kinds of habitats from grassland and upland and coastal forests to deserts and suburban areas. Summer night roosts are small and change frequently; populations in the north may hibernate in larger groups. Roost sites include tree holes, caves, buildings, even the ground. This bat flies in two main phases—from 10–11 p.m., and again from 1–2 a.m.—and hunts insects, both by hawking and gleaning. Mating takes place in the fall, and females store sperm over winter. Females form small maternity roosts, giving birth to a single pup in early summer.

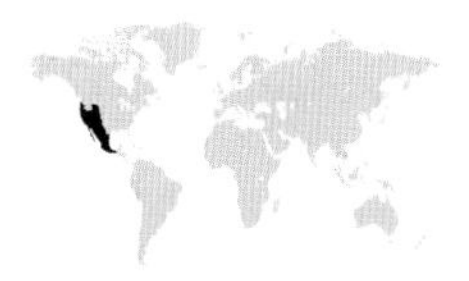

**IUCN STATUS** Least Concern
**LENGTH** 2¾–3¾in (7–9.4cm)
**WEIGHT** ¼oz (3.3–5.4g)

***Myotis ciliolabrum***

## Western Small-footed Myotis

**THIS BAT RANGES FROM** southeast Alberta and southwest Saskatchewan in Canada, then south to New Mexico and Kansas. It is named for its short hind feet (half the length of the lower leg). In general appearance it is very like *M. californicus* and *M. leibi*. Unlike *M. californicus*, its tail projects slightly (less than ⅛in/4mm) from the uropatagium. It hunts in woodland, farmland, and arid rocky countryside. It roosts and hibernates in mines, caves, and cracks in cliff faces, usually alone. Females rear young in small groups, often in erosion cavities and in spaces beneath large rocks. Both sexes swarm and roost together in early fall, the mating season. This bat is an insect-eater with a slow, maneuverable flight, often skimming close to rock faces or over water to catch beetles, moths, and other night-flying insects.

**IUCN STATUS** Least Concern
**LENGTH** 3–3⅜in (7.5–8.5cm)
**WEIGHT** ¼oz (3.5–6g)

*Myotis evotis*

# Long-eared Myotis

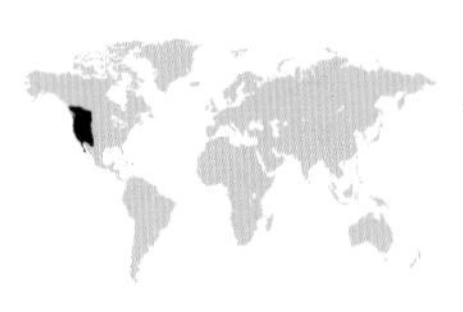

**IUCN STATUS** Least Concern
**LENGTH** 3⅜–3⅞in (8.7–10cm)
**WEIGHT** ¼oz (5–8g)

**THIS BAT OCCURS IN** the west half of North America, from Alberta and British Columbia, through to New Mexico, Arizona, California, and also Baja California, but not elsewhere in Mexico. It is a light tawny-yellow bat with a pale grayish underside, contrasting with its blackish face, ears, and membranes. The long, broad ears taper slightly to rounded tips. It occurs primarily in mixed coniferous forest, where it roosts in elevated tree hollows; other roosting and hibernation sites include derelict buildings, caves, and underneath loose bark. Studies show that in this species listening for prey-generated sounds is more important than echolocation. Although both techniques are used in the "search" phase, the bat does not use echolocation at all in its final attack. This bat has caught an underwing moth (*Catocala innubens*). Females gather in maternity roosts in spring, giving birth to a single pup.

*Myotis pilosus*

# Rickett's Big-footed Myotis

**IUCN STATUS** Near Threatened
**LENGTH** unknown
**WEIGHT** unknown

**THIS BAT OCCURS** in far east China, including Hainan island, and also in north Vietnam and Laos. It is a relatively large *Myotis* species with a longer jaw than most. It has large feet with powerful curved, laterally compressed claws on all toes. It occurs in varied habitats, but always near water, as it is a specialized fish-hunter, dipping its powerful feet into the water in flight to seize small fish and aquatic insects. The prey is then transferred to the mouth and carried to a night roost. The biology of these bats is otherwise not well known. A declining species, it is very sensitive to disturbance and also to degeneration in water quality; both affect prey abundance. Pollution is a serious issue on rivers and water bodies in China, though less so in Vietnam and Laos. This bat is also hunted in some areas.

***Myotis formosus***

# Hodgson's Bat

**IUCN STATUS** Least Concern
**LENGTH** 1¾–2¼in (4.3–5.7cm)
**WEIGHT** ½oz (15g)

**THIS BAT HAS BEEN** recorded in several parts of southeast Asia, from Pakistan east to China and Taiwan, and also parts of Malaysia, Indonesia, and the Philippines. Its range seems to be patchy, and different populations may represent separate species. It has pale, golden-yellowish fur, with pinkish-brown wing membranes. The ears and bare face are pink, the small black eyes standing out clearly. Its ears are triangular and medium-sized, its snout short and pointed. This is a forest species, occurring from lowland regions up to the Himalayan foothills. It will use primary forest, but also more disturbed secondary growth. It roosts in caves, among tree leaves and bushes, and in buildings, but tends to hibernate in caves. The species hunts insects by hawking and gleaning. In March, females form maternity roosts of up to fifty individuals, giving birth to a single pup in May or June. For the first two weeks after birth the females return to the roosts several times nightly to visit and groom their young. The pups start flying at about three weeks old. This is a widespread but rather rare species.

***Myotis grisescens***

## Gray Myotis

**THIS BAT IS ENDEMIC** to the eastern United States. Its range extends from Iowa in the northwest to West Virginia in the northeast, and south to Alabama and Florida. It is uniform dark gray in new molt, gradually bleaching to reddish brown during the nursery season with blackish wing membranes. It prefers to hunt over slow rivers and ponds in woodland. Summer roosting caves tend to be warm and close to water; in winter it uses cool, deep, well-aerated caves. Hibernating populations are large: 100,000 to over a million in a single cave. When hunting, this species may defend a small feeding territory from other bats. It declined severely from the 1960s to 1980s, but strict protection of key cave roosts is enabling rapid recovery.

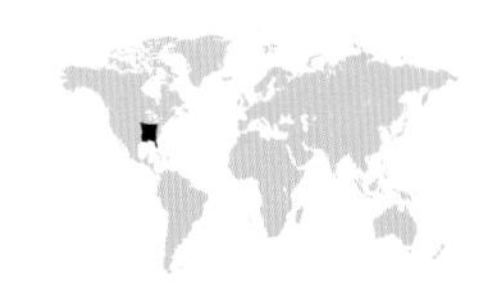

**IUCN STATUS** Near Threatened
**LENGTH** unknown
**WEIGHT** ¼–⅝oz (7–16g)

***Myotis horsfieldii***

## Horsfield's Myotis

**THIS BAT HAS POPULATIONS** scattered through the south and southeast of Asia, from India to the Philippines and Indonesia. It may prove to be more than one species; DNA studies are required on its various subpopulations to unravel its taxonomy. It is a robust *Myotis* with a thick coat of light gray-brown fur, fading to whitish on the belly, and a pink face. The ears are quite rounded, the snout medium-length and pointed. This bat is nearly always found close to water, but otherwise occurs in diverse habitats, from forests to town gardens. It is similarly comprehensive in its choice of roosting sites—anywhere offering shelter will do. It is an insectivore and catches its prey low over open rivers and streams. It is affected by habitat loss, but persists in some protected areas.

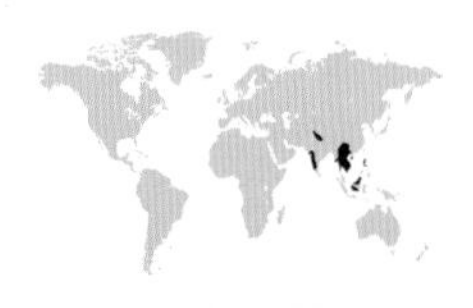

**IUCN STATUS** Least Concern
**LENGTH** unknown
**WEIGHT** unknown

***Myotis capaccinii***

## Long-fingered Myotis

**THIS IS AN UNCOMMON** bat that is found patchily in Spain and through the north of the Mediterranean to the Arabian Peninsula; it also occurs from Turkey to Iran and into northwest Africa. It is a drab brown *Myotis*, grayer on its underside, with a dark, pointed face and smallish ears; the feet are large for its size. It usually occurs close to rivers and lakes, roosting in nearby caves in colonies of up to 500. It mainly feeds on insects, which it catches over the water. The wide geographic range belies how sparse its distribution is suitable habitat is scarce, and water pollution is a major problem for this bat, as are development and drainage of waterways. The species is protected in many countries, as are some of its important maternity-roost caves.

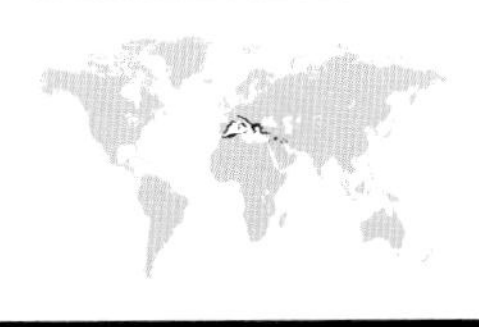

**IUCN STATUS** Vulnerable
**LENGTH** unknown
**WEIGHT** unknown

***Myotis chinensis***

## Large Myotis

**THIS BAT'S RANGE** encompasses central and southeast China, south to Thailand, Myanmar, and Vietnam; it is also likely to be present in Laos. Large for a *Myotis*, it has rather gray fur, paler on the belly and becoming sparse on the face to reveal pink skin below. The snout is robust and blunt, the ears small and narrow, with rounded tips. It is a bat of diverse habitats, from woodland and forest to open, rocky terrain and riversides; it roosts and hibernates in limestone caves. This species is usually found at low to mid altitudes. Little studied, it is probably mainly insectivorous, but could potentially take larger prey. Its wing shape suggests it is a slow and agile flier, hunting mainly by gleaning prey items from foliage.

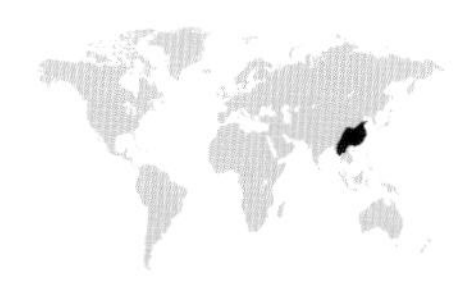

**IUCN STATUS** Least Concern
**LENGTH** unknown
**WEIGHT** unknown

***Myotis keenii***

## Keen's Myotis

**THIS RARE BAT** ranges from southeast Alaska south to Washington; it also occurs on Vancouver Island and the Queen Charlotte Islands. Its fluffy coat is orange-brown on the upperside, whitish below; its membranes, ears, and face are blackish. It prefers mature forest, but also visits towns and villages. It hibernates in caves within karst formations. The only known maternity roosts, which hold about forty individuals, are in a tree snag and in a rocky crevice warmed by geothermal activity. The species takes flying and non-flying prey, suggesting it gleans as well as hawking; it has been observed hunting over water. Mating occurs in the late fall. Females store sperm until spring, when fertilization occurs, and pregnancy may last up to sixty days. A single pup is then born, becoming independent after six weeks.

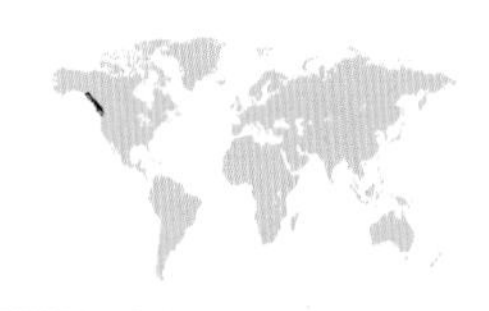

**IUCN STATUS** Least Concern
**LENGTH** 1⅝–2⅛in (4–5.5cm)
**WEIGHT** ⅛–⅜oz (4–9g)

***Myotis leibii***

## Eastern Small-footed Myotis

**THIS WIDESPREAD *MYOTIS*** species occurs in southeast Canada and the east and central United States. It is small with tiny feet, only ¼in (7–8mm) long. Its fluffy coat is yellowish brown with a dark undercoat. It has black wings, ears, and face. Its habitat is hardwood forest in both upland and lowland areas; in summer it roosts in tree holes, old buildings, rock crevices, and tunnels. In winter this bat hibernates inside very cold caves and mines, showing very high site fidelity. It has a slow, fluttering flight, and gleans as well as hawking. Maternity roosts are usually small (rarely more than thirty-five individuals). Post-breeding, the bats form larger mating "swarms"; each may mate with several partners. Pups, which at birth can be 35 percent of their mothers' weight, are born in summer.

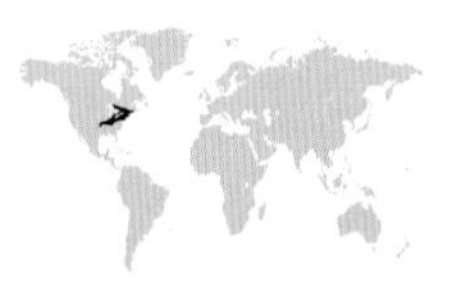

**IUCN STATUS** Least Concern
**LENGTH** 3–3⅜in (7.5–8.5cm)
**WEIGHT** ⅛–¼oz (3.5–6g)

***Myotis lucifugus***

# Little Brown Bat

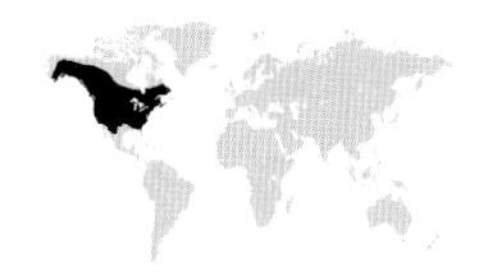

**IUCN STATUS** Least Concern
**LENGTH** 2⅜–4in (6–10.2cm)
**WEIGHT** ⅛–½oz (5–14g)

THIS IS A VERY WIDESPREAD North American bat, absent only from central southern states. It occurs in woodland habitats, but also in towns and more open, arid areas in some regions. It has a variable brown, yellowish or gray-brown coat with a paler underside. Its day roosts are very varied, ranging from buildings and tree holes to cracks under tree bark, under rocks, and in woodpiles, as well as bat boxes. In colder weather it forms hibernating populations of 200,000 or more individuals in caves and abandoned mines. These conditions unfortunately allow the spread of white-nose syndrome, which has devastated some colonies. This bat hunts mainly by open-air hawking. Mating occurs in pre-hibernation swarms; pups are born the following summer. Independent at just four weeks, they must quickly become effective foragers to reach a safe hibernating weight.

***Myotis myotis***

# Greater Mouse-eared Bat

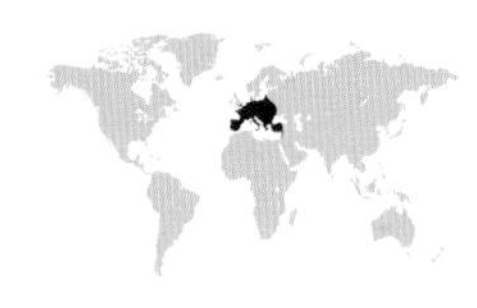

**IUCN STATUS** Least Concern
**LENGTH** 2½–3⅛in (6.5–8cm)
**WEIGHT** ¾–1⅝oz (20–45g)

A RELATIVELY LARGE *MYOTIS*, this species occurs across mainland Europe. The preferred habitat types have a mix of trees and open areas. Nursery roosts are in buildings and caves, whilst hibernation is only in caves. It is an adaptable insectivore that catches some of its prey on the ground, taking a large number of ground beetles and other insects including centipedes. It also catches prey by hawking and gleaning from foliage. It is polygamous, with dominant males guarding harems of about five females. Mating takes place in the fall, with young born in late winter and spring. Pregnancy lasts about two months, the same time that it takes pups to become independent. The species is of conservation concern and legally protected in some of the countries where it occurs.

***Myotis occultus***

# Arizona Myotis

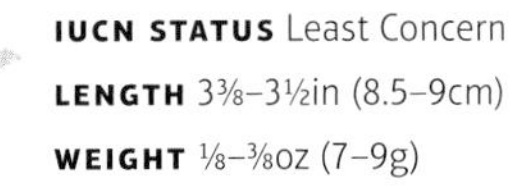

**IUCN STATUS** Least Concern
**LENGTH** 3⅜–3½in (8.5–9cm)
**WEIGHT** ⅛–⅜oz (7–9g)

**THIS BAT IS FOUND IN** the inland southwest of the United States, extending into central Mexico. Recently split from *M. lucifugus*, it closely resembles that species, but has only one premolar tooth behind each canine on the upper jaw, not two. It is found in woodland and more open country, and roosts in old tree hollows, under bridges, and sometimes loft spaces in summer. In winter it moves to caves and derelict buildings. This bat is closely associated with rivers and lakes. It can detect and target small prey at close range, so it often attacks swarms of newly emerged flying insects: it can eat 600 mosquitoes in an hour. Most females give birth in June to a single pup or occasionally twins. Maternity colonies may hold several hundred females.

***Myotis sodalis***

# Indiana Bat

**IUCN STATUS** Near Threatened
**LENGTH** 2¾–3⅝in (7.1–9.1cm)
**WEIGHT** ⅛–⅜oz (5–11g)

**THIS SPECIES RANGES** widely across the eastern United States. It is a grayish-brown bat with a whitish-gray underside, and dark wings and ears. The ears are fairly large, the snout short. It forages around riverside forest and open fields, forming small summer roosts in tree hollows or behind loose bark. Hibernation sites—typically cold, limestone caves with standing water on the floor—can hold hundreds or thousands of closely packed bats. The population seems to be distributed across relatively few hibernacula. In the fall, swarms develop in the vicinity of the hibernacula. Males roost separately by day, but visit the hibernacula at night to meet the females and mate. Females give birth to their pups in summer. They are attentive mothers, moving pups around in the roost as air temperature changes, and accompanying them on their first foraging flights. This species suffered severe decline in the last century, with the total known population falling from 883,720 to 387,300 in 2003 and consequently is classed as endangered by the U.S. Fish and Wildlife Service. Disturbance at hibernation roosts was believed to be a key cause. Protection of its major hibernacula has helped numbers to stabilize.

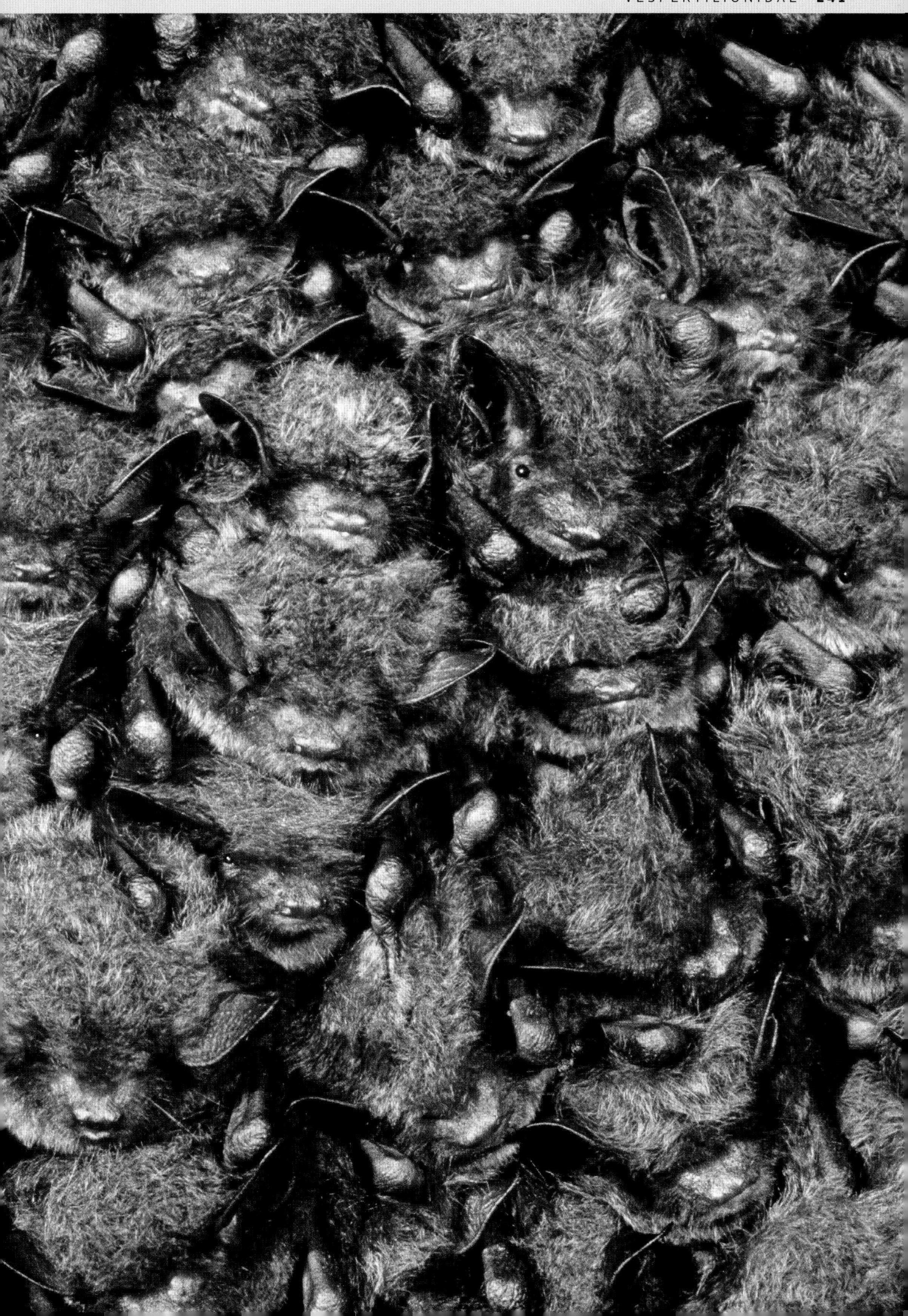

***Myotis nattereri***

## Natterer's Bat

**THIS BAT IS FOUND** over most of Europe including the British Isles, Spain, south and east Sweden (possibly also south Norway), the Baltic states, and south Finland. It is also present in parts of the Middle East and west and southwest Asia. It is grayish, paler below, with long but narrow ears that curve back at their tips with a long tragus, a short pink muzzle, and relatively large eyes. The broad wings are rather translucent gray. An adaptable bat, it occurs in woodland but also more open habitats with scattered trees. This species forms roosts in tree holes and may hibernate in buildings, squeezing into narrow crevices, alone or in groups of up to 150. Its quiet but wide-bandwidth echolocation calls enable it to find insects flying very close to foliage and near or on the ground.

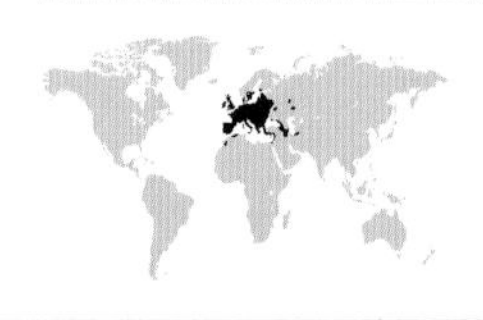

**IUCN STATUS** Least Concern
**LENGTH** 1¾–2in (4.4–5.1cm)
**WEIGHT** ⅛–⅜oz (5–9.5g)

***Myotis nigricans***

## Black Myotis

**THIS BAT IS FOUND** from the south of Mexico through most of Central America. Its range continues south through tropical South America, just reaching Uruguay and north Argentina. It is a small and very dark *Myotis*, often almost black, or very dark gray-brown, barely paler on the belly. The ears are rather small, the snout short and pointed. It uses diverse habitats, occurring in varied woodland, plantations, parks, and gardens. Roost sites similarly vary, including caves, crevices, buildings, tree holes, and spaces under loose bark. Large roosts may be shared with other *Myotis* bats. A female carries her pup for the first two or three days, then leaves it in the maternity roost while she forages; pups attain adult weight in as little as three weeks. This is an abundant species with a stable population.

**IUCN STATUS** Least Concern
**LENGTH** 2in (5cm)
**WEIGHT** ⅛oz (3–4g)

***Myotis ridleyi***

## Ridley's Myotis

**THIS BAT IS FOUND** in southeast Asia, in south Thailand and the Malay peninsula, and Borneo. It is a very small species, with a shorter forearm measurement than other *Myotis* species within its range. The fur is rather dark gray-brown, paler on the underside; its wing membranes are grayish. It has small, triangular ears, a short snout, and small eyes. This bat occurs only in lowland dipterocarp forest, and roosts in caves. However, there are no caves in the region where it occurs in Thailand, so alternative roosts must also be acceptable. It prefers areas with running water, and forages within the forest understory, its short broad wings giving it great agility on the wing. This rare species is poorly known and further study is needed. It does occur in some protected areas.

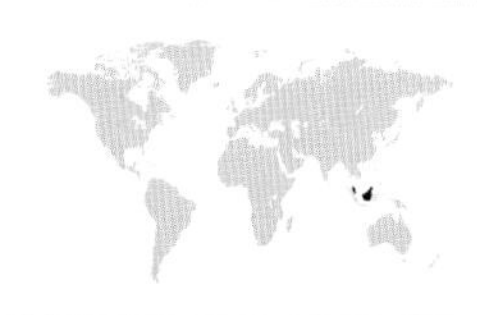

**IUCN STATUS** Near Threatened
**LENGTH** unknown
**WEIGHT** ⅛–¼oz (4–6g)

***Myotis riparius***

## Riparian Myotis

**THIS BAT'S EXTENSIVE** range extends south from Honduras in Central America to encompass nearly the whole of South America, reaching as far as the most northerly parts of Chile and Argentina. Dark golden-brown in color, with darker ears, face (especially at the tip of the snout), and membranes, it is a fairly typical *Myotis* in appearance. It is found most reliably in wet tropical forest, but also occurs in dry, deciduous forest and farmland, mainly in lowland areas, but occasionally up to elevations of 3,300ft (1,000m). It roosts in colonies in caves, flying out at dusk to hunt insects, often over rivers, ponds, and other fresh water. There is some uncertainty about its taxonomy, and molecular DNA research is needed to determine whether this bat actually comprises several species.

**IUCN STATUS** Least Concern
**LENGTH** 2⅞–3⅝in (7.3–9.1cm)
**WEIGHT** ⅛oz (4.4g)

***Myotis secundus***

## Taiwan Long-toed Myotis

**THIS SMALL, DELICATELY** built *Myotis* bat is found only on the island of Taiwan (China), and has only recently become known to science. It has shaggy, dark gray-brown fur, with a pinkish face, and brown ears and wing membranes. The snout is short, the ears relatively long. Its head is domed and rounded; it has rather long, delicately slender toes. This bat lives throughout the central, forested parts of Taiwan in both highland and lowland areas—mainly in forests, but it has also been observed in more open areas with rocky outcrops. It roosts in various caves and crevices. Although it is an island endemic with a naturally small range and small population, and so potentially rather vulnerable, it appears to be surviving well and occurs in several protected areas.

**IUCN STATUS** Least Concern
**LENGTH** unknown
**WEIGHT** ⅛oz (3–5g)

***Myotis septentrionalis***

## Northern Myotis

**THIS SPECIES OCCURS** over an extensive area of west Canada and the east and central United States, from Yukon to Alabama and Georgia. It has light brown fur, pale grayish on the underside, with dark brown membranes, face, and ears. The ears are fairly large with rounded tips. In summer, the population is mainly in the north of the range, in boreal forest. It roosts in trees and in buildings. Hibernation sites are small, often cold caves and crevices. Maternity roosts usually hold between twenty and sixty females; pups are independent at one month. This bat emerges soon after sunset and hunts primarily for moths, mostly picking them from foliage. Its echolocation calls are nearly inaudible to moths. The species has suffered declines from habitat loss.

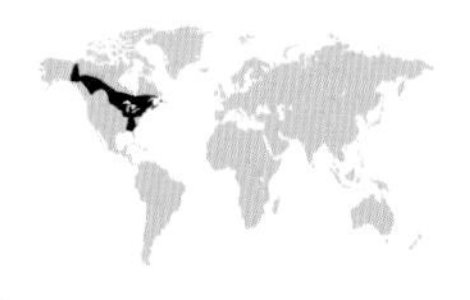

**IUCN STATUS** Least Concern
**LENGTH** 3⅛in (7.8cm)
**WEIGHT** ¼–⅜oz (6–9g)

***Myotis thysanodes***

## Fringed Myotis

**THIS BAT'S RANGE** extends from the southwest of Canada through the far west of the United States and well into the south of Mexico. Quite a large *Myotis*, it has light sandy-brown fur, contrasting well with the blackish-brown face, ears, and membranes. Its snout is short and pointed, its ears rather large and broad-based, with pointed tips. It inhabits oak and ponderosa pine woodlands in both highland and lowland areas, but also more open, arid regions; populations in the north are migratory. Roosts are inside large caves, buildings, and other sheltered places; they can hold several hundred animals. This bat hawks and gleans for prey around the canopy, taking moths and beetles, among other insects. Although the species has a stable population overall, the Mexican subspecies *aztecus* has suffered significant habitat loss and has probably declined.

**IUCN STATUS** Least Concern
**LENGTH** 3⅜in (8.5cm)
**WEIGHT** ⅛–⅜oz (5–9g)

***Myotis volans***

## Long-legged Myotis

**THIS BAT HAS A** broad distribution in the western North America, from the far south of Alaska through west Canada and the west and central United States, into central Mexico. It is a large *Myotis* with medium, gray-brown fur, whiter on its underside, and dark ears, face, and membranes. It has large ears with pointed tips, and relatively long legs with small feet. It is found in varied, but mainly quite arid and open habitats. This bat roosts in large caves and mines. Maternity roosts hold groups of a few hundred females with their young; the species hibernates in groups of several thousand. It leaves its roost well before dusk, but can consume all the prey it requires for the night in a single foraging flight lasting less than an hour.

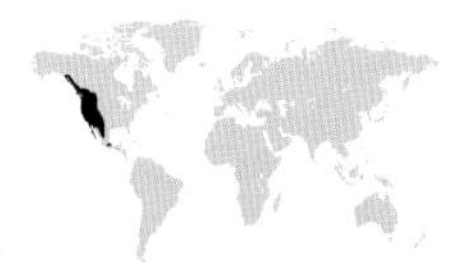

**IUCN STATUS** Least Concern
**LENGTH** 3¼–3⅞in (8.3–10cm)
**WEIGHT** ⅛–⅜oz (5–10g)

***Myotis velifer***

# Cave Myotis

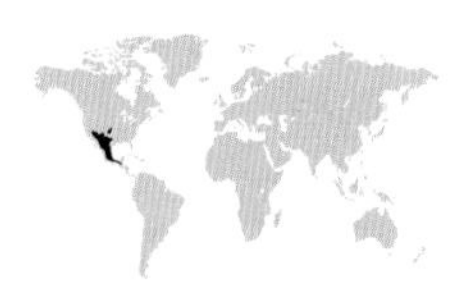

**IUCN STATUS** Least Concern
**LENGTH** 2¼in (5.6cm)
**WEIGHT** ¼–½oz (7–14g)

THIS BAT OCCURS IN southern United States; Nebraska marks its northern limit, and its range continues south through central Mexico and Central America to Honduras, though not reaching either coast. Its color varies, from paler and browner in the east of its range to much darker in the west. It has dark wing membranes and a dusky face and ears. It is a forest species found at a variety of elevations. Roosts, formed in caves, tunnels, old mines, and buildings, can hold up to 50,000 individuals, but may be much smaller. This bat, a relatively long-winged and fast-flying *Myotis*, preys on all kinds of insects. It hunts around vegetation, gleaning and hawking, and can consume 80 percent of its total daily requirement in its first two hours of foraging. However, its first action after leaving the roost is often to head for water and drink. Some populations migrate south in winter; others hibernate. Disturbance and vegetational overgrowth of cave entrances has caused serious decline in many areas.

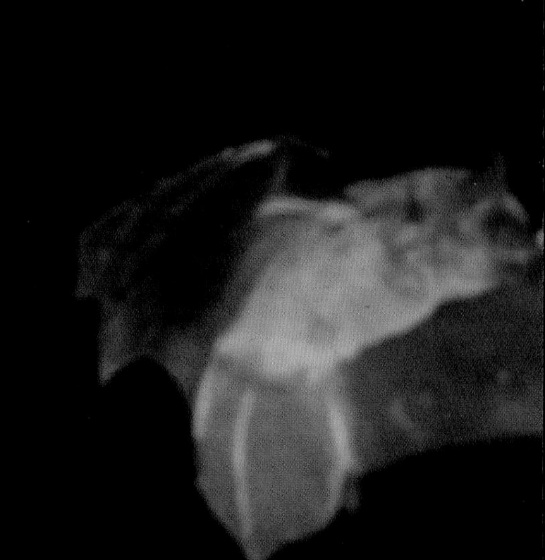

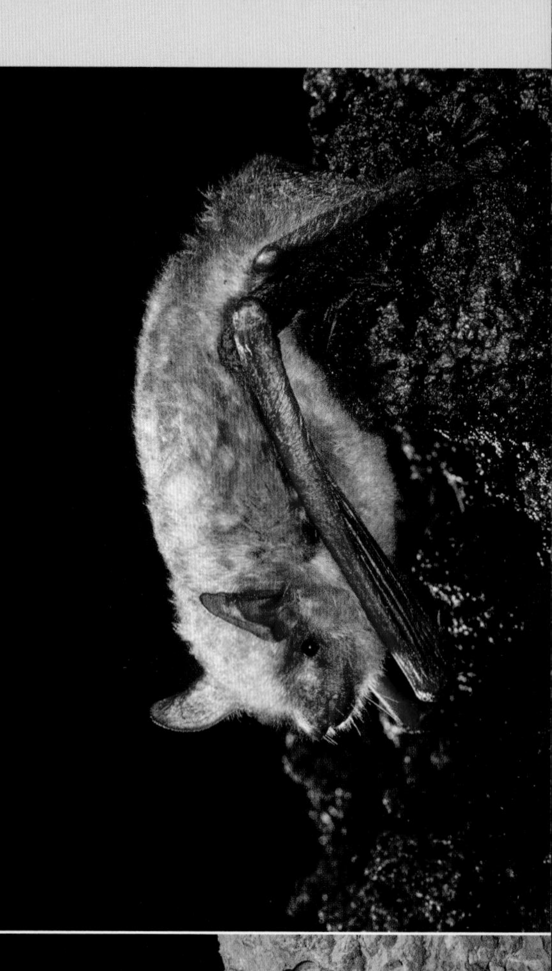

***Myotis tricolor***

## Cape Hairy Bat

**THIS IS AN AFRICAN** species which appears to have a wide but patchy distribution, mainly on the east side of the continent from Ethiopia to South Africa. It has pale tawny-golden fur, paler on the belly, while its ears, face, and wings are brownish pink. Its ears are slender. It is broad-winged with a light, agile flight. It uses varied habitats, such as scrubland and savanna, but also rainforest. It roosts in large, undisturbed caves and disused mines with pools of water at the bottom. Roosts are usually large, holding at least 1,500 bats; maternity roosts tend to be smaller. It hawks or gleans insects at the edge of vegetation. Timing of breeding season likely varies across its range; the singleton pup is weaned at six weeks.

**IUCN STATUS** Least Concern
**LENGTH** unknown
**WEIGHT** unknown

***Myotis vivesi***

## Fish-eating Bat

**THIS BAT IS RESTRICTED** to the Gulf of California, with populations along the coast of Sonora and Gulf islands, and also parts of the Pacific shore of Baja California. The largest *Myotis* in the Americas, it is a fast—though not agile—flier, with long, powerful wings. It is a fishing bat that hunts over sheltered sea water and roosts in coastal caves, rock crevices, and spaces between and under boulders. It catches prey by skimming its large, strong-clawed feet in the water; the claws are compressed laterally to reduce drag. As well as small fish, this bat feeds on marine crustaceans and can drink seawater, producing highly concentrated urine. Suitable roost sites for this species are scarce. Its strongest populations are on islands, and any introduction of non-native predators on those islands could be devastating.

**IUCN STATUS** Vulnerable
**LENGTH** 5¾in (14.5cm)
**WEIGHT** ⅞oz (25g)

*Myotis yumanensis*

# Yuma Myotis

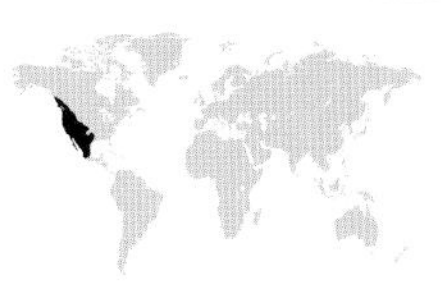

**IUCN STATUS** Least Concern
**LENGTH** 1½–1⅞in (3.9–4.8cm)
**WEIGHT** ¼oz (6g)

**THIS SMALL BAT OCCURS** in a broad band down the western side of North America, from British Columbia in the north down to the southern half of Mexico, and east across to Oklahoma. Some populations in the north likely migrate south in winter. The species is strongly associated with water. Its roosts are never far from rivers or lakes, though they may be in any habitats from juniper woodlands and thorny scrub to deserts. It roosts in caves, attics, abandoned buildings, mines, and underneath bridges, in large colonies. It hunts flying insects very efficiently. This bat is common and widespread, but suffered its first recorded cases of white-nose syndrome in May 2017. Seen here are a male, a pup and a mother whose fur has been bleached orange by ammonia fumes.

## Rhogeessa

This book recognizes eleven *Rhogeessa* species, a group of attractive, golden-furred pipistrelles of which *R. tumida* is described on page 257. The other species within this American genus include the endangered *R. genowaysi*, known from only two patches of mature forest in south Chiapas, Mexico; it is fast losing its habitat. *R. minutilla* occurs in Colombia and Venezuela and *R. mira* is a very rare species but with an apparently stable population—known only from a particular river basin in Michoacán, Mexico. Three of the remaining six species are little known. They are *R. hussoni* (an apparently rare, recently split species known from just three records in Brazil and Suriname), *R. menchuae* (a newly described species from humid forest in Central America), and *R. velilla* (another newly described species, from Ecuador). The other species are *R. aeneus* (found on the Yucatán peninsula, Mexico), *R. bickhami* (Central America), *R. io* (Central and South America), and *R. parvula* (the south of Mexico).

***Rhogeessa io***

## Hesperoptenus

Bats of the Asian genus *Hesperoptenus* are also known as the "false serotines." They are blunt-faced, pipistrelle-like bats with large teeth. Detailed accounts for *H. blandfordi* and *H. tickelli* are on pages 253 and 252, respectively. *H. doriae* of Malaysia and *H. gaskelli* of Sulawesi, Indonesia, are both rare and little known species. The remaining species is *H. tomesi*, which is present on mainland Malaysia and Borneo, possibly also in Thailand; it is a declining, forest-dependent species.

## Aeorestes

The American genus *Aeorestes* was formerly part of *Lasiurus*, the "tree bats" or "yellow bats." However, its four species have now been placed in a new genus (albeit with an old name, originally used in 1870), based on DNA studies that show they are distinct from other *Lasiurus*. The group is sometimes known as the "gray-haired bats" because of the distinctive, frosted appearance of their fur, the result of whitish tips to the longest, coarsest hairs in the coat (the guard hairs). *A. cinereus* is described on page 256. The other species are *A. egregius*, a rare, species of Panama and possibly Colombia, the small

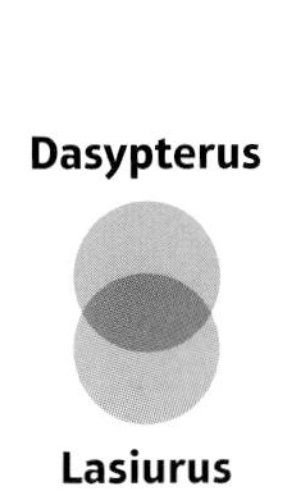

*A. semotus* of Hawaii—a recent split from *A. cinereus* (yet to be assessed by the IUCN but very likely to be placed in a threatened category)—and *A. villosissimus* (until recently considered a South American subspecies of *L. cinereus*).

## Dasypterus

*Dasypterus* is another American genus of "yellow bats," which has been separated from *Lasiurus* in recent years. *D. ega* is described on page 255, *D. intermedius* on page 254, and *D. xanthinus* on page 253. The fourth and final species is *D. insularis*, which is endemic to Cuba. It is rare with a patchy distribution. Like other related species, it roosts not in caves or other shelters, but in the open, usually among palm tree fronds. This habit gives versatility, but also places this bat at risk of heavy losses in extreme weather events, which can be frequent in the Caribbean.

## Lasiurus

Although several of its species have recently been assigned to new genera as modern-day understanding of genetic relatedness grows, the American genus *Lasiurus* remains fairly large, with eleven species. Traits shared by this and other closely related genera include the habit of roosting among foliage in the open. For warmth, some species (especially from regions in the north) have particularly thick coats. The fur is often reddish or yellowish in color, sometimes with silvery tips to the hairs. Another unusual trait is that these bats possess two pairs of teats (most bats have only one pair of true teats, though may have accessory "false teats"). Members of this genus regularly have twins and triplets, occasionally even quadruplets.

Three species of *Lasiurus* are covered in detail. They are *L. blossevillii* on page 257, *L. borealis* on page 258, and *L. seminolus* on page 259. Of the rest, *L. atratus* is a species found widely in the Guianas, the south of Venezuela, and Bolivia. *L. castaneus* occurs in elfin forest in Costa Rica and in Panama, and is little known. *L. degelidus* is endemic to lowland parts of Jamaica; it has only six known sites. *L. ebenus* is known from a single specimen, taken in 1999 in Brazil. *L. frantzi* is a recent split from *L. blossevillii*. It is known from Mexico. *L. minor* is of uncertain taxonomy, and occurs on Puerto Rico, the Bahamas, and Hispaniola. *L. pfeifferi* is a poorly known but declining species endemic to Cuba. The final species is *L. varius*, a little-known species found in Argentina and Chile.

***Hesperoptenus tickelli***

# Tickell's Bat

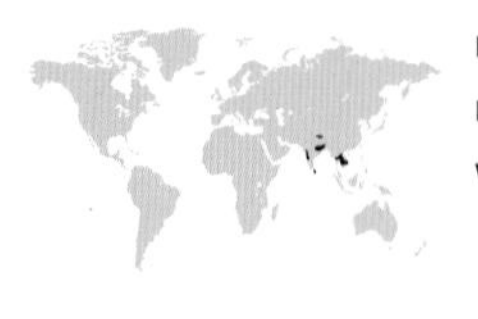

**IUCN STATUS** Least Concern
**LENGTH** 2¾in (7cm)
**WEIGHT** ¼–¾oz (6–10g)

THIS BAT HAS A PATCHY distribution, including the west, south and northeast of India, Sri Lanka, and parts of Nepal, Bhutan, and Bangladesh; it also occurs in southeast Asia, from the extreme east of Myanmar to north Thailand, Laos, Vietnam, and Cambodia. It is a pale yellowish bat with pinkish-brown membranes (darkening toward the outer edge), ears, and face. The ears are small and rounded; the snout short and pointed. It hunts along forest edges and in grassland, vegetated dunes, and other open areas, and roosts alone or in small groups in foliage in the canopy. This bat's hunting flight is steady and agile; it mainly feeds on slow-flying insects such as beetles and termites. Females give birth once a year, to a single pup. This species is elusive, and may be more widespread than present records indicate.

***Hesperoptenus blanfordi***

## Blanford's Bat

**THIS SMALL, SOUTHEAST** Asian bat ranges from south Myanmar and Thailand through the Malay peninsula; it is also recorded in Laos, Vietnam, and Borneo. It has darkish chestnut-brown fur on upper and underside, and rather short, broad wings. There are flat pads on the bases of its feet and thumbs. The face is short and blunt, the ears smallish and rounded with kidney-shaped tragi. Its habitat is varied, including lowland wet forest and dry dipterocarp forest in lowlands and uplands, usually close to small rivers and streams. It roosts in the mouths of limestone caves in small colonies of fewer than ten individuals; possibly also in bamboo clumps. This species is neither very common nor well studied, but its adaptable ways mean it is unlikely to be under any particular threat.

**IUCN STATUS** Vulnerable
**LENGTH** unknown
**WEIGHT** ¼oz (6.1–6.4g)

***Dasypterus xanthinus***

## Western Yellow Bat

**THIS SPECIES, A CLOSE** relative of *D. ega* (and sometimes considered a subspecies of it), is found in the far southwest United States and into north and central Mexico. It is a medium-sized vesper bat with fluffy, pale yellowish fur and a pinkish face. The wing membranes are dark gray, contrasting with the pink limbs and digits; the pink ears are smallish and low-set with broad tragi. The face is pointed, and the teeth large for its size. It occurs in light woodland and more open grassland and farmland with scattered trees, often close to water, and tends to roost alone, hanging from the midrib of a large leaf (especially palm leaves). The typical litter is two pups, but up to four may be born. It is common with a stable population.

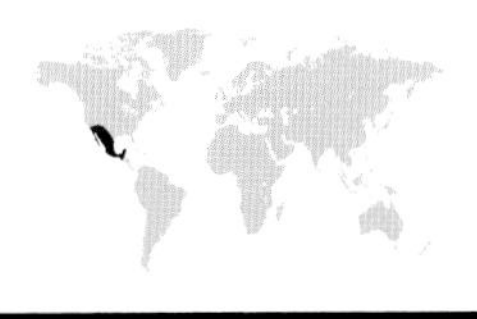

**IUCN STATUS** Least Concern
**LENGTH** 4–4¾in (10.2–11.8cm)
**WEIGHT** ⅜–½oz (10–15g)

***Dasypterus intermedius***

# Northern Yellow Bat

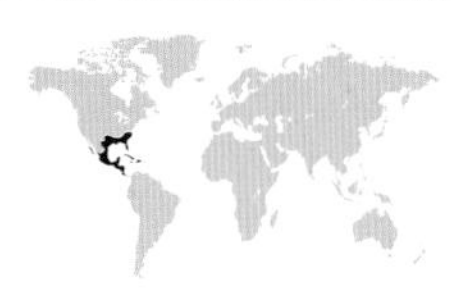

**IUCN STATUS** Least Concern
**LENGTH** 4⅞–5¼in (12.1–13.2cm)
**WEIGHT** ⅝–¾oz (17–22g)

**THIS IS A RATHER UNCOMMON** bat of the southeast United States through Mexico (mainly east coast) to Guatemala and Honduras. In appearance it is similar to the Southern Yellow Bat. It is found in woodland and dry scrub, usually roosting by itself in dead foliage, such as these dead palm fronds. It does not hibernate as such, but becomes torpid in prolonged cold spells. Females give birth in early summer, usually to twins, within small maternity roosts; if these are disturbed, females carry their young away to a new site. By late summer, adult females and pups form feeding colonies, hunting night-flying insects around open areas. Males tend to be solitary year-round, apart from the mating season in late fall. This bat is uncommon in most of its range, and is threatened by habitat loss from deforestation.

***Dasypterus ega***

# Southern Yellow Bat

**IUCN STATUS** Least Concern
**LENGTH** 4–4¾in (10.2–11.8cm)
**WEIGHT** ⅜–⅝oz (10–18g)

**THIS BAT RANGES** from the far southeast of the United States through east Mexico and throughout Central America; it also occurs across much of east and central South America to north Argentina. At least some northern hemisphere populations migrate south in winter (particularly males), taking a coastal route and making short sea crossings. This species occurs in woodland and forest (seen here among Spanish moss). In the United States it particularly favors ornamental palm fronds for roosting, which provide ideal camouflage; the prevalence of ornamental palm trees here is thought to be facilitating its northward spread. It hunts well above ground level. It is usually solitary when roosting, but can form gatherings of up to twenty. Most females produce twins or triplets.

***Aeorestes cinereus***

# Hoary Bat

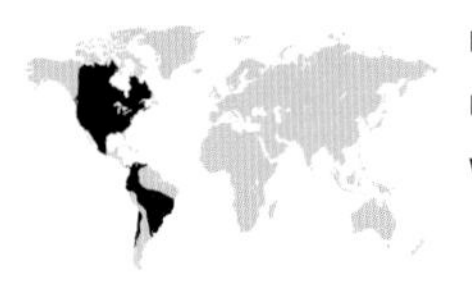

**IUCN STATUS** Least Concern
**LENGTH** 5⅛–6in (13–15cm)
**WEIGHT** ¾–¼oz (20–35g)

**THIS SPECIES HAS AN** extensive range, covering most of North America. Some migrate, wintering in states in the south through to Central America. It also occurs over large parts of South America. It is a dark gray-brown, largish bat (females significantly larger), with white hair tips that give a frosty appearance. It occurs in varied wooded habitats. In most of North America the sexes occupy different areas except during mating season. Females are more abundant in the west, and also in lowland regions; males seem to favor uplands. They roost alone, usually in tree foliage. Fast, powerful fliers, these bats catch insect prey in flight, producing calls audible to human ears. Migrants travel in groups; they can suffer significant mortality from wind turbines.

***Rhogeessa tumida***

## Black-winged Little Yellow Bat

**THIS VERY SMALL BAT** is present in southeast Mexico and through Guatemala, Honduras, Nicaragua, and just into Costa Rica. It has quite bright orangish fur with contrasting blackish wings and ears. The face is short and rather blunt, with a dark muzzle and small eyes; the ears are quite large. It occurs in wooded habitats at up to 5,000ft (1,500m), especially slightly disturbed deciduous forest, roosting in large groups in caves and buildings. The bats leave their roost at sunset and hunt close to ground or over water, following a set patrolling route as they search for flying insects. Activity is highest just after sunset and just before dawn. Twin pups are born during the rainy season. Though generally uncommon, this species is not considered to be declining or threatened.

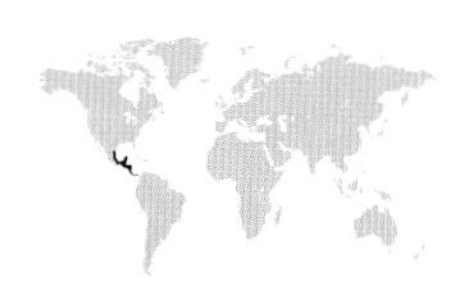

**IUCN STATUS** Least Concern
**LENGTH** 1⅝in (4.1–4.2cm)
**WEIGHT** ⅛oz (3.5–4g)

***Lasiurus blossevillii***

## Western Red Bat

**THIS BAT IS FOUND** widely in the west of North America, down through Mexico and across South America, including in Bolivia, north Argentina, Uruguay, Brazil, and Trinidad and Tobago; it is also found on the Galápagos islands and Cozumel island. It has bright ginger-red fur on its head and upperside, and is grizzled silver-gray below. The muzzle is short and blunt, and the ears small and narrow. It occurs in forests, but also parks and gardens, plantations, and other habitats with plenty of lush vegetation; it usually roosts in foliage, on its own or in small clusters. The species seems to breed in the austral summer, having litters of three or four pups. At least some populations probably migrate in the austral winter. Its ability to live in varied habitats will help protect it (to some extent) from the effects of ongoing deforestation.

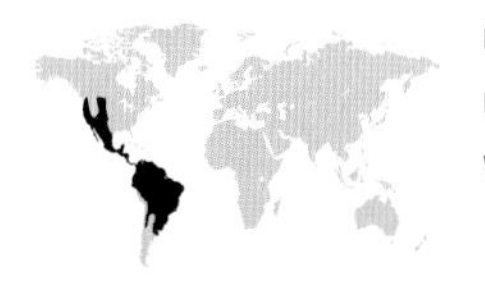

**IUCN STATUS** Least Concern
**LENGTH** 4–4⅛in (10–10.5cm)
**WEIGHT** ¼–⅜oz (6–10g)

***Lasiurus borealis***

# Eastern Red Bat

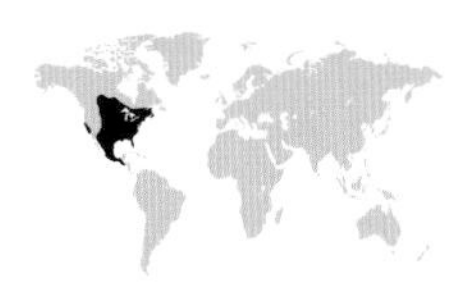

**IUCN STATUS** Least Concern

**LENGTH** 3⅝–4⅝in (9.3–11.7cm)

**WEIGHT** ¼–½oz (7–13g)

**THIS BAT OCCURS IN** eastern North America, including southeast Canada. Some populations in the north migrate to southern United States in winter, burrowing into bunch grass and leaf-litter to hibernate. Its body fur is light gray-brown on the upperside and whitish on the underside, with a light orange-red face; females tend to be paler. It avoids populated areas, occurring mainly in remote forest, and roosts by day alone among foliage. A fast-flying bat, it mainly catches moths in flight. Often the prey can detect the bat's echolocation calls and tries to evade capture. A rapid rising and falling chase ensues, with prey escaping 60 percent of the time. Females give birth in summer, usually to twins. This species is still considered common, but appears to have declined over the past 150 years.

***Lasiurus seminolus***

# Seminole Yellow Bat

**IUCN STATUS** Least Concern
**LENGTH** 1¾–2in (4.4–5.2cm)
**WEIGHT** ¼–½oz (7–14g)

**NAMED FOR AN INDIGENOUS** people in Florida, this bat occurs in the southeast of the United States and just into Mexico. The coat is dark, glossy nut-brown with a white underside. It occurs in forest and woodland, roosting alone close to the forest edge. They may hang up within foliage, or often within clumps of Spanish moss on pine tree branches between ca. 5 and 20ft (1.5 and 6.1m) in height. They emerge at dusk and fly rapidly in pursuit of insects around treetops, over water, and around artificial light sources. These bats mate in the fall, coupling in flight. Females produce three or four pups, born in early summer; the young become independent at about one month old. They become torpid on the coldest winter days; those living furthest north migrate southward.

Arielulus

Eptesicus

## Arielulus

The pipistrelle-like bats of the Asian genus *Arielulus* are also known as "sprites"; they tend to have golden-tipped, sparkly looking fur. *A. aureocollaris* is particularly pretty, with its yellow collar and facial markings. This species is found in Laos, Vietnam, Thailand, and possibly in China. *A. circumdatus* occurs widely though patchily distributed across south and southeast Asia. *A. torquatus* is endemic to Taiwan (China). The other two *Arielulus* species are both considered vulnerable. *A. cuprosus* is a rare bat found only in Borneo, while *A. societatis* occurs on the Malaysian peninsula.

## Eptesicus

With thirty-five species, *Eptesicus* is one of the larger vespertilionid genera. The group are known as "house bats," or sometimes as serotines. This genus is distributed on both sides of the Atlantic, and includes some widespread and familiar species. Among the *Eptesicus* bats is the well-known species *E. fuscus* (see page 262), possibly the most often-observed bat in North America. This book also presents full accounts for *E. chiriquinus* and *E. furinalis* on page 264.

Four *Eptesicus* species are considered to be under threat. *E. innoxius* (Ecuador and Peru) and *E. laephotis* (Peru, Bolivia, Argentina), the former because of a small and fragmented distribution, and the latter because of rapid habitat loss. *E. guadeloupensis* (Guadeloupe, Lesser Antilles) is rare on the single small island where it occurs, as is the almost unknown montane species *E. japonensis* (Japan).

Several *Eptesicus* bats are poorly known mainly due to a paucity of records. They are: *E. alienus* (south Brazil, Uruguay, east and central Argentina), *E. dimissus* (Thailand, Nepal), *E. humboldti* (Colombia and Venezuela), *E. kobayashii* (Korea), *E. platyops* (Nigeria and Senegal), *E. taddeii* (south and southeast Brazil), *E. tatei* (India), and *E. velatus* (Brazil, Bolivia, Argentina, Paraguay, Peru). The following species are yet to be assessed by the IUCN: *E. anatolicus* (Turkey and Syria), *E. bottae* (southeast Europe and Asia), *E. diaphanopterus* (northeast Brazil), *E. lobatus* (Ukraine), *E. ognevi* (India), and *E. pachyomus* (east Asia).

***Eptesicus brasiliensis***

The remaining species of *Eptesicus* are: *E. andinus* (Colombia, Ecuador, Peru, Bolivia, and Venezuela), *E. bobrinskoi* (Kazakhstan), *E. brasiliensis* (Mexico, Central America, and widely in South America), *E. diminutus* (Venezuela, Brazil, Paraguay, Uruguay, the north of Argentina), *E. floweri* (Mali, Sudan, Mauritania), *E. gobiensis* (central and east Asia), *E. hottentotus* (east and south Africa), *E. isabellinus* (Iberia and north Africa), *E. macrotus* (Chile, Argentina, Paraguay), *E. magellanicus* (Argentina, Chile), *E. montanus* (extensively across South America), *E. nilssonii* (western Europe across Asia to Japan), *E. pachyotis* (south and southeast Asia), and *E. serotinus* (Europe, north Africa, the Middle East).

## Glauconycteris

The genus *Glauconycteris*, the "butterfly bats," are mostly pale bats with round ears that have lower folds curving round below the eyes. They have a light, fluttering, butterfly like flight, and are often on the wing before sunset. The genus includes some strikingly attractive species, such as the black-and-white patterned *G. superba*, from West Africa, and *G. variegata*, found in southern Africa, which is pale with a bold, reticulated pattern on its wing membranes.

The widespread *G. argentata* is described on page 265. There are another ten *Glauconycteris* species; the majority are poorly known forest species from west and central Africa. These include *G. curryae* (found in Democratic Republic of Congo and Cameroon), *G. egeria* (Cameroon and Central African Republic), *G. gleni* (Cameroon), *G. humeralis* (the east of the Democratic Republic of Congo and Uganda), *G. kenyacola* (Kenya), and *G. machadoi* (Angola).

Of the remainder, *G. alboguttata is* of the Democratic Republic of Congo and Cameroon. *G. beatrix*, which occurs over most of west Africa, and *G. poensis*, which ranges from west Africa across to central Africa, possibly reaching as far east as Tanzania. The final species, *G. atra*, is newly described (2018) from the Democratic Republic of Congo.

***Eptesicus fuscus***

# Big Brown Bat

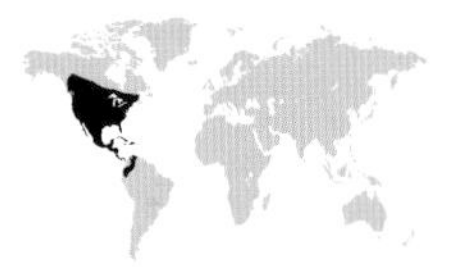

**IUCN STATUS** Least Concern
**LENGTH** 3⅜–5⅜in (8.7–13.8cm)
**WEIGHT** ½–¾oz (14–20g)

**THIS BAT'S HUGE RANGE** covers much of the south of Canada, virtually the entire United States and Mexico, central areas of Central America, and parts of Venezuela and Columbia; it also occurs in much of the Caribbean. It has mid-brown fur with a darker face, ears, and membranes. The ears are smallish, the face blunt, and eyes very small. Often observed in urban and suburban areas, this bat roosts in various artificial structures, and is also common in wider countryside, especially forests. It is extremely tolerant of cold conditions in hibernation roosts, provided it has sufficient fat stores, so uses spaces quite exposed to the air, even hibernating among icicles. The diet consists of large, often hard-bodied flying insects, especially beetles. Mating occurs at any point through winter. The following summer, females give birth in single-sex day roosts. In some areas one pup is normal, but in the east of the United States twins are usually born. Females leave pups packed together in a cluster to forage, locating their offspring on return by scent. This widespread species has a stable population.

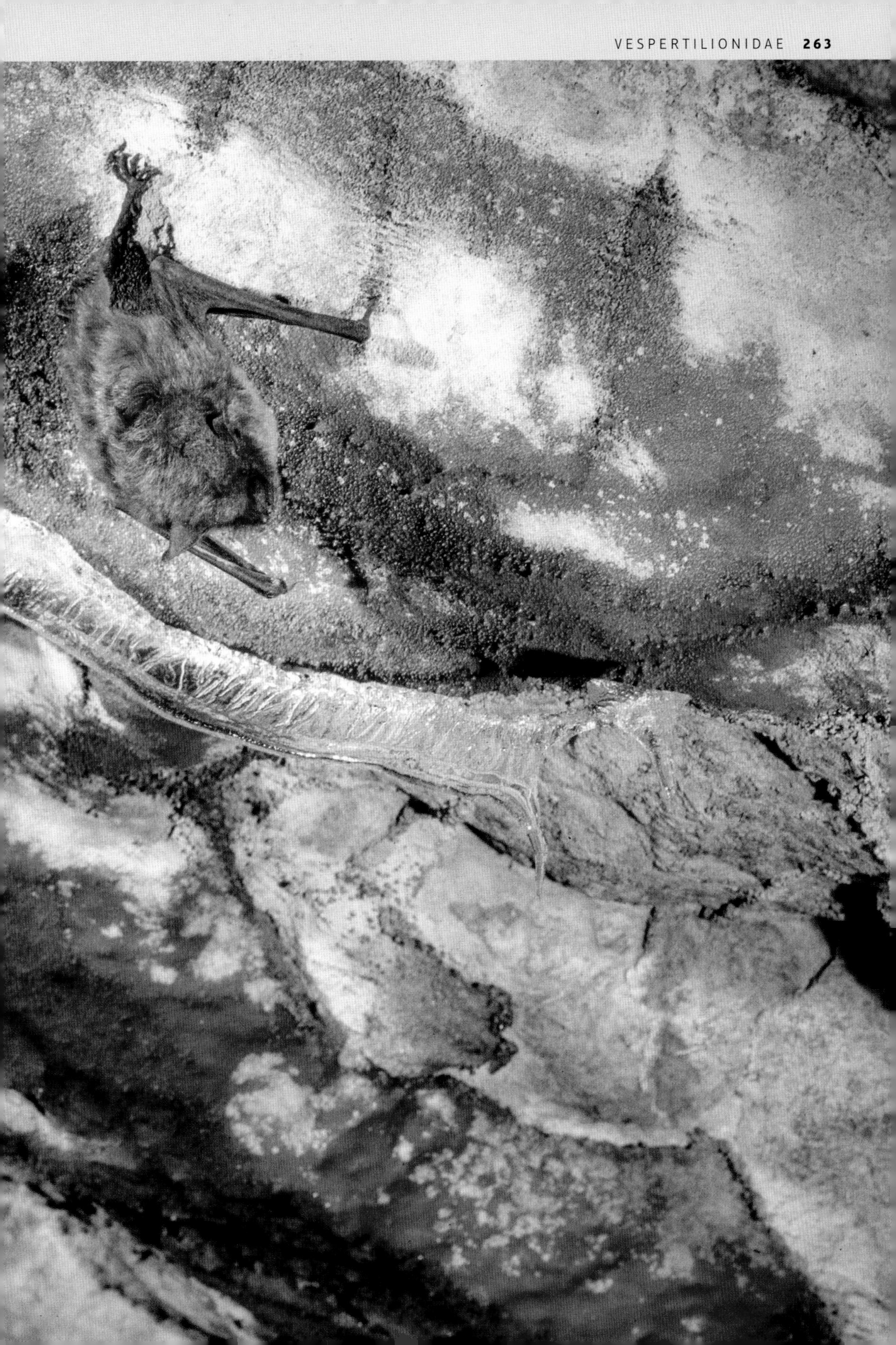

*Eptesicus chiriquinus*

## Chiriquinan Serotine

**THIS SPECIES IS PRESENT** in Panama and patchily in northwestern South America, to as far east as Guyana and southwest as Ecuador. It is a dark vesper bat with long, blackish-brown fur, becoming sparse on the dark pinkish face, and blackish-gray membranes. It has small eyes and rather small, low-set narrow ears with rounded tips. The snout is short and blunt, the chin prominent. It occurs in the lowlands and highlands up to 10,000ft (3,000m), in moist evergreen forest and along forest edges, tending to hunt along paths, in clearings, and in other, more open situations close to tree cover. Its roosting habits are not known, but it most likely uses small tree hollows or similar sheltered spots. It is not very common nor well studied; habitat loss probably affects it in some areas.

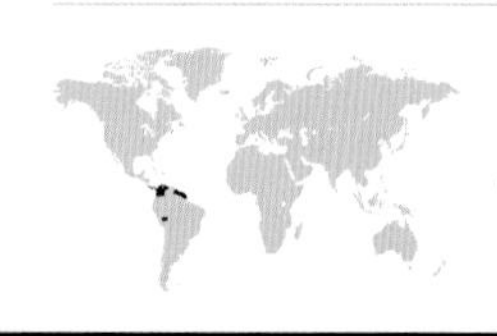

**IUCN STATUS** Least Concern
**LENGTH** 2½–2¾in (6.2–6.9cm)
**WEIGHT** ⅜–½oz (9–13g)

*Eptesicus furinalis*

## Argentine Brown Bat

**THIS BAT OCCURS FROM** south Mexico through Central America and then across most of South America, as far south as central Argentina; it is absent from Chile and west Peru. Over this range it is represented by several subspecies; these require further study, as some may be full species. It has mid-gray-brown, rather woolly looking fur and a sparsely furred, pinkish face with a round snout. The eyes are small but prominent, the ears dark, smallish, and narrow, with forward-curved tragi. It is a forest species, most common at higher altitudes (usually above 3,300ft (1,000m) and up to about 5,200ft (1,600m), and in moist forest. It also occurs in clearings, along roadsides, in drier thorn-scrub, and other, more open areas, often hunting over water. One or two pups may be born per annual litter.

**IUCN STATUS** Least Concern
**LENGTH** 2in (5.2cm)
**WEIGHT** ¼–⅜oz (7.5–9g)

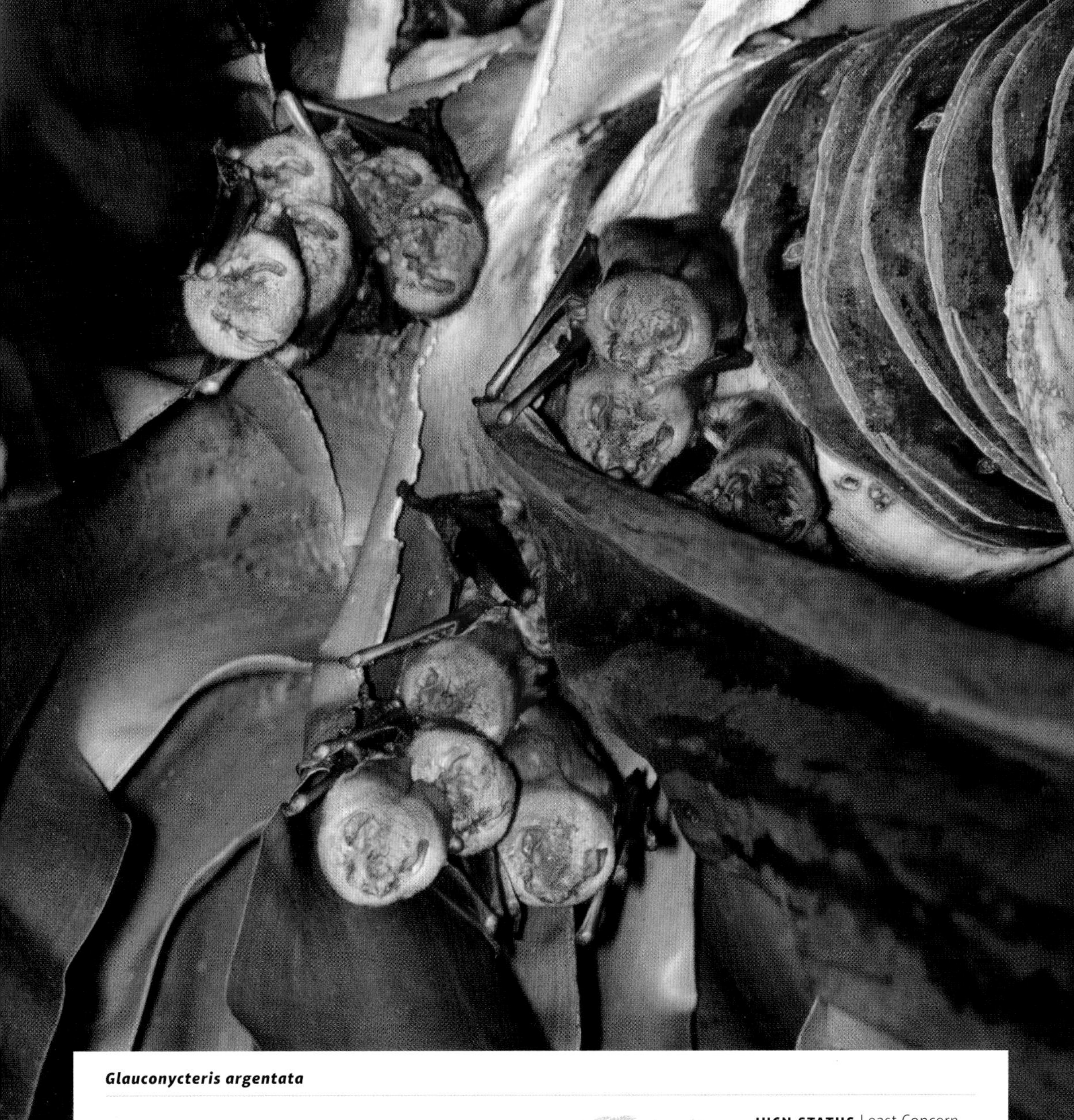

***Glauconycteris argentata***

# Silvered Bat

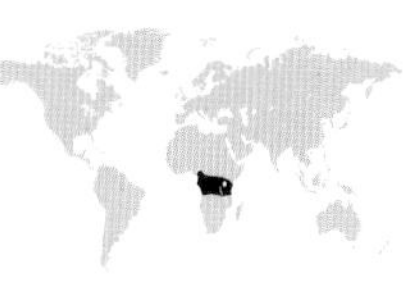

**IUCN STATUS** Least Concern
**LENGTH** unknown
**WEIGHT** unknown

THIS SPECIES, ONE OF the "butterfly bats"—so-called because some in the genus are boldly patterned—occurs across central Africa, from Cameroon and Angola in the west to Kenya, Tanzania, and north Malawi in the east. It has a drab, yellowish-gray coat with distinct bands of paler, silvery fur where the wings meet the body. The wings are pinkish gray, darkening away from the body. The face is rather square and boxy, the eyes very small, and the outer folds of the small, rounded ears curve round and down to the jawline. Very little is known of this bat and records are infrequent. The species has been observed in moist, lowland, tropical forest and savanna, and it roosts within trees. Deforestation threatens its survival throughout its range, but its population trend is not currently known.

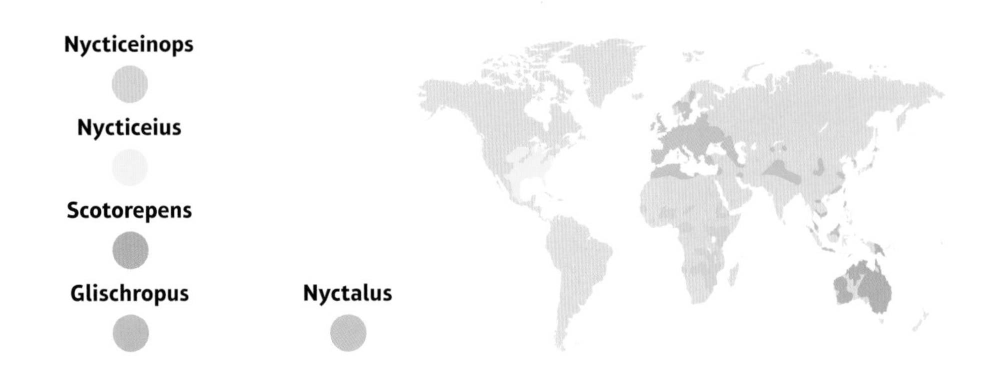


## Nycticeinops & Nycticeius

*Nycticeinops* and *Nycticeius* are two small, closely related (and sometimes lumped together) genera of small, pipistrelle-like bats that occur in Africa and the Americas respectively. *Nycticeinops* holds two species: *N. eisentrauti* and *N. schlieffeni*. The former is only known from Cameroon, while the latter occurs extensively (though discontinuously) in Africa. Of the three *Nycticeius* species, *N. humeralis* is a well-known North American species; it is described on page 269. The other two are *N. aenobarbus* and *N. cubanus*. The former is known from only one specimen of unknown origin (possibly South America). The latter is a bat that occurs only on Cuba.

## Scotorepens

*Scotorepens* are the "broad-nosed bats"—small species named for their well-spaced, puffy nostrils. They are mainly found in Australia. This book recognizes four species, of which *S. orion* is described on page 269. The other three are *S. balstoni*, which is widespread in inland Australia; *S. greyii*, which occurs in north, east and central Australia; and *S. sanborni*, which is found in parts of north Australia. The latter's range also extends to New Guinea, and some of the islands of east Indonesia.

## Glischropus

The species that make up the small genus *Glischropus* are known as the "thick-thumbed bats." They are very small, with pointed, shrew-like faces, enlarged nostrils, and thickened, mobile pads on their thumbs and ankles—probably to provide mechanical adhesion, enabling them to stick to surfaces such as large leaves. *G. aquilus* is newly discovered (2017) from Sumatra. The same goes for *G. bucephalus*, found in Cambodia in 2011. The other species are the little-known *G. javanus*, known from just two specimens taken on Java, and *G. tylopus,* found widely across southeast Asia.

## Nyctalus

The noctules, of the genus *Nyctalus*, are large, fast-flying vespertilionids. They are adept hunters, and several species even prey regularly on small birds. They occur only in the Old World, and have sleek, dark fur, blackish bare parts, and short, pug-like faces with wide mouths. The group includes the well-known *N. noctula* (see page 270) and a further seven species. *N. aviator* (Japan, Korea, and parts of east China) is a rare species. *N. azoreum* occurs only on the Azores. *N. furvus* is also vulnerable, with a small range on Honshu, Japan. *N. lasiopterus* is the only bat known to hunt and catch birds in flight (rather than at their roosts). It has a very patchy distribution in the south of Europe, north Africa, the Middle East, and the west of Asia. The other three species are not considered to be vulnerable, they are *N. leisleri*, present over most of Europe, also northwest Africa, the Middle East, and separately in

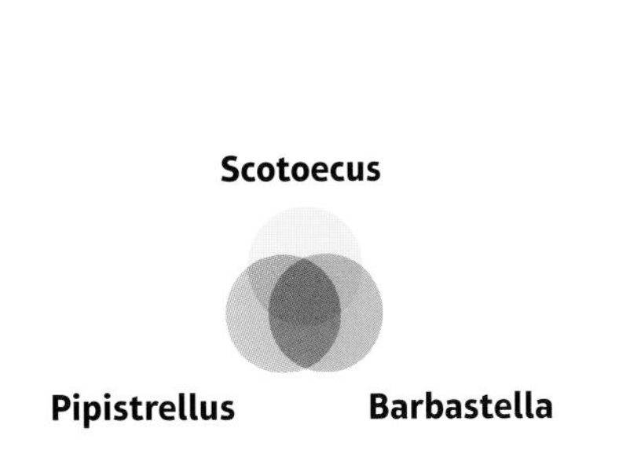


Afghanistan, Pakistan, and India. *N. montanus* occurs in South Asia, and *N. plancyi* is endemic to China, where it is widespread.

### Scotoecus

The *Scotoecus* bats of Africa (and in one case Asia) are sometimes known as "lesser house bats." There are five species, though only three are recognized by the IUCN. These are *S. hirundo* and *S. pallidus* (the former found in parts of sub-Saharan Africa, the latter in India and Pakistan), and *S. albofuscus*, recorded sparsely from west and east Africa. The other two species—*S. hindei* and *S. albigula*, both of east Africa—have not yet been assessed by the IUCN, as both are recent splits from *S. hirundo*.

### Barbastella

The barbastelles, genus *Barbastella*, comprise four species. They are named for the white-tipped hairs on the face of *B. barbastellus* (see page 270) —the name deriving from the words *barba* (meaning "beard"), and *stella* (meaning "starry"). These medium-sized vespertilionid bats are dark with large, tall, and close-set ears; their faces are short and rather flat. *B. leucomelas* occurs widely in Asia and also north Africa. *B. beijingensis* is a recently discovered species found in China. *B. darjelingensis* is another recently discovered species, occurring in the Middle East and west Asia.

### Pipistrellus

The genus *Pipistrellus* might be defined as the "true pipistrelles." The genus has undergone considerable revision since the turn of the century, with several species reassigned to new genera, but it remains one of the larger vespertilionid genera, with thirty-three species. These are small bats with simple, pointed faces, medium-sized ears, and, typically, an agile, fluttering flight. They occur in the Old World only, as all New World pipistrelles have now been separated into other genera.

Four species are described in detail: *P. abramus* (see page 271), *P. kuhlii* (page 271), *P. raceyi* (page 272), and *P. rueppellii* (page 272). The others are *P. adamsi* (north Australia), *P. angulatus* (Indonesia, New Guinea, the Solomon Islands), *P. ceylonicus* (India, Sri Lanka, Pakistan, Bangladesh, Myanmar), *P. collinus* (New Guinea), *P. coromandra* (south and southeast Asia), *P. hesperidus* (most of sub-Saharan Africa), *P. javanicus* (north and central south Asia, southeast Asia), *P. nanulus* (west, central, and east Africa), *P. nathusii* (Europe, west Asia), *P. papuanus* (Indonesia, New Guinea), *P. paterculus* (patchily in mainland southeast Asia), *P. pipistrellus* (Europe, northwest Africa, and through the Middle East and parts of Asia), *P. pygmaeus* (patchily through Europe and the Middle East), *P. rusticus* (west, central, and south Africa), *P. stenopterus* (Thailand, Malaysia,

**Otonycteris**

**Corynorhinus**

Indonesia, Philippines), *P. tenuis* (south and southeast Asia), *P. wattsi* (southeast Papua New Guinea), and *P. westralis* (north Australia).

Just two species are threatened. They are: *P. maderensis*, which has a restricted range in Madeira and the west of the Canary Islands, and *P. endoi*, an endangered species occurring in Japan. Several more are poorly known, mainly because few specimens exist. They are *P. grandidieri* (central Africa), *P. aero* (Kenya), *P. hanaki* (Libya), *P. inexspectatus* (west and central Africa), *P. minahassae* (Sulawesi), *P. permixtus* (Tanzania), and *P. sturdeei* (possibly Bonin islands, Japan). Two more are *P. dhofarensis* (a newly discovered Middle Eastern species) and *P. aladdin* (a recent split from *P. pipistrellus*, found in central Asia).

***Pipistrellus stenopterus***

## Corynorhinus

The "big-eared bats" of the genus *Corynorhinus* occur in the Americas. These striking bats are named for their huge, long, and tall ears—the faces are flattened, with suggestions of a nose-leaf. *C. mexicanus* has a small range in Mexico, and a small and declining population. The other two species, *C. rafinesquii* and *C. townsendii*, are covered on page 273 and page 274, respectively.

## Otonycteris

*Otonycteris* is another small genus of large-eared bats. They are pale-furred and occur in desert areas, where they mainly take prey from the ground, detecting them by the sounds these creatures make as they move around. *O. hemprichii* is found patchily in arid areas in north Africa and the Middle East. *O. leucophaea* is known from deserts in central Asia.

***Nycticeius humeralis***

## Evening Bat

**THIS WIDESPREAD NORTH AMERICAN** bat ranges across the eastern half of the United States, from Nebraska and Iowa east to New Jersey, south to Florida, and across to northeast Mexico. The most northerly populations are migratory, moving south in winter. This is a *Myotis*-sized bat with dark, rufous-brown fur and blackish ears, snout, and membranes. The ears are triangular, with rounded tips. Its broad, rectangular tragus immediately sets it apart from *Myotis*. It has a short, delicately pointed snout. This species occurs in various forest habitats, and roosts in groups of ten to thirty in tree holes or beneath loose bark; it is not known to roost in caves. It hunts insect prey, using a slow, agile, searching flight. Females bear twins, occasionally triplets; the litter at birth weighs 50 percent of their mother's body weight.

**IUCN STATUS** Least Concern
**LENGTH** 3⅝in (9.3cm)
**WEIGHT** ⅛–½oz (5–14g)

***Scotorepens orion***

## Eastern Broad-nosed Bat

**THIS AUSTRALIAN SPECIES** has a rather localized distribution. It occurs on the east coast from Melbourne to Sydney, with a smaller, separate population in the Cairns area in north Queensland. It is a medium-sized vesper bat with mid-brown fur, a little paler on the belly. It has small, triangular ears, and a short snout with slightly puffy nostrils. It occurs in mature, wet forest and more open woodland with some tall trees; here it roosts in tree holes, sometimes in buildings, usually alone but sometimes in pairs or small groups. After emerging at sunset, the bats hunt around trees, using echolocation to find flying insects of all kinds. Females bear one or two pups per litter, and reportedly carry them on early flights for the first ten days.

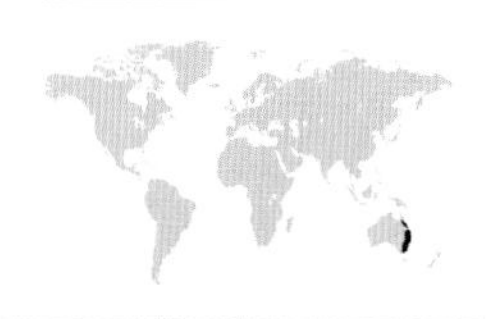

**IUCN STATUS** Least Concern
**LENGTH** unknown
**WEIGHT** ⅜–½oz (9–15g)

***Nyctalus noctula***

## Noctule

**THIS RATHER LARGE BAT** occurs extensively in Europe, including most of Great Britain and south Scandinavia, and into western Asia; from France it extends broadly east into Russia. The most northerly populations migrate south for winter. A separate population exists in central Asia, reaching western China. It has glossy chestnut fur. The wings are long and slim, affording fast, powerful flight. It hunts in woodland, gardens, farmland with trees, and other habitats. Summer roosts are in tree holes, with females forming maternity roosts of up to fifty bats. In winter it hibernates in large groups (up to 10,000) using caves and buildings. It captures flying insects in the open air, and is occasionally seen hunting in broad daylight. A modest population decline is occurring, due to habitat loss and overuse of pesticides.

**IUCN STATUS** Least Concern
**LENGTH** 2–4in (5–10cm)
**WEIGHT** ⅝–1¾oz (16–49g)

***Barbastella barbastellus***

## Western Barbastelle

**THIS BAT IS MAINLY** confined to Europe, ranging from north Spain and the south of Britain to Latvia, Belarus, and Ukraine. It also occurs in Georgia, just reaches Iran, and is recorded in Morocco and on the Canary Islands. This is a dark-furred bat with large, upright, round-tipped ears, set close together, with outer folds that reach below the eyes. It has a blunt, broad muzzle with a white-haired "beard" and a very large uropatagium. The species occurs in woodland, forming summer and hibernation roosts in hollow trees, more rarely in buildings; it hunts around foliage near the canopy, hawking and gleaning small insects. Mating occurs in the fall, with females having single pups (occasionally twins) the following spring. Though widespread, it is rare and generally declining due to loss of suitable roost trees.

**IUCN STATUS** Near Threatened
**LENGTH** 1¾–2⅜in (4.5–6cm)
**WEIGHT** ¼–⅜oz (6–10g)

***Pipistrellus abramus***

## Japanese Pipistrelle

**THIS SMALL BAT** is found in southeast Asia, from south Japan across to North and South Korea, east and south China and into Myanmar, Laos, and Vietnam. It is a rather plain pipistrelle with drab, gray-brown fur and barely contrasting pinkish-brown bare parts. The ears are smallish and rather narrow, the eyes small, and the snout short and pointed. Often known as the Chinese or Japanese House Bat, it seems to avoid natural habitats, roosting instead in roof spaces in buildings, even in city centers. It commonly hunts around built-up areas, where artificial light attracts insects. Roosts can hold up to one hundred individuals, but may be much smaller, depending on available space. This is a very common species and not considered to be threatened, though despite its abundance its ecology is little studied.

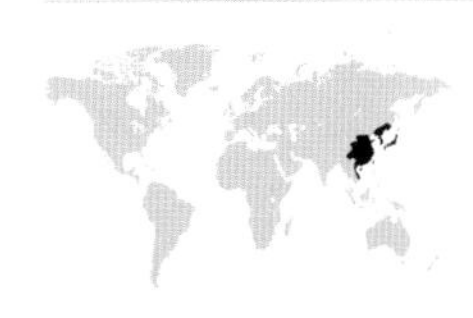

**IUCN STATUS** Least Concern
**LENGTH** 1½–2⅜in (3.8–6cm)
**WEIGHT** ⅛–¼oz (3.9–6.5g)

***Pipistrellus kuhlii***

## Kuhl's Pipistrelle

**THIS BAT HAS AN** extensive Mediterranean and west Asian range. It occurs in the south of Europe from Spain eastward, in north Africa, and the Middle East, extending to Russia, Pakistan, and the far northwest of India. It is also present on many larger Mediterranean islands. A dark, rather large pipistrelle, it has dense, chestnut-brown fur and brownish-black bare parts, but tends to become paler with age; populations in desert areas are also paler. It is found in woodland, scrub and maquis, semi-desert, grassland, parks, and gardens—including in towns, where it can be very common. Many larger colonies use buildings. The bat often hunts for insects around artificial lights. Despite heavy pesticide use in some areas, overall the species is faring well, and in many areas is on the increase.

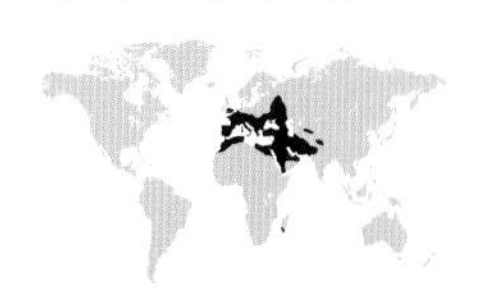

**IUCN STATUS** Least Concern
**LENGTH** 3–3⅝in (7.5–9.2cm)
**WEIGHT** ⅛–⅜oz (5–8.5g)

***Pipistrellus raceyi***

## Racey's Pipistrelle

**THIS RECENTLY DISCOVERED** bat (described in 2006) is endemic to Madagascar. It has been recorded on both east and west coasts, from just four locations, but may be more widespread. It has reddish-brown fur, tinted paler yellowish on the underside, and blackish wings and bare parts. The ears are dark and medium-sized. The few existing specimens were trapped in varied habitats, including in farmland close to a town, and in dry forest; all were below 260ft (80m) elevation. The only known roost was in a hollow space within a house wall; here a male and several females were caught, suggesting the species could form harems when breeding. The bat's behavior and ecology are otherwise unknown. It is more similar to Asian than African *Pipistrellus*, unlike most Malagasy bats, which are of likely African origin.

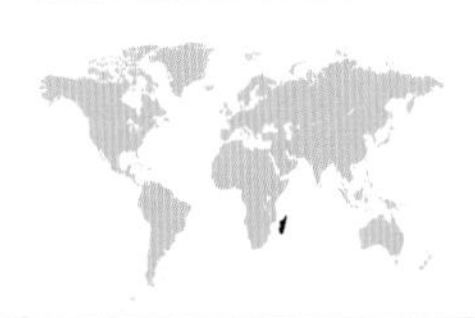

**IUCN STATUS** Data Deficient
**LENGTH** unknown
**WEIGHT** unknown

***Pipistrellus rueppellii***

## Rüppel's Pipistrelle

**THIS BAT IS RECORDED** widely but patchily across Africa, especially from Sudan south to Zambia, but also in parts of the north and west. It is also known from north Egypt, and Yemen, Iraq, and Iran. It is a darkish pipistrelle (though becomes paler with age) with a white underside, the latter making it distinctive among pipistrelles. It has dark limbs with paler membranes. The dark ears have prominent, blade-shaped tragi. It favors open, arid countryside with some bodies of water and scattered trees. This species roosts within rocky crevices, forming colonies of up to one hundred individuals. It catches insects on the wing, usually hunting close to ground level and chasing prey with agile, fluttering flight. Females give birth to twins in late summer. It is affected by overuse of pesticides in some areas.

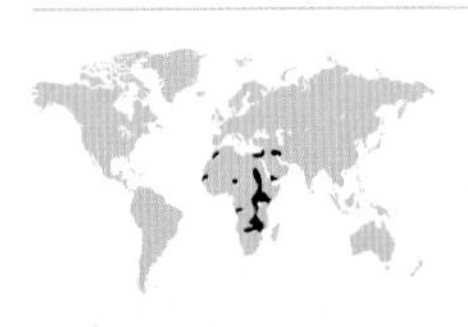

**IUCN STATUS** Near Threatened
**LENGTH** 3–3⅜in (7.7–8.5cm)
**WEIGHT** ¼oz (7g)

*Corynorhinus rafinesquii*

# Rafinesque's Big-eared Bat

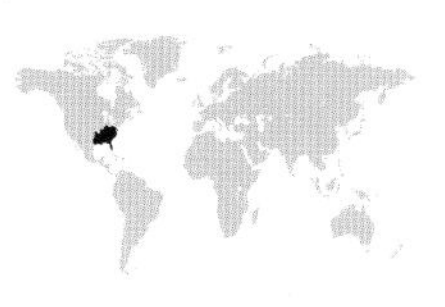

**IUCN STATUS** Least Concern
**LENGTH** 3–4⅛in (7.5–10.5cm)
**WEIGHT** ¼–½oz (7–13g)

**THIS BAT OCCURS SPARSELY** across much of the southeast of the United States. Its range extends north to Indiana and west into Texas. Males have the largest testes relative to total body mass of any mammal, suggesting a highly competitive mating system. Its habitat is undisturbed, deciduous forest. This bat roosts in deep tree holes, and also in caves, wells, and derelict buildings. Most colonies are small, but hibernation roosts can hold two hundred individuals. It hunts by gleaning from foliage and hawking. This species seems resistant to white-nose syndrome, perhaps because it is a "shallow" hibernator, often active on winter nights. However, it is sensitive to disturbance, and has lost much habitat over the last century. The elective removal of decaying trees is a particular problem.

***Corynorhinus townsendii***

# Townsend's Big-eared Bat

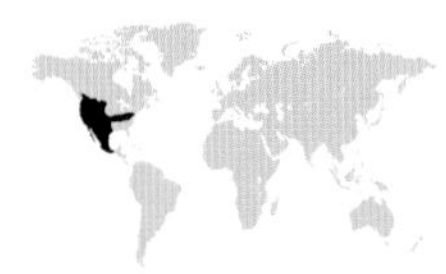

**IUCN STATUS** Least Concern
**LENGTH** 3½–4⅝in (8.9–11.6cm)
**WEIGHT** ⅜oz (9–12g)

**THIS SPECIES OCCURS FROM** southeast Canada down the western half of the United States into Mexico; two smaller, well-separated populations occur in the eastern United States. A close relative of Rafinesque's Big-eared Bat, it resembles that species in appearance, and has similar hunting behavior. Habitat varies over the range, but most occur in dense, upland forest, roosting in deep but well-aerated caves and rock crevices. Mating occurs in the fall, following a courtship display; if accepted, the male marks the female's body with scent from his facial glands. She is usually already in a state of torpor when mating occurs, and stores sperm until the following spring, when fertilization occurs. She joins an all-female nursery colony and gives birth to her pup between fifty-six and one hundred days later (depending on spring temperatures).

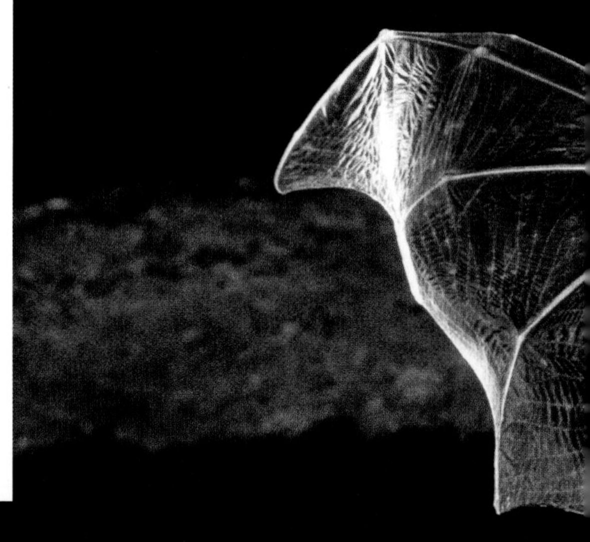

## Plecotus

The bats in the Old World genus *Plecotus* are also exceptionally large-eared. At rest the ears curl back, recalling ram's horns, and may be tucked under the wings, leaving the tragi sticking up—though in these bats the tragus itself is large enough to be mistaken for the entire ear. Known as "long-eared bats" because of their outrageously oversized ears, they are agile and slow-flying, taking much of their prey by gleaning from foliage or even the ground. Like *Corynorhinus* bats, they mainly detect their prey by the mechanical sounds it makes. See page 280 for accounts on *P. auritus* and *P. austriacus*.

Several *Plecotus* species are newly discovered or newly split. They are *P. homochrous* (Ethiopia), *P. kozlovi* (China and Mongolia), *P. strelkovi* (central Asia), *P. turkmenicus* (central Asia), and *P. wardii* (Europe). A further three are little known—*P. ariel* of China, *P. balensis* of the Ethiopian highlands, and *P. christiei* of northeast Africa. *P. taivanus* is a declining and threatened species present on Taiwan. The six remaining species are: *P. kolombatovici* (south and southeast Mediterranean regions), *P. macrobullaris* (south Europe and the Middle East), *P. ognevi* (central and east Asia), and *P. sacrimontis* (Japan). *P. sardus* is endemic to Sardinia, and *P. teneriffae* of the Canary Islands.

***Plecotus taivanus***

## Scotophilus

*Scotophilus*, the "yellow bats," or "house bats," is a fairly large genus of pale brownish, yellowish-bellied bats that are distributed across Africa and the south of Asia. They are medium-sized with stout, robust snouts and wide-based, rounded ears. Identification is challenging within this group. Many occur around human habitation, roosting in buildings and hunting insects attracted by artificial light sources. They tend to be harem-forming, and females usually give birth to twins.

Four species are described in detail: *S. dinganii* (page 278), *S. kuhlii* (page 278), *S. nigrita* (page 279), and *S. viridis* (page 279). Of the remaining fourteen species, several of them are relatively new to science, having been described since the year 2000. *S. andrewreborii* is a recently discovered species that appears to

**Chalinolobus**

be widespread in Kenya. *S. collinus* is a little-known species that is present on various Indonesian islands, and also occurs on Borneo. *S. ejetai* is another newly described species (2014), found in the Rift Valley in Ethiopia. *S. heathii* occurs widely in south and southeast Asia. *S. leucogaster* is present in much of sub-Saharan Africa, especially regions in the west. *S. livingstoni* is a newly described species, so far known only from Ghana and Kenya, but likely also to occur in between. *S. marovaza* is a Malagasy species, present on the west side of the island. And finally, *S. nux* occurs in west and central Africa, *S. robustus* in Madagascar, and *S. trujilloi* in coastal Kenya.

The final four species in the genus are poorly known *S. borbonicus* (known from only two specimens, one from Madagascar, the other from Réunion island), *S. celebensis* (found on Sulawesi and possibly a subspecies of *S. heathii*), *S. nucella* (found apparently patchily in sub-Saharan Africa, but few specimens ever taken), and *S. tandrefana* (known only from two specimens, both taken on Madagascar).

## Chalinolobus

The forest-dwelling Australian bats within the genus *Chalinolobus* are variously known as "pied bats" (because many have white markings on their dark coats, around where the wings meet the body), "wattled bats" (because of fleshy lobes on the face below the ears and on the bottom lip), or "long-tailed bats" (because they have long tails, albeit fully enclosed in the uropatagium). They are round-headed bats with small, low-set ears, and short snouts. *C. gouldii* and *C. tuberculatus* are described on page 281. There are five other species: *C. morio* (south Australia and also some central areas) and *C. nigrogriseus* (north and northeast Australia), *C. dwyeri* (coastal east Australia) and *C. picatus* (east Australia), and *C. neocaledonicus* (Grand Terre island, New Caledonia), with a population of probably no more than 1,500 individuals.

*Scotophilus dinganii*

## African Yellow House Bat

**THIS BAT OCCURS** extensively in sub-Saharan Africa, from Guinea-Bissau in the west to Ethiopia and Somalia in the east, and south to southeastern South Africa; it is absent from desert and heavily forested regions. A fairly large vesper bat, it has drab, brownish straw-colored fur, more yellowish on the underside, and darker bare parts; the snout is pointed with puffy, slightly tubular nostrils and the ears are smallish. It is found in savanna (dry and moist) and often roosts in buildings in groups of around twenty to thirty, also sometimes singly; it is very quiet at its roost and often goes unnoticed by householders. This bat emerges well after dark and feeds efficiently, often spending just two hours actively foraging each night. Females give birth to twins, which attain independence in a few weeks.

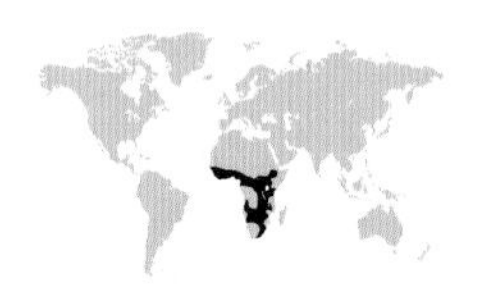

**IUCN STATUS** Least Concern
**LENGTH** 5⅛in (13cm)
**WEIGHT** 1oz (27g)

*Scotophilus kuhlii*

## Lesser Asiatic Yellow House Bat

**THIS WIDESPREAD BAT** occurs in south and southeast Asia, from Pakistan and India across to east China, and reaching south through Myanmar, Thailand, Laos, Cambodia, Vietnam, the Malay peninsula, Sumatra, and Borneo to many Indonesian islands and throughout the Philippines. It has light-brown to yellowish fur with a paler underside. The membranes are grayish. It has small eyes and ears, and a broad snout with a deep lower lip and slightly tubular nostrils. This highly adaptable species occurs in both primary and secondary forest, and also in parks and gardens. It roosts in buildings of all kinds and also in the open among dead foliage, in colonies of up to several hundred. It leaves its roost to hunt at around sunset. Females give birth to single or twin pups.

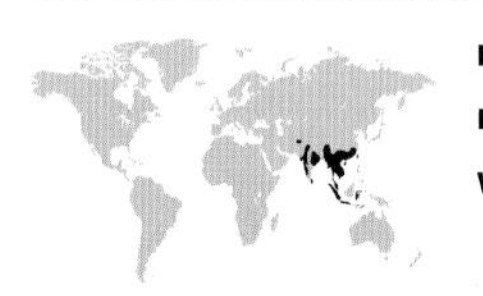

**IUCN STATUS** Least Concern
**LENGTH** 4½–5xin (11.5–12.7cm)
**WEIGHT** ⅝–¾oz (17–22g)

***Scotophilus nigrita***

## Giant House Bat

THIS BAT OCCURS in Africa, in the west (Senegal to Ghana) and also southeast (Tanzania, Zambia, and Zimbabwe). It is a very large species, by far the largest in its genus. The sleek fur is mid-brown, more yellowish on the underside, with darker, brownish membranes, ears, and face. The ears have forward-curved, pointed tragi; the snout is blunt with puffy nostrils. It is poorly known, with infrequent records. It has been found in forest areas and in both moist and dry savanna habitats. Roosting individuals have been found in buildings and in tree hollows, and also in purpose-built bat houses. The diet and hunting behavior have not been studied, though it is likely to be capable of handling the largest insect prey. The species is rare and possibly declining due to habitat loss.

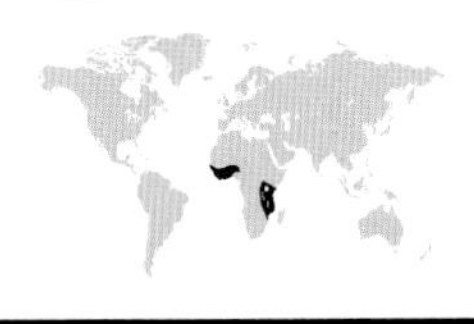

**IUCN STATUS** Least Concern
**LENGTH** 6⅞in (17.5cm)
**WEIGHT** 1⅞oz (53g)

***Scotophilus viridis***

## Green House Bat

THIS AFRICAN BAT'S range extends from Senegal narrowly across west and central Africa to Ethiopia, then south through the east side of the continent to Zambia, Zimbabwe, and northeastern South Africa. It has olive-brown upperparts. The belly is bright yellowish with a green tint, sometimes strongly orange-tinted around the vent. The ears and membranes are dark. This bat is mostly found in warm, lowland, wooded savanna (moist and dry) and wooded river valleys; it avoids more open grassland, and has been observed roosting in tree hollows—and also in buildings—in small colonies. Its ecology is otherwise almost entirely unstudied. It seems to be quite common over most of its range, and faces no particular threats at present. There are records of the species from several protected areas.

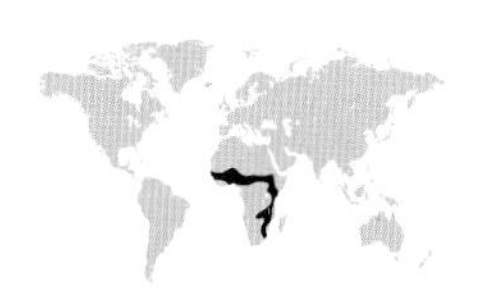

**IUCN STATUS** Least Concern
**LENGTH** 4⅜–5in (11.1–12.5cm)
**WEIGHT** ½–⅝oz (13–18g)

***Plecotus auritus***

## Brown Big-eared Bat

**THIS BAT OCCURS WIDELY** in Europe, from Ireland east to the Ural mountains. It is small, light brown, with small eyes and a short muzzle. The ears are long and broad; they often curl back at rest, exposing the long, slender tragi. It has wide wings, giving a slow and agile fluttering flight, and occurs in wooded habitats, including parks and gardens. Roosts (day and hibernation) may be in buildings or tree holes; this bat usually roosts alone or in small groups. Unusually, the sexes roost together all year, with both remaining in the same colony long-term. It listens for movements to glean prey from foliage, and also hawks; its quiet echolocation calls are not easily detected by their prey. This species is long-lived, some reaching thirty years old.

**IUCN STATUS** Least Concern
**LENGTH** unknown
**WEIGHT** ¼–⅜oz (6–12g)

***Plecotus austriacus***

## Gray Big-eared Bat

**THIS IS A EUROPEAN BAT** that ranges from Spain across to Ukraine and north Greece. It is slightly larger and grayer than *P. auritus*, but was only distinguished from it in the 1960s; identification is difficult even with the bat in hand. It may be found around farmland, riversides, and woodland edges. This species roosts and hibernates in tree holes, caves, and buildings, sometimes alongside *P. auritus*. It roosts alone, but breeding females form maternity roosts of up to thirty. Its hunting behavior resembles that of *P. auritus*; it is also a specialized moth-catcher with very quiet ("whispering") echolocation calls. In Britain and parts of central Europe this bat is rare and of conservation concern, most likely because of agricultural changes, but in Mediterranean regions it remains quite common.

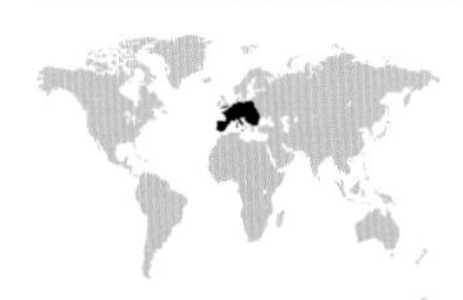

**IUCN STATUS** Least Concern
**LENGTH** 1¾–2¾in (4.5–7cm)
**WEIGHT** ⅛–¾oz (5–20g)

***Chalinolobus gouldii***

## Gould's Wattled Bat

**THIS BAT OCCURS** in Australia and is widespread, being absent only from north Queensland and south Tasmania. It is a small bat with shiny, velvety-looking blackish fur and similarly dark bare parts. Its smallish ears are very wide-based, tapering to rounded tips. The snout is very short, the nostrils slightly tube-shaped, and the dark eyes fairly large. The species occurs in varied habitats including forest, scrubland, and more open areas; it roosts in buildings and tree hollows. Its colonies usually hold some thirty individuals but sometimes up to 200; males often roost alone. It may hibernate briefly in the north of its range. The diet is varied, but moths are the most frequently taken insects. Twin pups are born in October, and are independent by January. This is a common species, facing no immediate threats.

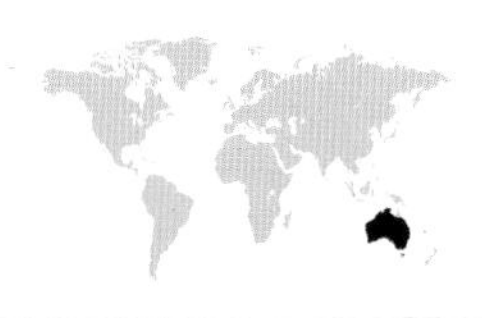

**IUCN STATUS** Least Concern
**LENGTH** 2¾in (7cm)
**WEIGHT** ½oz (14g)

***Chalinolobus tuberculatus***

## New Zealand Long-tailed Bat

**ONE OF ONLY TWO** extant species native to New Zealand—this bat occurs over much of North Island, and on South Island's west and central-east coast. It has rich dark brown or red-brown fur, and a long tail, enclosed within its large uropatagium. The head is highly domed, with a very short, blunt snout; the ears are rounded and low-set. It occurs in native forest, though may venture into more open countryside to hunt. It usually roosts alone in small tree hollows, and hibernates in winter (duration depending on latitude). Emerging at dusk, it hunts on the wing, capturing various flying insects (possibly using the uropatagium as a scoop). Females produce a single pup in December. It has declined because of habitat loss, and predation by non-native species.

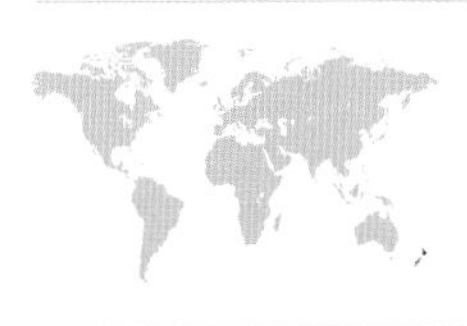

**IUCN STATUS** Vulnerable
**LENGTH** unknown
**WEIGHT** ⅛–⅜oz (8–12g)

## Falsistrellus

*Falsistrellus* is a small genus of pipistrelle-like bats—sometimes called the "false pipistrelles"—that occurs in Asia and Australia. *F. affinis* is found patchily in south and southeast Asia and *F. tasmaniensis* in east and southeast Australia. *F. mordax* (a barely known species from Java) and *F. petersi* (recorded from Borneo, Sulawesi, the Philippines, and Moluccan islands) are both little known. The final species, *F. mackenziei* occurs only in far southwest Australia.

## Hypsugo

*Hypsugo* bats are allies of the pipistrelles. They often have a pale coat that contrasts with the almost black membranes, limbs, face, and ears. The genus holds eighteen species, though by some other taxonomies several are placed in the genus *Pipistrellus*; they occur in Africa and Asia. A large proportion of the species are little known, they are: *H. arabicus* (known from a few specimens from Oman and Iran), *H. ariel* (Israel, Jordan, Egypt, Saudi Arabia, Yemen, and Sudan), *H. anthonyi* (known from one specimen only, taken in Myanmar), *H. joffrei* (recorded from a few sites in Vietnam, Myanmar, Nepal, and India), *H. kitcheneri* (Borneo), *H. lophurus* (Myanmar), *H. macrotis* (west Malaysia), *H. musciculus* (west Africa), and *H. vordermanni* (Borneo and some nearby islands). The other species are *H. alaschanicus* (east Asia), *H. bemainty* (central west Madagascar), *H. cadornae* (northeast south Asia, mainland southeast Asia), *H. crassulus* (west, central and east Africa), *H. imbricatus* (Indonesia, Borneo), *H. pulveratus* (China and Myanmar), and *H. savii* (widespread in Eurasia). *H. dolichodon* was described in 2015, and is found in forest habitats in Vietnam and Laos. Its most distinctive trait is its impressive mouthful of large teeth; the upper canines are particularly long and alarming in appearance. The similar *H. lanzai* was described in 2011 from a specimen taken in Yemen.

## Laephotis

There is much disagreement over which species should be placed in the genus *Laephotis*. This book treats twenty-one African and Malagasy species as *Laephotis*, but the IUCN recognizes just four, placing the rest in *Neoromicia*, or *Pipistrellus*. Traditionally

***Hypsugo savii***

Nyctophilus

recognized *Laephotis* have large ears. The following species are covered in detail: *L. botswanae* (see page 285), *L. capensis* (page 285), *L. nana* (page 286), *L. tenuipinnis* (page 284), and *L. zuluensis* (page 284).

There are a number of *Laephotis* bats that have reasonably or very large ranges. They are as follows: *L. anchietae* (the south of Africa), *L. guineensis* (west, central, and east Africa), *L. matroka* (the east highlands of Madagascar), *L. namibensis* (Namibia and South Africa), *L. rendalli* (widespread in sub-Saharan Africa), *L. somalica* (west, central, and east Africa), and *L. wintoni* (much of sub-Saharan Africa).

Several poorly known *Laephotis* species are *L. angolensis* (known from two sites in Angola and one in Democratic Republic of Congo), *L. flavescens* (central and south Africa), *L. helios* (east Africa), *L. isabella* (Guinea and Liberia), and *L. robertsi* (central highlands of Madagascar). The species *L. stanleyi* is not yet assessed; it was split from *L. capensis* in 2017.

Three species are considered to be in danger. *L. brunnea* occurs in west and central Africa. It is a forest species in a region with heavy deforestation, and has a patchy distribution. *L. malagasyensis* is found at just four sites in the Isalo Massif region of Madagascar. *L. roseveari* is known from Liberia and adjacent Guinea, but seems to be very rare and limited to the dwindling areas of primary rainforest.

## Nyctophilus

*Nyctophilus* is another genus of large-eared, rather flat-faced vespertilionids, present in Australia, New Guinea, and east Indonesia. These species are foliage-gleaners and roost in tree hollows. *N. gouldi* is described on page 290. The genus includes two species that are considered endangered. *N. howensis* occurs or occurred on Lord Howe Island, Australia; it is known from a single skull, found in 1972. No other sign of it has ever been found, but islanders still report seeing bats that would be (by size) attributable to this species, rather than the much smaller *Vespadelus darlingtoni*—the only other bat species known to occur on the island. The other endangered species is *N. nebulosus*, which occurs only around Mount Koghis on the island of New Caledonia. There are also four little-known species: *N. heran* (known only from its type specimen, taken in the Lesser Sundas), *N. microdon* (a few locations in Papua New Guinea), *N. sherrini* (Tasmania), and *N. shirleyae* (Papua New Guinea). The following three species are all splits from what was known as *N. timoriensis*. *N. corbeni* occurs in southeast Australia, *N. daedalus* in north Australia, and *N. major* in southwest Australia. The remaining *Nyctophilus* species are: *N. arnhemensis* (coastal northwest and north Australia), *N. bifax* (New Guinea and north Australia), *N. geoffroyi* (widespread in Australia, including Tasmania), *N. microtis* (New Guinea), and *N. walkeri* (north Australia).

***Laephotis tenuipinnis***

## White-winged Bat

**THIS SPECIES OCCURS** in west and central Africa, from the Gambia across to Ethiopia and Kenya, and south as far as west Angola. It has brown upperside fur and a white belly; the hair on the flanks at the base of the wings is particularly silvery, long, and silky. The wing membranes are also whitish, contrasting with the brown body when seen from above. Males in breeding condition can show an orange tint, from glandular secretions. This bat has rounded ears and a short snout. It occurs in rainforest in lowland areas, and roosts in tree hollows and under bark, either alone or in small groups of up to twenty. It is rather common and occurs in several protected areas. However, its dependence on rainforest may have caused declines where deforestation is an issue.

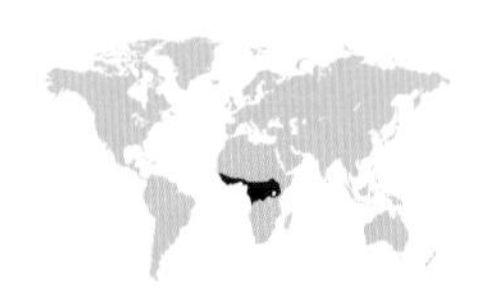

**IUCN STATUS** Least Concern
**LENGTH** 2½–3⅜in (6.4–8.7cm)
**WEIGHT** ⅛oz (3g)

***Laephotis zuluensis***

## Aloe Bat

**THIS SPECIES IS** found in the south of Africa, including parts of Zambia, Zimbabwe, South Africa, and Namibia (though not the far southwest of the continent) and separately in east Africa (Kenya, Tanzania. and Ethiopia). It has mid-brown, yellow-tinted fur, slightly paler on the belly, and blackish wing membranes. The wings are relatively short and broad. The grayish ears are rather small and narrow, the eyes inconspicuous, and the snout short and neatly pointed. It prefers open, lightly wooded habitats, including wet and dry savanna with rivers, and also semi-desert if standing water is available. A slow and active flier, it catches insects on the wing, and particularly targets beetles. Although the species is uncommon and probably under-recorded, it is present in some protected areas and not considered under threat.

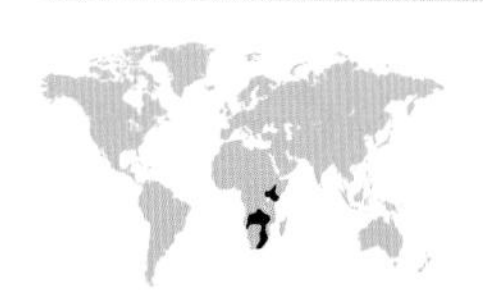

**IUCN STATUS** Least Concern
**LENGTH** 2¾–3⅜in (7–8.6cm)
**WEIGHT** ⅛oz (4–4.9g)

***Laephotis botswanae***

## Botswanan Long-eared Bat

**THIS BAT IS FOUND** in southern Africa, from the far south of Democratic Republic of Congo to the north of South Africa, and seems to exist in several fairly small, separate populations. It has fluffy, light brown fur, whitish on its underside, with a rather sharply pointed snout. The striking ears are long, curved backward, and round-ended, with prominent tragi. Most records are from open woodland and savanna (dry and moist), especially in the uplands and close to rivers. It may roost under loose tree bark, but little is known of its habits—most records are of animals netted by night while they foraged. It probably hunts mostly by gleaning insects from foliage.

**IUCN STATUS** Least Concern
**LENGTH** 3½–3⅞in (9–10cm)
**WEIGHT** ⅛–¼oz (5–7.2g)

***Laephotis capensis***

## Cape Bat

**THIS IS A VERY WIDESPREAD** African species, found south of the Sahara from Senegal across to Ethiopia and almost continuously to the south of South Africa. It is absent from the Horn of Africa, the Namib desert, and parts of the east coast. It has sleek, light-brown fur and darker bare parts, rather small, well-spaced, triangular ears, and a short, delicately pointed snout. It occurs in almost all natural habitats, from forest to savanna, and also in bushveld and light woodland. Roosts are usually in tree hollows, under loose bark, or in rural buildings, in groups of usually twenty or fewer. An agile and lively flier, this bat consumes all kinds of small flying insects. Because of its extensive range and adaptable habits, the species is under no particular threat.

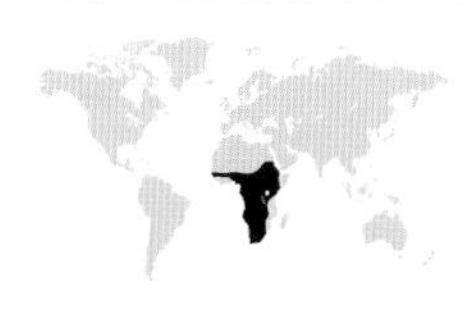

**IUCN STATUS** Least Concern
**LENGTH** 2⅜–4¼in (6–10.7cm)
**WEIGHT** ⅛–⅜oz (3.4–10.1g)

***Laephotis nana***

# Banana Bat

**IUCN STATUS** Least Concern
**LENGTH** 2⅜–3¾in (6.1–9.4cm)
**WEIGHT** ⅛–¼oz (2.5–6.4g)

THIS BAT OCCURS extensively in sub-Saharan Africa, though it is absent from the southwest and present only patchily in east Africa. It has mid-brown fur and darker grayish membranes, limbs and face. The ears are triangular with rounded tips and tragi shaped like hatchets. There are thick, gland-rich pads on the thumbs and feet; these help the bat to roost on smooth surfaces such as banana leaves, through adhesion or suction. It occurs in forest and plantations, and roosts in large, rolled-up leaves, especially of banana plants—usually alone, but sometimes in groups of one hundred or more. Dominant males defend roosts from other males, while females frequently change roosts, showing no attachment to particular roosts or males. This species hunts fairly close to the ground, using echolocation. It is also highly vocal, with pairs engaging in "duels" that can last up to twenty minutes. Females bear one or two pups at the start of the rainy season. These are weaned at two months old, and fully grown and independent at five months.

Vespadelus

Tylonycteris

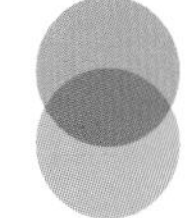

Vespertilio

## Tylonycteris

*Tylonycteris* bats are extremely small. They possess gripping pads on the thumbs and feet, enabling them to stick onto bamboo stalks (hence their common English name of "bamboo bats"). There are just three species: *T. pachypus*, which is described on page 295, *T. robustula*, a widespread species found in south and southeast Asia, and *T. pygmaeus* of southwest China, only discovered in 2007.

## Vespadelus

*Vespadelus* bats are also very small. They occur in Australia and are sometimes known as "cave bats" or "forest bats." They have delicate, short, snub snouts and relatively small ears; the fur is usually much paler on the underside than the upperside. They are: *V. baverstocki* (central Australia), *V. caurinus* (north Australia), *V. darlingtoni* (southeast mainland Australia and Tasmania), *V. douglasorum* (Kimberley region, west Australia), *V. finlaysoni* (widespread across north, west, and central Australia), *V. pumilus* (Queensland and New South Wales), *V. regulus* (south Australia, including Tasmania), *V. troughtoni* (east Australia), and *V. vulturnus* (southeast Australia, including Tasmania).

***Vespadelus pumilus***

## Vespertilio

There are just two species in the genus *Vespertilio*. These are the "particolored bats," *V. murinus* and *V. sinensis*. They are striking, medium-sized bats with silvery-tipped, gray-brown fur, paler on the underside, and contrastingly dark bare parts. The dark snout is blocky and robust, the ears medium-sized. *V. murinus* ranges from central Europe across through central Asia to east Asia, while *V. sinensis* has a more of a northeast distribution, being present patchily in east Siberia, east China, Korea, and Japan.

## Monotypic genera

The family Vespertilionidae also includes seventeen monotypic genera; several of these bats are highly distinctive in both appearance and habits. Six of the species in monotypic genera are described in full. They are: *Antrozous pallidus* (see page 294), *Idionycteris phyllotis* (page 290), *Lasionycteris noctivagans* (page 291), *Parastrellus hesperus* (page 292),

*Perimyotis subflavus* (page 293), and *Euderma maculatum* (page 296).

## Eudiscopus

Of the remaining eleven species within monotypic genera, *Eudiscopus denticulus* is a small bat with flattened fleshy disks on its feet and thumbs. These enable the bat to attach itself to leaves and the inner parts of bamboo stems for roosting. It hunts in open air and around foliage, and occurs in various kinds of forest where bamboo grows. This species was formerly (2008) classed as Data Deficient by the IUCN due to a shortage of observations and lack of information about its population and distribution. However, more populations have been found through additional surveys since that time and the bat has now (2016) been reclassified as Least Concern. It is known from a number of sites in Thailand and Laos.

## Bauerus

*Bauerus dubiaquercus* is a curious-looking, largish vespertilionid. It possesses a rather long, robust, and squarish snout that bears puffy growths near the tip. This bat preys on relatively large insects, which it captures by gleaning. The fur is variably grayish or rufous. It is present from Mexico south through Guatemala, Belize, and Honduras. It is quite rare and seems to be quite dependent on forest, which, within its range, is subject to deforestation.

## Submyotodon

*Submyotodon latirostris* is a rather *Myotis*-like species, but it has been shown to be distinct from them in terms of DNA, as well as in details of the skull and the dentition. It is a very small, delicate-looking bat with a short, simple snout and rather small ears set low on its domed head; its fur is rich, dark brown. The species is endemic to Taiwan (China), where it is quite common in the uplands.

## Scotozous

*Scotozous dormeri* is a south Asian species, occurring in Pakistan, northern India, and Bangladesh. It was formerly classed as a *Pipistrellus* species, but is now recognized as distinct. It is widespread, and highly adaptable in its habitat usage.

*Nyctophilus gouldi*

## Gould's Long-eared Bat

**THIS BAT OCCURS** in Australia, mainly on the east coast from the Cairns area south to Melbourne, but also in the southwest around Perth. A smallish bat, it has light grayish-yellow fur and brownish ears and membranes. The ears are curved backward and broad. Its snout is short, wide at the tip and a little upturned, with some warty bumps. It occurs in tall forest with a rich understory, roosting in tree hollows, spaces in buildings, and other small crevices. Up to twenty females gather in maternity roosts to have their young (usually twins); males roost alone. In the south of its range this bat hibernates through the austral winter. Although fairly common, this species is affected by forest loss and selective removal of tall, old trees.

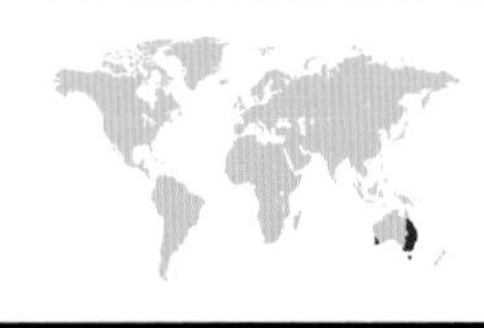

**IUCN STATUS** Least Concern
**LENGTH** 2⅛–2½in (5.5–6.5cm)
**WEIGHT** ⅛–¼oz (9–13g)

*Idionycteris phyllotis*

## Allen's Big-eared Bat

**THIS SPECIES OCCURS** in the southwest United States and south through central Mexico. It is an attractive bat with mid-brown, long, and silky fur, and strongly pink-toned limbs, ears, and snout; the membranes are darker gray. It has long, prominent ears with large tragi, and distinctive paired lappets over the nose. It occurs mainly in wooded areas in the uplands, but also in river valleys and semi-desert scrubland. Roost sites are varied, including crevices in limestone rocks, mineshafts, and beneath the loose bark of mature pine trees. Maternity roosts can hold up to 150 females, each of which bears a single pup in early summer. This bat is an extremely agile flier that gleans prey from foliage and rock surfaces. The population has declined slightly over recent decades, mainly through disturbance and destruction of roosts.

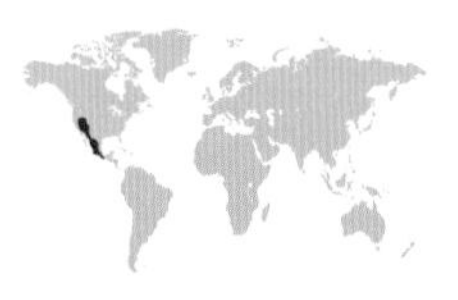

**IUCN STATUS** Least Concern
**LENGTH** 4⅛–4⅝in (10.3–11.8cm)
**WEIGHT** ¼–⅝oz (8–16g)

***Lasionycteris noctivagans***

# Silver-haired Bat

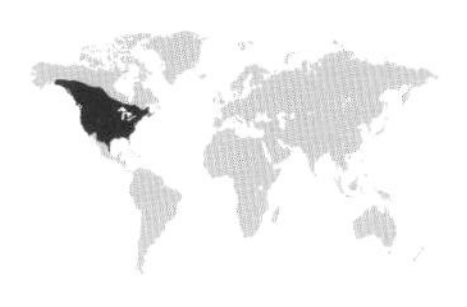

**IUCN STATUS** Least Concern
**LENGTH** 3⅝–4½in (9.2–11.5cm)
**WEIGHT** ¼–⅜oz (8–11g)

**THIS BAT OCCURS** across almost the whole of the United States. At least some in the north are migratory, but evidence suggests not all individuals in the same population migrate. It has blackish fur with silver-tipped guard hairs on the body, giving a frosty appearance. The muzzle is short and rather broad. It feeds on a variety of flying insects. It occurs most often in hardwood forest with rivers, forming small nursery colonies in tree cavities or under loose bark, and hibernating alone or in small groups in cliff-face crevices and tree cavities. Females store sperm through winter, and gestation begins in early spring; twin pups are born in early summer. Time of emergence seems to vary considerably, depending on which competing bat species are present, as well as varying risk of predation.

***Parastrellus hesperus***

# Canyon Bat

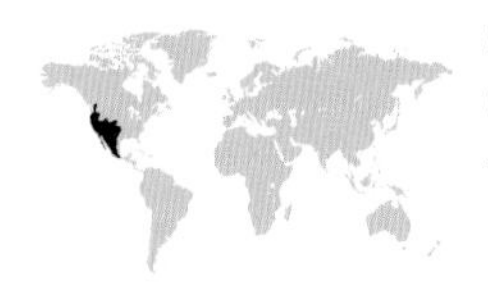

**IUCN STATUS** Least Concern
**LENGTH** 2½–3⅛in (6.2–8cm)
**WEIGHT** ⅛oz (3–6g)

**THIS SMALL BAT**, still placed in the genus *Pipistrellus* by some authorities, occurs in the west United States and in west and central Mexico. To the north, its limit is the extreme south of Washington; its range extends eastward up to Texas. It has light sandy fur with dark, bare parts, a short snout, and small ears. Its broad, short wings permit slow, fluttering, and agile flight. The species occurs in open, arid habitats such as grassland and semi-desert, as well as light woodland. It roosts and hibernates in rocky crevices, abandoned mines, and old buildings. Roost sites are invariably close to water. The bat often emerges well before sunset and may be seen in full daylight. Females usually give birth to twins, which are adult-sized at only one month old.

***Perimyotis subflavus***

# Tricolored Bat

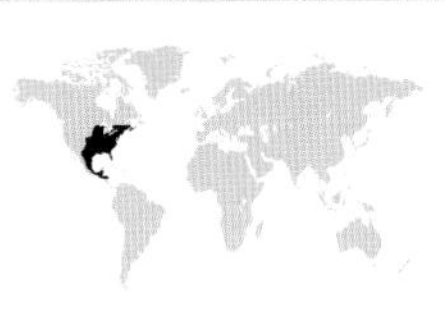

**IUCN STATUS** Least Concern
**LENGTH** 3–3½in (7.7–8.9cm)
**WEIGHT** ⅛–¼oz (4.6–7.9g)

**THIS BAT RANGES** from southeast Canada through the eastern half of the United States, and through east Mexico to Honduras. It has recently been placed in a new monotypic genus, having formerly been a *Pipistrellus*. A smallish vesper bat, it has yellow-brown fur, a short face, and relatively large eyes and ears. The species tends to occur in open woodland or woodland edges near water. Nursery roosts are formed in foliage, and caves used for summer roosts and winter hibernation. It preys on all kinds of flying insects. In the fall it forms mating "swarms," and females give birth the following spring, usually to twins. The pups are born well-developed and are able to fly at three weeks old. Though abundant, this species has suffered some mass die-offs due to white-nose syndrome.

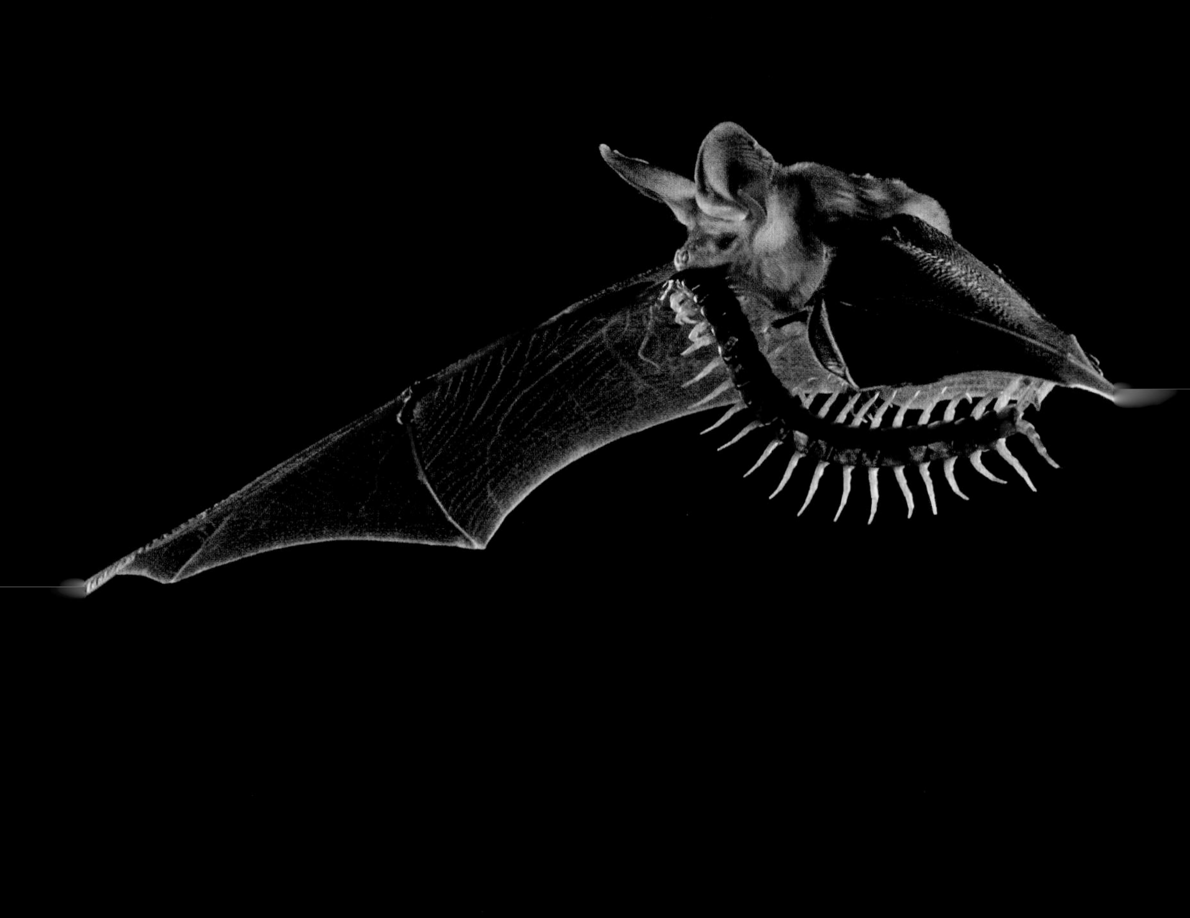

***Antrozous pallidus***

# Pallid Bat

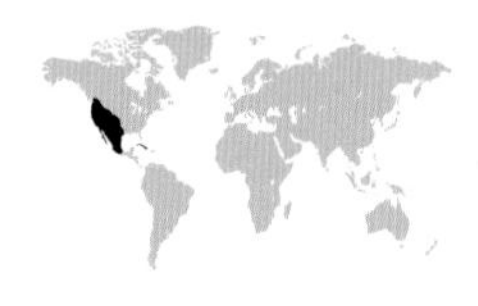

**IUCN STATUS** Least Concern
**LENGTH** 2⅜–3⅜in (6–8.5cm)
**WEIGHT** ⅝–1oz (17–28g)

**THIS LARGE VERSPERTILIONID** ranges from southwest Canada through the west of the United States to Mexico, and also Cuba. It has pale yellowish fur, whitish below, prominent eyes and long, broad ears. It occurs in arid, sparsely vegetated rocky habitats, roosting mainly in rock crevices, and also tree holes and buildings. It forages over grassland, woodland, and riversides, mainly gleaning from the ground, but also landing and crawling after prey. This species feeds on large insects, centipedes, scorpions (apparently resistant to their venom), and some small vertebrates, detected by sound. It is also an important pollinator of giant desert cacti. Roosts usually hold at least twenty bats, which communicate with an individually unique, consistent series of calls, allowing roost-mates to recognize one another.

*Tylonycteris pachypus*

# Lesser Bamboo Bat

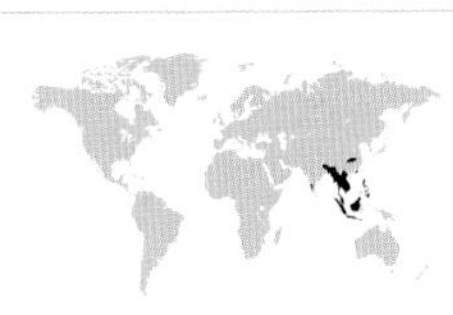

**IUCN STATUS** Least Concern
**LENGTH** 1⅜–2in (3.5–5cm)
**WEIGHT** ⅛–¼oz (2–5.8g)

**THIS TINY SPECIES** occurs in southwest India, and then from Nepal, Bangladesh, and central and eastern China south through Myanmar, Thailand, Laos, Cambodia, Vietnam, and Malaysia. It is also present in west Indonesia, the whole of Borneo, and most likely the Philippines. It is a brown or reddish-brown bat with dark bare parts. The face is compressed, with small eyes and ears. Its feet bear large, flat pads enabling it to attach to surfaces. The species uses various habitats where bamboo grows, including forest and farmland, and may also roost inside buildings. It eats insects of various kinds. Roosting groups tend to contain multiple females with a single male; other males roost alone. Females often bear twin pups, which take six weeks to reach independence.

*Euderma maculatum*

# Spotted Bat

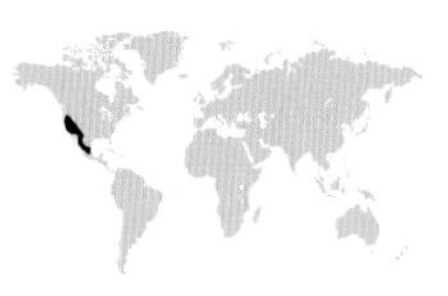

**IUCN STATUS** Least Concern
**LENGTH** 4¼–5in (10.7–12.5cm)
**WEIGHT** ⅝–¾oz (16–20g)

THIS BAT OCCURS in the west of North America, from southwest Canada south to northern Mexico. Strikingly beautiful, its dark upperside is marked with three large, white spots, one on the rump and one on each shoulder. The ears are very large. It occurs in a wide variety of both arid and forested habitats, and needs large areas of open water to drink. It roosts by day in small, tight cracks in rock faces, and occurs at up to 10,000ft (3,000m) altitude. Shortage of open water sites are disappearing rapidly throughout its range. This bat preys mainly on moths, especially species able to hear most bat echolocation calls; its calls are at an unusually low frequency (about 10kHz) so it can approach prey undetected. However, its calls only allow navigation through quite open environments.

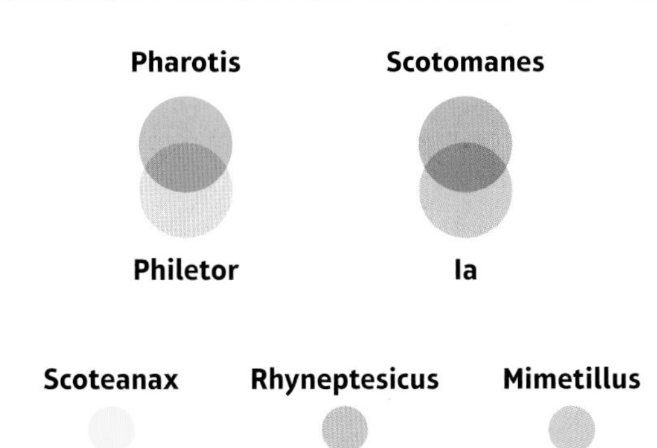

## Rhyneptesicus

*Rhyneptesicus nasutus* is often classed within *Eptesicus*, and is similar to those species. It is a bat of open, arid habitats close to rivers, with rocky areas or buildings for roosting. It occurs in Afghanistan, Iran, Iraq, Oman, Pakistan, Saudi Arabia, and Yemen.

## Ia

*Ia io* has the distinction of being the animal with the world's shortest scientific name (along with a dinosaur species, *Yi qi*). It is also one of the largest vespertilionids, with a body length of up to 4⅛in (10.5cm). This bat is able to prey on birds as well as a wide range of larger insects, tending to leave its roost relatively early in the evening to hunt small birds as they go to roost. It occurs in south and southeast Asia, and roosts in caves in groups of up to one hundred.

## Scoteanax

*Scoteanax rueppellii* occurs in Australia, and is related to the Australian genus *Nyctophilus*. Its range extends down the east coast from north Queensland to southeast New South Wales. It is found in all kinds of forested habitat.

## Scotomanes

*Scotomanes ornatus*, known in English as "harlequin bat," is a highly distinctive east Asian species—one of the most boldly patterned of all bats, with various conspicuous white or silvery markings on its mid-brown fur. The underside is mostly white, but with a neatly delineated brown chin, collar, and central stripe; the upperside shows a white line down the spine and patches around the shoulders.

## Pharotis

*Pharotis imogene* is another striking bat, very dark-furred with a blunt face and large, long ears that can curl back like rams' horns. It has bumpy projections on the face above the nostrils. Native to Papua New Guinea, this species has not been observed with certainty since 1890, though there is anecdotal evidence that it survives. The forest terrain where it may still occur is very difficult to survey accurately.

## Mimetillus

*Mimetillus moloneyi* is a distinctive, small vespertilionid. It has a wedge-shaped snout and rounded ears with pronounced lower folds, resulting in a face rather reminiscent of a molossid bat. For its size its wings are unusually short. It occurs over a broad swathe of south Africa, in wooded savanna habitats.

## Philetor

*Philetor brachypterus* is a small, dark, blunt-faced bat that occurs in Nepal and patchily in insular southeast Asia to New Guinea. On the mainland it is an upland species, hunting and roosting in montane forest, while on the islands it occurs in both lowland and upland tropical forest and forest edges.

# YINPTEROCHIROPTERA

## PTEROPODIDAE

The family Pteropodidae, also known as the Old World fruit bats and as megabats, contains 200 species, in forty-five genera and eight subfamilies. In their appearance and way of life they are distinct and different to all other bat families, and were long classified as a suborder, Megachiroptera; all the other bat families were placed in a second suborder, Microchiroptera. The *mega* and *micro* prefixes referenced an obvious difference between the two groups: megabats were (mostly) big, microbats (mostly) small.

Today, biologists have powerful new tools in their armory to help classify animals properly. With advances in molecular biology, there is no longer a need to rely just on anatomical and behavioral traits. Instead, an animal's DNA can be examined and compared to that of others. This way, the evolutionary pathways of different species can be traced, and how far back in time they shared a common ancestor can be worked out. Insights from this kind of research have helped to find a place for many animals that had hitherto proved difficult to classify. However, it has also shaken up understanding of the relationships within and between many different animal groups that had been thought to have been already figured out—and bats are no exception.

Despite their physical distinctiveness, we now know that the megabats do not form a suborder all on their own. At suborder level, they must be grouped together with several microbat families, specifically Craseonycteridae, Hipposeridae, Megadermatidae, Rhinolophidae, Rhinonycteridae, and Rhinopomatidae. This grouping has been named Yinpterochiroptera (or Pteropodiformes). The remaining bat families form the suborder Yangochiroptera (or Vespertilioniformes).

### Evolution and anatomy

The lesson to be learned from this is that the characteristics most obvious to us (with our human perspective) are not necessarily the "deepest," or most significant. Nevertheless, a typical pteropodid does look drastically different to a typical microbat—so much so that, for many years, biologists seriously entertained the idea that the two groups were not related at all. They believed that pteropodids were a lineage of primates, and had evolved their wings quite independently to the microbat lineage (see Introduction, page 15). Side by side, a generic pteropodid is much larger than a generic microbat. It has no tail and shows a space between its feet when it flies. Its face is totally different: wide-eyed and foxy, with simple, pointed ears, quite unlike the microbat's squashed and folded face, smaller, sometimes tiny eyes, and complex, convoluted ears. And in habits the two differ hugely as well. The microbat roosts in caves or other hollows, catches insects, and finds both its prey and its way about through echolocation. The pteropodid roosts hanging from tree branches, feeds on fruit that it locates by sight and smell, and does not echolocate.

The devil is in the detail, though, and there are exceptions that blur the lines. The largest microbat is larger than the smallest pteropodid. Some pteropodids have tails, some microbats do not. There are many microbats with long snouts and big eyes. Feeding and roosting habits overlap too, and a few pteropodids echolocate (albeit through a completely different method to that employed by microbats). In general, pteropodids have claws on the second digits of their hands and microbats do not, but a few pteropodids also lack the extra claw. Studies on how bats move on the ground and on vertical surfaces reveal that pteropodids excel at climbing in a head-up position but are very clumsy on the ground. Many microbats are surprisingly agile moving on flat ground, however, and locomote in a manner recalling land mammals.

The family Pteropodidae occurs mostly in tropical regions. It is found across most of Africa, south and southeast Asia, and east Australia, with south Asia home to the greatest diversity of species and most likely the region where the earliest members of the family evolved. There are no species in the Americas or in the north of Asia, and only one species, *Rousettus aegyptiacus* (see page 325), just edges into Europe (Cyprus). Larger pteropodids are strong fliers, and in the Pacific many species will readily commute between remote islands, wherever suitable habitat exists. Some smaller species stick to just one or a handful of islands, though, and are not great travelers. Most species are associated with forest habitats, but some live in more open country and alongside human settlements.

Most pteropodids feed mainly or entirely on fruit and nectar, and this diet requires particular anatomical adaptations. The bats locate their food by sight and scent. Unlike insectivorous bats, locked into a constant evolutionary "arms race" against prey that does not want to be caught, fruit-eating bats have a mutually beneficial relationship with the plants whose fruit and nectar they eat. Because the plants rely on bats for seed dispersal or pollination, their fruits and flowers have developed bat-friendly qualities. Plants that "want" to be visited by bats are often pale or located where they are easier to find in the darkness of the night. They grow at points on the plant that bats can easily access, and they are highly fragrant. Fruit-eating bats, accordingly, have very large eyes with acute low-light sensitivity; they also possess a highly developed sense of smell. It is believed these bats lost the ability to echolocate, although some cave-dwelling species, which must navigate in very dark conditions, have "re-evolved" simpler forms of echolocation using tongue clicks. It has also recently been discovered that at least three species of pteropodids—*Cynopterus brachyotis*, *Eonycteris spelaea* (see page 337), and *Macroglossus sobrinus* (page 322)—have evolved another basic form of echolocation, involving wing sounds. All pteropodids still have good hearing, as evidenced by their complex systems of vocal communication.

Pteropodids have longer snouts than most microbats, giving the fox-like look that has led to the alternative name of "flying fox" (for species in the genus *Pteropus* in particular). They lack the elaborate nose-leaves that many microbats possess, though the actual nostrils are often quite flared and prominent. They have strong jaws, and sharp front teeth for gripping and biting into fruit; at the back of the jaw are broader and flatter teeth for chewing. The palate on the roof of the mouth has a variable number of prominent hard ridges. These are used, along with the teeth, to crush mouthfuls of fruit to a pulp. The bat swallows the juice and spits out the fibrous leftovers, compressed into a pellet. The digestive tract is relatively short and simple, with little else besides sugary fruit juice to process. The fruits these bats take often contain very small seeds; these are eaten along with the fruit pulp, but are excreted intact. Their bodies can function on a lower protein intake than most other mammals and a fruit diet can meet their needs. However, many pteropodids also consume some pollen, which is relatively rich in protein, and some species have been observed to catch and eat insects occasionally.

A few pteropodids take most of their food from flowers, consuming nectar and pollen. Among them are the "long-tongued bats" (*Macroglossus sobrinus* and *M. minimus*), several species known as "blossom bats" (*Syconycteris australis*, page 324, *S. hobbit*, and *S. carolinae*, and also the genus *Eonycteris*), and those of the genus *Melonycteris* (*M. fardoulisi*, *M. melanops*, *and M. woodfordi*). These bats are typically smaller than the species that eat a higher proportion of fruit; with less need to chew their food, they have longer, more delicate snouts which can probe into flowers. They have small teeth but very long tongues, with brush-like projections (papillae) that help to "mop up" nectar.

When pteropodids need to drink, they swoop low over water and scoop up a mouthful (as do other bats). They may also lick rainwater from their fur. Some coast-dwelling fruit bats have been observed drinking seawater. This may be because they require more sodium than their diet can provide.

## Pteropodids and people

As fruit-eaters and nectar-drinkers, pteropodids are important parts of their ecosystem as they disperse seeds and pollinate flowers. They cover more ground than most other pollinators and seed dispersers. These bats are also very important agents in maintaining genetic diversity in plant populations and in kickstarting reforestation. Many of the wild plants that depend on bats are of economic importance, and bats' pollination services are of great value to agriculture. They can, of course, also do economic harm by raiding fruit crops. However, keeping bats away from orchards is accomplished relatively easily with the use of netting, and there is no need to use lethal methods.

## Social and breeding behavior

Like the microbats, all but a few island species of pteropodids are nocturnal and do their foraging and feeding after dark. The majority of species roost in the open, in tall trees, and could be mistaken for large dead leaves as they hang by their feet, with wings folded around their bodies. They may roost in groups or alone. Some are able to enter a torpid state to cope with cold weather, though most live in regions where temperature fluctuations are not too significant. In areas where several different pteropodids coexist, there are often marked consistencies in the position each species takes up within the levels of the forest; some stick to the canopy, and others live much lower down.

Some species roost inside large caves. Among them are the genus *Rousettus*, which has evolved a simple system of echolocation (using tongue clicks) to help bats find their way around the dark interior of their large roosting caves.

The gregarious species of pteropodids can show elaborate social systems. Bats do not sleep solidly through the daylight hours, but are regularly awake and interacting with one another. Social behaviors between adults include sniffing and grooming one another (this being particularly common between a mother and full-grown pup), and antagonistic encounters involving "boxing" with the wings, accompanied by loud calls. In many species, males are polygynous, guarding a harem of several females from the attentions of other males. There is a social hierarchy between unpaired males as well, with lower-status individuals forced to roost in suboptimal positions, lower in the tree. Mornings at the roost often see a surge in sexual activity, which in some species at least includes licking of the genitals as well as copulation (male licking female and vice versa, and sometimes male-to-male too, though this probably serves a different, social function). Female groups tend to stay together regardless of any change to which male is guarding them. They also can and do reject unwanted mating attempts from males.

Male pteropodids have various ways to attract the attention of females. The epauletted fruit bats (genus *Epomophorus*) have shoulder pouches that, when opened, extrude a clump of pale, elongated hairs—the so-called epaulette. This provides not just a visual display, but also an olfactory one, as the hairs help disperse pheromones. The bat beats its wings to spread the scent around or waft it toward a female. The genus *Epomops* also show pale shoulder patches, and some other species have pale markings around the face to help draw attention.

One pteropodid species, *Hypsignathus monstrosus* (see page 323), forms leks. In this bat, male–female associations are short: males display together to attract females, which then select a mate from the group. Each male holds a territory within the displaying group, and the most successful males are clumped together—just 6 percent of the group accounts for 79 percent of all matings. This species shows notably marked sexual dimorphism, a trait observed in other lekking species in nature, for example, in birds such as grouse. In the case of these bats, it is manifested in the males being much larger, with a very different facial structure. Their hugely enlarged noses, lips, and cheek pouches enable them to produce the loud, guttural croaking and honking calls they give when displaying. The lekking occurs for a short period just after dark and just before dawn, with the males tending to establish their territories in the early morning session. Mating then occurs in the evening.

Monogamous pair formation is known in some individuals of at least one species of pteropodid bat (*Pteropus conspicillatus*). Some others are solitary, only forming groups for a brief mating season—among them the "blossom bats" (genus *Syconycteris*).

Timing of breeding is tied into food availability. For most of these bats, births occur in the rainy season when food supplies are high. As with other bats, pteropodids typically have lengthy gestation periods, often four months or longer. They do not normally show delayed fertilization, as is the norm with microbats that live in temperate climates and hibernate over winter. Females of the species that breed twice a year will usually mate again soon after giving birth; the female will be nursing one pup while pregnant with the next. Relatively few species have more than one young in a year, however. In many cases, females of these species will not associate with males during their pregnancies, but roost in single-sex groups. A few pteropodids sometimes give birth to twins, but singleton pups are usual.

Cooperative maternal behavior between females has been observed in captive pteropodids. A female *Pteropus rodricensis* was seen to be struggling to birth her pup and was not in the correct position (pteropodids give birth in a head-up position, but this individual was hanging head-down). Another female seemed to encourage her to change position, and also licked the struggling female's genitals until the birth process began. Several other species have been observed to suckle pups that were not their own, while the mother was away.

The pup is born furry and with open eyes. It is unable to fly, but able to hang onto its mother's belly, using its teeth as well as its wings and feet. The female will carry it with her on each of her foraging trips from birth until it is a few weeks old, which is much safer than leaving it alone in the roost. By the time the pup is too large to carry it is not so helpless, is better able to regulate its own body temperature, and will probably come to no harm if left on its own.

As it grows, the pup will begin to climb around the roosting area, and to start to build up its wing strength before attempting its first flight. It will not become independent until it is three months old at the youngest. Even when it is flying confidently and is weaned, the pup will continue to associate with its mother, sometimes for longer than a year. Sexual maturity may take several years to attain in larger species; most are capable of breeding at two years old.

Like other bats, pteropids can live to an impressive age, with forty-year-old individuals known in some species. Most, of course, will not reach such a great age. They face a variety of predators, from large birds of prey to tree-dwelling and cave-invading snakes. Some species that live on remote islands enjoy a predator-free existence—or did, until humans arrived on the scene. People are direct predators of pteropodid bats. The larger species in particular are hunted in vast numbers for food in many countries, and also killed as crop pests. For the island endemics, humans themselves are not necessarily the only, or the worst, threat; non-native predators introduced (deliberately or accidentally) by humans can also cause devastation. On the island of Guam in the Pacific, a combination of overhunting by people and predation by introduced Brown Tree Snakes have almost wiped out the two species of *Pteropus* bats that occur there. On Christmas Island, another *Pteropus* species is threatened by the tiniest non-native enemy—the ferociously predatory Yellow Crazy Ant, which has established "super-colonies" on the island, and has devastated numbers of many of the island's unique endemic animals.

Pteropodids may also die as a result of disease, or by collision with overhead power lines or wind turbines. Island species may be killed directly or lose their roosts when violent weather strikes, or even volcanic eruption. When roosting spots are lost and new ones cannot be found, the bats will be forced to roost in unsuitable places. For example, many of the two million *Rousettus amplexicaudatus* (see page 326) roosting in the Monfort Bat Sanctuary cave on Samal island in the Philippines cannot find a safe place on the cave roof. Instead, they have to roost on lower slopes where predators such as rats can reach them and their pups. The cave remains a vital shelter for the species, but is vastly over-occupied because disturbance has driven the bats from their other roosting caves.

From the accounts that follow, it is clear that even within the relatively easily observed family Pteropodidae there are many poorly known and under-researched species. It is therefore likely that the number of recognized species will grow, as these more obscure bats are studied and ways in which their separate populations differ are discovered. Research is also needed urgently to work out how best to safeguard the many threatened species. Even those that are endemic to a single island are likely to fulfill important ecological roles; their loss may well have a domino effect on a wide range of other species.

***Pteropus alecto***

## Black Flying Fox

**THIS BAT OCCURS** along the north coast of Australia and parts of New Guinea and Indonesia. It is large and mainly black, showing a dark reddish undercoat on the upperside. The large eyes are dark brown, the snout long and robust. It occurs in coastal areas, roosting in mangrove swamps as well as tropical forests and wooded savannas, in groups and sometimes with other *Pteropus* species. It may fly long distances to forage; nightly movements of 12 miles (20 km) are not unusual, and it will commute between islands. The species is thus a key long-distance seed disperser for many plant species. It feeds on fruit and nectar, visiting flowers of *Eucalyptus, Banksia, Ficus* (seen here) and *Melaleuca,* and taking native and non-native fruits. Females have one pup a year, after a twenty-seven-week pregnancy.

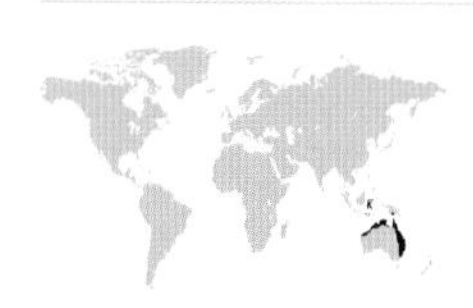

**IUCN STATUS** Least Concern
**LENGTH** unknown
**WEIGHT** 1lb 7oz (660g)

***Pteropus mariannus***

## Marianas Flying Fox

**THIS BAT OCCURS** in Micronesia, on the Northern Marianas Islands, Guam, and the Ulithi Atoll. It is a medium-sized, dark gray-brown bat, with pale silvery-golden shoulders and sides of the neck. Its habitat is primary tropical forest, with a sparse understory, a canopy reaching 26–50ft (8–15m), and taller trees for roosting. The diet is mainly forest fruits, with some nectar, from plants such as this coral tree. Males live with up to fifteen females; nonbreeding males form bachelor groups. Females have a single pup after six months' gestation. Its severe decline is due to habitat loss, illegal hunting, volcanic eruption (on Anatahan Island), and (on Guam) predation of pups by the non-native Brown Tree Snake—only around one hundred survive there. Urgent strict protection of the estimated 4,200 bats and their habitat on the Marianas Islands is needed.

**IUCN STATUS** Endangered
**LENGTH** 7½–9½in (19–24cm)
**WEIGHT** 12oz–1lb 4oz (330–577g)

**Pteropus**

## Pteropus

The largest genus within Pteropodidae is *Pteropus*: the flying foxes. These are the classic fruit bats, mostly large or very large, with attractive, fox-like faces. Most have dark, rather dense and soft fur, and often a paler area on the upper back. Others are pale brown, silvery, or bright rufous. The genus includes the world's largest bats: *P. neohibernicus* and *P. vampyrus* (see page 314). Members of both species can have a wingspan greater than 5ft (1.5m) and can weigh over 2lb 3oz (1kg). The smallest species, *P. personatus*, *P. temminckii*, *P. tokudae*, and *P. woodfordi*, all typically weigh less than 6oz (170g). A few species are very common and widespread, but many more have a restricted distribution, being limited to just one or a few islands. Of these a high proportion are considered to be threatened with extinction.

These bats are fruit-eaters and can dextrously manipulate their food with one foot while hanging from the other; the lack of a complete membrane between the legs makes this an easier task for them. They can also grip food items with their thumbs. They will travel long distances in search of food, at least 37 miles (60km) in many cases. For those living on islands, sea crossings are no obstacle to their foraging flights. When these bats locate fruit they make a rather clumsy crash-landing into nearby foliage, then scramble their way to the fruit.

Among those species classed as at risk of extinction is *P. aldabrensis*, endemic to the Aldabra Atoll in the Seychelles. The islands that hold its small population (just a few hundred individuals) are under strong UNESCO environmental protection, and its numbers are stable at present. However, climate change and a resultant rise in sea level could prove catastrophic for this species. Another vulnerable species is *P. anetianus* of Vanuatu. Although known to exist on at least twenty-one islands and considered to be common, it is under heavy pressure from hunting and is in decline. This species does not appear to be able to recolonize islands from which it has been eradicated.

*P. aruensis* of the Aru Islands, Indonesia, is probably already extinct. It has not been seen alive since the late nineteenth century, and the islands have suffered substantial deforestation since then. *P. rennelli* occurs only on the island of Rennell in the Solomons, which is just 380 square miles (984 km$^2$). Available habitat on the island has dwindled severely since the 1950s and 1960s, and a single weather event could easily wipe out the species. It is a similar story for *P. fundatus* of the Banks Islands in Vanuatu. While this small species occurs on two islands rather than one, those islands have a total area of less than 135 square miles (350km$^2$). The bat is still hunted for food, and neither it nor its habitat has any kind of legal protection. *P. tuberculatus* was known only from the Solomon Islands and is now probably extinct, as are *P. allenorum*, *P. brunneus*, *P. coxi*, *P. pilosus*, *P. subniger*, and *P. tokudae.*

Many species of *Pteropus* are little known, while others have had no kind of assessment at all. There is no doubt that some of these are in dire straits; they may be extinct or very close to it already, before researchers have managed to find out anything useful about their population or ecology. These include *P. gilliardorum*, known only from three specimens taken on New Britain and New Ireland of Papua New Guinea, and *P. keyensis*, of the Kei Islands in Indonesia. This bat is almost completely unknown, but its home islands suffer uncontrolled forest destruction because of the slash-and-burn agriculture practiced there.

The following species are treated with individual accounts; *P. alecto* (see page 305), *P. conspicillatus* (page 308), *P. lylei* (page 310), *P. mariannus* (page 305), *P. poliocephalus* (page 312), *P. scapulatus* (page 313), and *P. tonganus* (page 311). The other *Pteropus* species this book recognizes are as follows: *P. admiralitatum* (Admiralty and Bismarck islands), *P. caniceps* (Indonesia), *P. capistratus* (Papua New Guinea), *P. chrysoproctus* (Indonesia), *P. cognatus* (Solomon Islands), *P. dasymallus* (Japan, Philippines, and Taiwan (China)), *P. ennisae* (Bismarck Islands), *P. faunulus* (Andaman and Nicobar Islands), *P. griseus* (Indonesia), *P. howensis* (Solomon Islands), *P. hypomelanus* (south and southeast Asia), *P. intermedius* (Myanmar and Thailand), *P. livingstonii* (Comoros Islands), *P. lombocensis* (Indonesia), *P. loochoensis* (Japan), *P. macrotis* (Australia, Indonesia, and Papua New Guinea), *P. mahaganus* (Papua New Guinea and Solomon Islands), *P. medius* (south Asia), *P. melanopogon* (Indonesia), *P. melanotus* (Andaman and Nicobar Islands), *P. molossinus* (Micronesia), *P. niger* (Mauritius), *P. nitendiensis* (Solomon Islands), *P. ocularis* (Indonesia), *P. ornatus* (New Caledonia), *P. pelagicus* (Micronesia), *P. pelewensis* (Micronesia), *P. personatus* (Moluccan Islands), *P. pohlei* (Papua New Guinea), *P. pselaphon* (Japan), *P. pumilus* (Philippines and Indonesia), *P. rayneri* (Papua New Guinea and Solomon Islands), *P. rodricensis* (Rodrigues island), *P. rufus* (Madagascar), *P. samoensis* (Samoan Islands), *P. seychellensis* (Seychelles), *P. speciosus* (Philippines), *P. temminckii* (Indonesia), *P. ualanus* (Micronesia), *P. vetulus* (New Caledonia), and *P. voeltzkowi* (Pemba Island off Tanzania).

***Pteropus conspicillatus***

# Spectacled Flying Fox

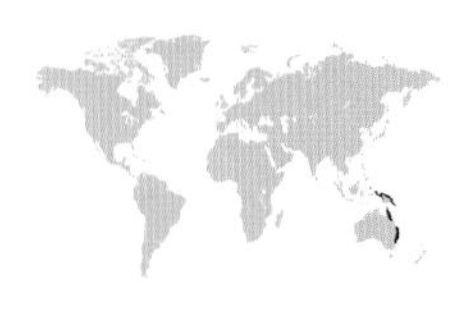

**IUCN STATUS** Least Concern
**LENGTH** 8¾–9½in (22–24cm)
**WEIGHT** 1lb 2oz–2lb 3oz (510–1,000g)

**THIS BEAUTIFUL BAT** occurs around much of the New Guinea coastline and nearby Indonesia, and also parts of coastal Queensland, Australia. It has black fur and membranes with a pale back, and spectacle markings on its face that extend down to the muzzle. The large eyes are very dark and the ears small. Males are significantly larger than females. It occurs in lowland swamp, mangrove forest, and tropical, moist forests, and increasingly is colonizing wooded areas near towns. It favors pale-colored fruits from the forest that it can find easily in the dark, and may also eat fruit from orchards during droughts. Groups ("camps") roost high in the trees in full sunshine by day, and forage at night. Roosting colonies are single-sex until pups are born, when males join the colonies and form territories, defending one or more females. This bat is hunted for food and persecuted as a crop pest in parts of its range. Farmers in Queensland may legally kill small numbers, even though the species is listed as threatened under the Commonwealth Environmental Protection and Biodiversity Conservation Act 1999.

***Pteropus lylei***

# Lyle's Flying Fox

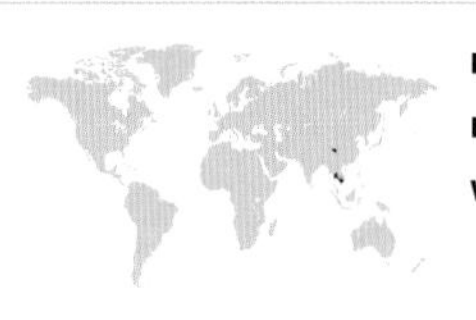

**IUCN STATUS** Vulnerable
**LENGTH** unknown
**WEIGHT** unknown

**THIS IS A DECLINING** species with a scattered distribution in southeast Asia, from south Vietnam through Cambodia to Thailand. It is also present in Yunnan, China. A large bat, it has bright reddish-golden fur with a blackish face and ears. It occurs in damp wooded areas and may visit orchards for food, as well as forage in the forest. It is hunted for bushmeat in Thailand and Cambodia, and, like other flying foxes, is widely persecuted as a crop pest, although colonies within the grounds of Buddhist monasteries in Thailand enjoy strict protection. Only twelve colonies are known in Thailand, the largest of which holds 3,000 individuals. Just three colonies are known in Vietnam. It appears to have undergone dramatic though poorly documented decline.

***Pteropus tonganus***

# Pacific Flying Fox

**IUCN STATUS** Least Concern
**LENGTH** 5⅞–10¼in (15–26cm)
**WEIGHT** 7oz–2lb 7oz (200–1,100g)

**THIS WIDESPREAD PACIFIC** species occurs on many islands and island groups. It is medium-sized and dark, with pale yellowish shoulders. It prefers primary forest, roosting in tall trees, but may forage in cultivated and suburban areas. It breeds once a year, males joining female groups after the singleton pups are born. The diet includes fruit, pollen, and nectar; individuals may defend food plants from other bats. Like other fruit bats, it crushes mouthfuls of fruit against its palate, spitting out pellets of fiber and swallowing the juice. Assessing this bat's population is difficult, as it migrates between islands. It is declining, mainly due to habitat loss, but is also hunted for food. It is also susceptible to natural disasters such as cyclones. Fiji holds about 50 percent of its total population.

***Pteropus poliocephalus***

# Gray-headed Flying Fox

**IUCN STATUS** Vulnerable
**LENGTH** 9⅛–11½in (23–29cm)
**WEIGHT** 1lb 15oz–2lb 3oz (600–1,000g)

THIS DISTINCTIVE, LARGE flying fox is found in coastal east Australia. It occurs mainly in lowland tropical and swampy forest, and roosts in large trees. Because of habitat loss, it is increasingly observed in the suburbs. Although roosts can persist for many years, at times large parts of the population migrate long distances to find food or new habitat. It forms monogamous pairs until the single pup becomes independent, the male defending his mate and territory with smelly secretions and loud calls. This bat is sometimes considered a nuisance. The bats' presence around houses produces noise and mess. A coordinated national plan is needed to manage these situations while protecting and improving natural habitat to halt this ecologically essential species' alarming decline.

***Pteropus scapulatus***

# Little Red Flying Fox

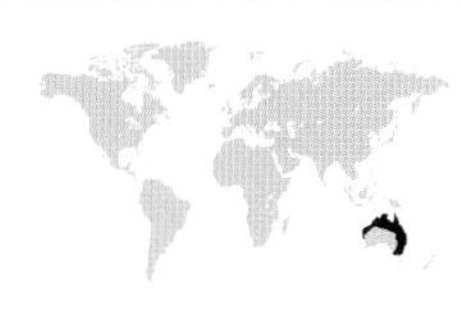

**IUCN STATUS** Least Concern
**LENGTH** 9½–10¼in (24–26cm)
**WEIGHT** 1lb 3oz (550g)

**THIS SMALL FLYING** fox occurs around the coast of Australia, except for the southwest quarter; it also occurs on some islands in the Torres Strait. It has dark red-brown fur with pale hair tips. It forms unusually dense roosts in large trees, typically alongside watercourses in varied habitats including mangrove swamps, bamboo clumps, and forest. The diet is mostly nectar and pollen from eucalyptus flowers (seen here); when drought affects supply of these, it undertakes lengthy migrations to find new food sources. It forms harems that separate when females are pregnant; males rejoin after pups are born. Speculation about these bats as potential sources of disease has harmed public perception of this and other flying fox species. Additional threats come from climate change and habitat loss.

***Pteropus vampyrus***

# Large Flying Fox

**IUCN STATUS** Least Concern

**LENGTH** 10⅝–13½in (27–34cm)

**WEIGHT** 1lb 5oz–2lb 3oz (600–1,000g)

**THIS GRAND BAT IS FOUND** over much of southeast Asia, from south Vietnam, Thailand, Cambodia, and the Malay peninsula through most of the Indonesian islands, Borneo, and the Philippines. They have wingspans up to 5½ft (1.6m), the largest of any bat. It has rufous fur, dark membranes, and a dark face. The eyes are reddish, the ears quite small and pointed. It roosts in trees in both primary and secondary forest, and forages at night in both forest and cultivated areas, preferring the former. This species is polygynous, with successful males defending up to ten females in harems. Each female produces one pup a year, in late winter or spring, depending on region. This species, like other *Pteropus* bats in south Asia, is declining, due to overhunting for bushmeat; it is also persecuted as a crop pest. Much current research on *Pteropus* bats in the region is focused on trying to establish a link between them and viral diseases that can affect humans. However, studies on their population dynamics are needed urgently, as is education for local people on the bats' economic importance as pollinators and seed dispersers for many fruit crops.

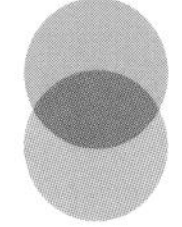


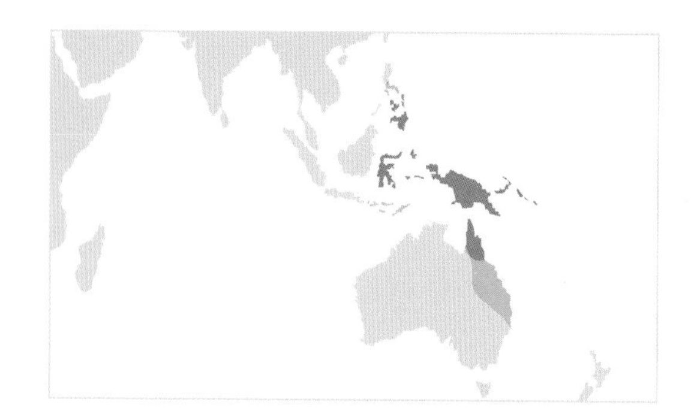

## Nyctimene

Another fairly large genus is *Nyctimene*, of southeast Asia and Australia, with sixteen species. These distinctive animals are known as the "tube-nosed bats," or "tube-nosed fruit bats." They can be recognized by their unusual nostrils, which are extended into fleshy tubes that stick out sideways, projecting about 3⁄16in (0.5cm) from the snout. They are medium-sized fruit bats with rather plain, light brown fur, a short tail, small narrow ears, and distinctively cat-like eyes with pointed inner corners. Many also have beautiful golden spots on their wings and ears.

*N. vizcaccia* is found in Papua New Guinea and the Solomon Islands. Another well-known species is the widespread *N. albiventer*, which occurs in the Halmahera, Banda and Aru Islands, Bismarck Archipelago, Moluccan Islands, New Guinea, Admiralty and Solomon Islands, and the Cape York peninsula of Australia. This species has dark wing membranes with bright yellow spotted markings on the upperside; the exact pattern varies between individuals. These markings probably help to provide camouflage for the bats at their solitary tree roosts.

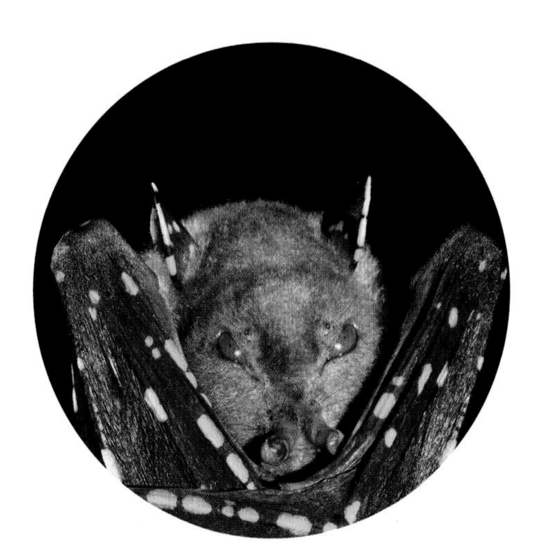

***Nyctimene robinsoni***

The other *Nyctimene* species this book recognizes are: *N. aello* (New Guinea), *N. cephalotes* (Indonesia and New Guinea), *N. certans* (New Guinea), *N. cyclotis* (New Guinea), *N. draconilla* (New Guinea), *N. keasti* (Moluccan Islands), *N. major* (islands around New Guinea), *N. malaitensis* (Solomon Islands), *N. masalai* (Bismarck Islands), *N. minutus* (Indonesia), *N. rabori* (Philippines and Indonesia), *N. robinsoni* (Queensland in Australia), *N. sanctacrucis* (Santa Cruz Islands), and *N. wrightae* (New Guinea).

## Dobsonia

The genus *Dobsonia* holds fourteen species. They are known as the "naked-backed fruit bats" because their wing membranes meet along the midline of the back, so the entire back appears covered with naked skin. The fur is light-colored, with a greenish tint in the case of *D. viridis*. These are large pteropodids with sturdy, dog-like faces, small ears, and

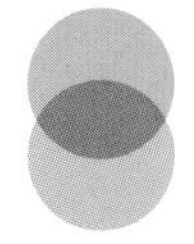


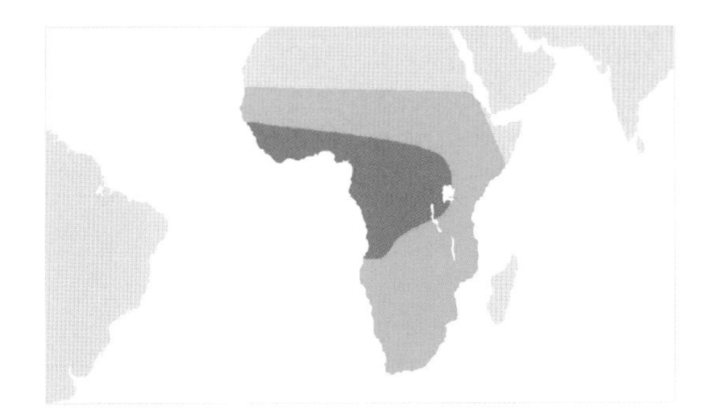

prominent, well-separated nostrils. They are gregarious and roost in caves.

The best-known species is *D. chapmani*, found in the Philippines (Negros and Cebu islands). With the last sighting in 1964, the species was thought to have become extinct in the 1970s. However, a small population was rediscovered on Cebu in 2001, and another on Negros in 2003. The species continues to be hunted and its habitat destroyed: unless immediate measures are taken, its status as an extinct species will soon be restored for real. Most other *Dobsonia* bats have restricted distributions on various Pacific islands, but are not under serious threat of extinction. The other species this book recognizes are *D. anderseni* (Bismarck Archipelago), *D. beauforti* (New Guinea), *D. crenulata* (Sulawesi and adjacent islands), *D. emersa* (Indonesia), *D. exoleta* (Indonesia), *D. inermis* (Solomon Islands), *D. magna* (New Guinea to north Queensland), *D. minor* (New Guinea and Indonesia), *D. moluccensis* (Moluccan Islands) *D. pannietensis* (Papua New Guinea), *D. peronii* (Indonesia), and *D. praedatrix* (Bismarck Islands).

## Epomophorus & Epomops

The epauletted fruit bats form the genus *Epomophorus*. They are rather small and broad-winged for pteropodids. As well as the shoulder epaulettes that give them their name, they show white markings on the face at the base of the ears. Both the epaulettes and the facial markings are associated with scent glands. These bats have large eyes, and a short tail that extends just beyond the narrow uropatagium.

The *Epomophorus* bats are found in Africa, where most species are quite widespread. They live in relatively small groups and shift their roost sites around in response to changes in food availability. The diet is mainly fruit, but sometimes supplemented by leaves. The species this book recognizes are: *E. labiatus* (see page 319), *E. wahlbergi* (page 318), and *E. angolensis* (Angola and Namibia), *E. anselli* (Malawi), *E. crypturus* (southern Africa), *E. dobsonii* (southern Africa), *E. gambianus* (west, central, and southeast Africa), *E. grandis* (Angola and Congo), *E. minor* (central and eastern Africa), and *E. minimus* (east Africa). The genus *Epomops*, holding the west African species *E. buettikoferi* and also *E. franqueti* (see page 318), is a closely related and similar genus.

***Epomops franqueti***

## Franquet's Epauletted Fruit Bat

**THIS BAT OCCURS** across central and west Africa in a broad band. It is fairly large (males are substantially bigger), with brown or reddish fur and darker membranes. It has large, bulging, brown eyes, and white markings in front of its medium-sized ears. The male's epaulettes are tufts of long white fur at the shoulders, concealed within pouches when not in use. It occurs in forests and forest edges, both primary and secondary, roosting either alone or in small groups. The diet is mainly fruit, but other plant parts are also eaten. Males attract females by calling loudly and insistently while displaying their epaulettes and beating their wings. There is no fixed breeding season. The species is hunted for bushmeat.

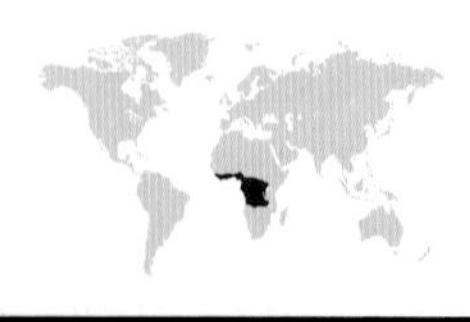

**IUCN STATUS** Least Concern
**LENGTH** 5½–7in (14–17.8cm)
**WEIGHT** 2¾–5⅝oz (78–158g)

***Epomophorus wahlbergi***

## Wahlberg's Epauletted Fruit Bat

**THIS BAT IS WIDESPREAD** in sub-Saharan Africa, from Gabon east to Kenya and Tanzania, and south to South Africa. It has white markings at the base of its ears that may provide disruptive camouflage when roosting. The male has white epaulettes and cheek pouches. It occurs in wooded savanna, suburbs, and farmland, usually roosting in tree canopies in small groups; it changes roosts frequently to avoid predators. Like other *Epomophorus* bats, it uses tongue-clicking echolocation to navigate, but finds fruit by scent. It sometimes plucks fruits while hovering, or eats them while holding onto the plant. This species also feeds on nectar, and is an important baobab pollinator. Males court females by calling and displaying the epaulettes. Common in some areas, its an invaluable seed disperser and pollinator.

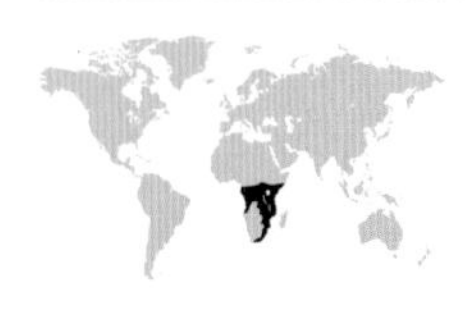

**IUCN STATUS** Least Concern
**LENGTH** 5⅛in (13.2cm)
**WEIGHT** 1⅜–4¼oz (40–120g)

***Epomophorus labiatus***

# Ethiopian Epauletted Fruit Bat

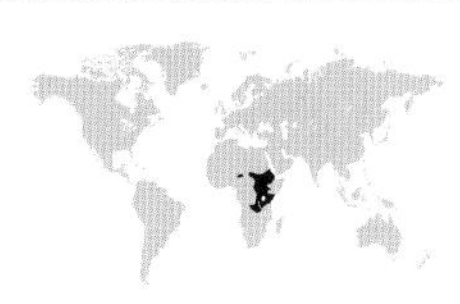

**IUCN STATUS** Least Concern
**LENGTH** unknown
**WEIGHT** unknown

**THIS SMALL FRUIT BAT** has a patchy distribution in east Africa, including parts of Ethiopia, Sudan, South Sudan, Kenya, Uganda, and Tanzania. It has light gray-brown fur and membranes, large, brown eyes, and a fox-like snout. It has a white spot above each eye, and males have white epaulettes of long hairs in their shoulder pouches; males also possess inflatable cheek pouches. This bat forages in savanna, woodland, mangrove forest, and mixed habitats, occurring at altitudes of up to 7,200ft (2,200m). It usually roosts in trees in groups of about twelve, but occasionally uses buildings or caves. It is an important pollinator and vital seed disperser for fruits, some of which only germinate after passing through a bat's digestive tract. They locate food by scent, and may settle on the tree to take fruits, or bite at them while hovering. The species is widespread, often common. However, it is hunted for bushmeat, at rates which may be unsustainable in some areas. This mother is carrying its half-grown pup.

**Micropteropus**

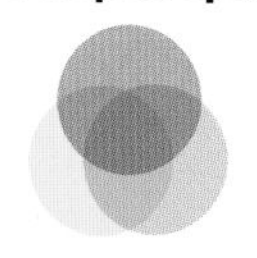

**Megaloglossus** **Myonycteris**

**Macroglossus**

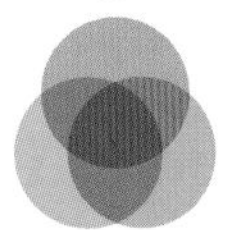

**Syconycteris** **Melonycteris**

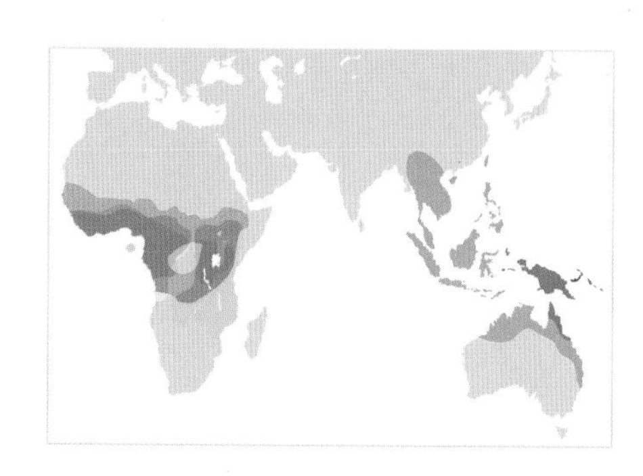

## Micropteropus

The small bats that make up the genus *Micropteropus* are known as the "dwarf epauletted fruit bats." There are two species: *M. intermedius* and *M. pusillus* (described on page 322). These pale, bug-eyed, little bats have erectable white ear-tufts. *M. intermedius* is known only from a few specimens, taken in Angola, and is poorly known.

## Myonycteris

Another African genus is *Myonycteris.* These bats have a contrastingly colored band of fur around the neck, and so are known as the "collared fruit bats." This book recognizes five species: *M. angolensis*, *M. brachycephala*, *M. leptodon*, *M. relicta*, and *M. torquata*. *M. brachycephala* has a very limited range, being endemic to the islands of São Tomé and Príncipe, where it is known from a single location—an area of forest just 290 sq. mi. (750km$^2$). Every known specimen has shown an unusual trait that is not seen in any other mammal: an asymmetrical dental formula, with one incisor tooth missing from the lower jaw. It is believed that this trait is not adaptive but neutral. It would have arisen from a genetic mutation, but because it did not impact negatively on survival it was able to persist in the population. With a presumably very small population, *M. brachycephala* is probably highly inbred, with many genetic traits shared by all surviving individuals. *M. angolensis*, sometimes placed in the monotypic genus *Lissonycteris*, occurs very widely in sub-Saharan Africa. *M. leptodon* is of west Africa, *M. relicta* of east Africa, and *M. torquata* of central Africa.

## Megaloglossus & Macroglossus

*Megaloglossus* is a small, central African genus of just two species. This was formerly only one, but *M. woermanni* was split in 2013, its west African population being renamed *M. azagnyi* (see page 328). These are medium-sized, nectar-feeding bats with long, slim snouts and long tongues. Another nectar-feeding genus is *Macroglossus*, the "long-tongued bats" (*M. sobrinus*, described on page 322, and *M. minimus,* a widespread species of mainland and insular southeast Asia and northern Australia).

## Melonycteris & Syconycteris

A third nectar-feeding genus is *Melonycteris,* comprising three species (*M. fardoulisi* of the Solomon Islands, *M. melanops* of Papua New Guinea, and *M. woodfordi*, which occurs both in Papua New Guinea and on the Solomon Islands). The genus *Syconycteris* is also similar—this holds *S. australis* (see page 324), *S. hobbit* of Papua New Guinea, and *S. carolinae*, which occurs in the north Moluccas in Indonesia.

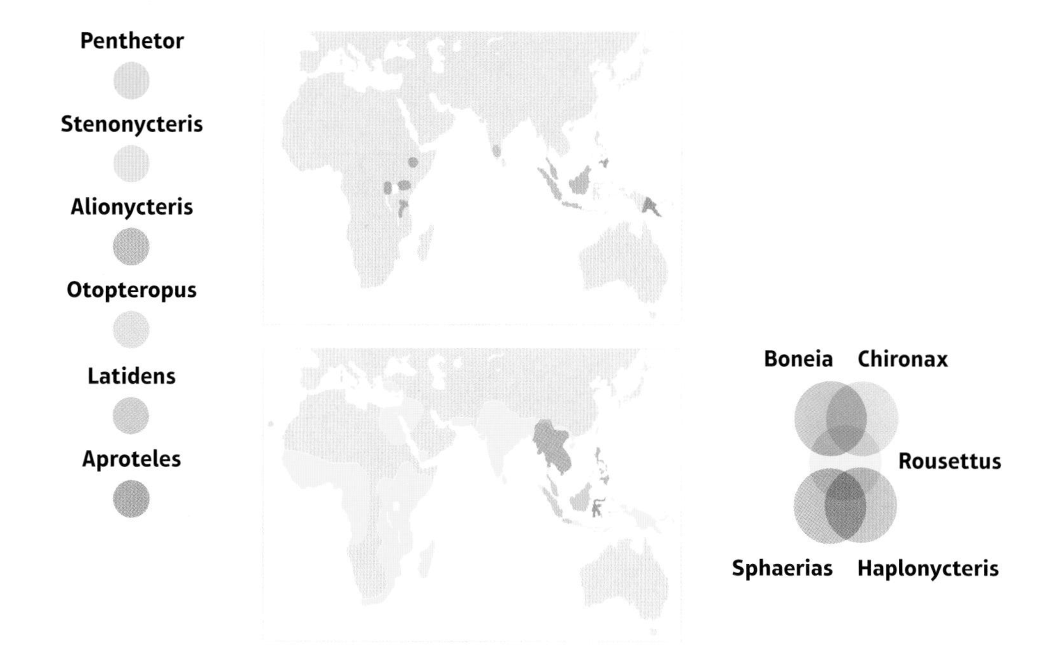


### Rousettus

Bats of the genus *Rousettus* are African and south Asian in distribution. These cave-roosting species are best known for their ability to echolocate, using tongue clicks. They are mostly dark and rather plain-colored, medium-sized pteropodids. This book recognizes *R. aegyptiacus* (see page 325), *R. amplexicaudatus* (page 326), and *R. celebensis* (Sulawesi and nearby islands in Indonesia), *R. leschenaultii* (south and southeast Asia), *R. linduensis* (Indonesia), *R. madagascariensis*, *R. obliviosus* (Comoros islands), and *R. spinalatus* (Malaysia and Indonesia).

The remaining genera of pteropodid bats all contain fewer than eight species each.

### Monotypic genera

Fifteen of the genera are monotypic—they hold just one species. Two are described in detail: *Hypsignathus monstrosus* (see page 323) and *Nanonycteris veldkampii* (page 329). Others include: *Alionycteris paucidentata* (Philippines), *Chironax melanocephalus* (Thailand, Malaysia, and Indonesia), *Haplonycteris fischeri* (Philippines), *Otopteropus cartilagonodus* (Philippines), *Penthetor lucasi* (Malaysia, Borneo, and Sumatra), *Sphaerias blanfordi* (south and southeast Asia), *Boneia bidens* (Sulawesi), and *Stenonycteris lanosus* (east and central Africa).

### Latidens & Aproteles

Another monotypic genus is the Indian upland cave-roosting species *Latidens salimalii*, endangered because of its restricted distribution, habitat loss (including tree-felling to make way for coffee plantations), disturbance, and hunting pressure, both for food and to make folk medicines. This is the only Indian pteropodid to have the highest available level of state protection. *Aproteles bulmerae* of New Guinea is also endangered, with a population of no more than 160 animals, 90 percent of which are found at a single locality that is currently unprotected.

***Micropteropus pusillus***

## Peter's Dwarf Epauletted Fruit Bat

**THIS BAT IS FOUND** in sub-Saharan Africa, from Senegal to Ethiopia and south into Angola. It has light gray-brown fur and pinkish-gray membranes; adult males possess white shoulder epaulettes. It occurs in wooded and more open habitats, and roosts in trees—usually alone, but sometimes in small groups (up to ten). It leaves its roost just after dusk to forage, taking fruit and some nectar. It covers considerable distances on foraging flights and, like other small fruit bats, is a very important disperser of forest tree seeds; it is also a key pollinator of the commercially and culturally important *Kigelia pinnata* ("sausage tree"). Males court females with loud calls and display of the epaulettes. Breeding may occur at any time of year, with births most frequent in spring and fall.

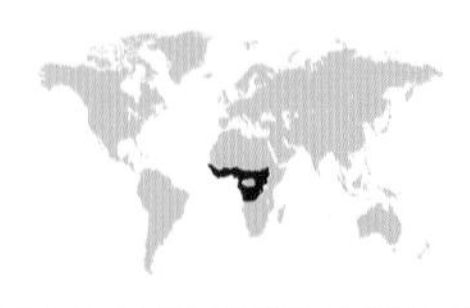

**IUCN STATUS** Least Concern
**LENGTH** 3–3⅛in (7.7–8cm)
**WEIGHT** ⅞–1¼oz (24–34g)

***Macroglossus sobrinus***

## Greater Long-tongued Fruit Bat

**THIS BAT IS WIDESPREAD** in southeast Asia, from east Nepal down through Thailand, Cambodia, Vietnam, Myanmar, and the Malay peninsula. A small nectar bat, it has light-brown fur and a strikingly long, slim muzzle to accommodate its very long, brush-tipped tongue. It lives in forests in both uplands and lowlands, and in mangrove swamps; it roosts in palm trees or banana plants. Its diet is nectar and pollen plus some fruit. This bat is a key pollinator of various wild species of banana, including *Musa itinerans*—a "pioneer" plant, important for regenerating forests in southwest China. These bats can breed year-round, with each female bearing one or two pups per year. Otherwise these bats tend to be solitary and may be territorial—their home ranges show little overlap.

**IUCN STATUS** Least Concern
**LENGTH** 3⅛–3½in (7.8–8.9cm)
**WEIGHT** ⅝–¾oz (18.5–23g)

***Hypsignathus monstrosus***

# Hammer-headed Fruit Bat

**IUCN STATUS** Least Concern
**LENGTH** 7¾–11¼in (19.5–28.5cm)
**WEIGHT** 7¾–11¼oz (218–450g)

**AFRICA'S LARGEST BAT** species ranges from Guinea Bissau to Angola, and east across central Africa to Ethiopia. Its name aptly describes its broad, blocky muzzle. Males, such as this one, are substantially larger than females, with enlarged lower lips and fleshy flaps around the muzzle tip. Mainly found in forested habitats, it is a key seed disperser; this bat also feeds on fig, guava, and banana crops. It roosts in small groups, usually fewer than ten. In some parts of its range males also form leks of more than one hundred, displaying to attract females: just 6 percent of males account for nearly 80 percent of all matings. Some local communities regard the males' loud calls when lekking as a nuisance. It has been speculated to be a potential source of Ebola virus, but substantial evidence is lacking.

***Syconycteris australis***

# Common Blossom Bat

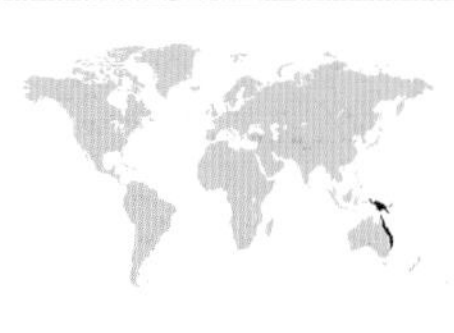

**IUCN STATUS** Least Concern
**LENGTH** 1⅝–2⅜in (4–6cm)
**WEIGHT** ⅝–¾oz (18.9–20.5g)

THIS CHARMING BAT, the smallest pteropodid, is found along the northeast coast of Australia and much of New Guinea, and on nearby island groups. Its tongue is very long and brush-tipped—uniquely among Pteropodidae, this species is believed to feed only on nectar and pollen. It roosts alone in forest habitats, using different roosting spots each night, and forages in more open ground, visiting many types of flower, including this *Banskia robur*. The sexes come together briefly each spring and again in the fall to mate; they then separate. Females give birth after a three-month pregnancy and pups become independent after about two months. Because this species roosts alone, it is susceptible to temperature changes, and can enter a torpid state in cold conditions.

***Rousettus aegyptiacus***

# Egyptian Fruit Bat

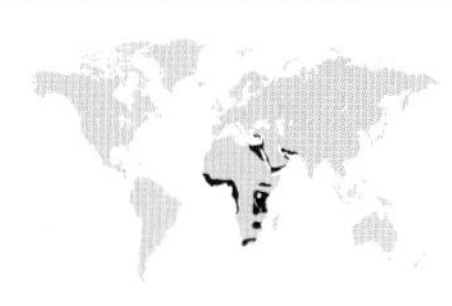

**IUCN STATUS** Least Concern
**LENGTH** 4¾–7⅝in (12.1–19.2cm)
**WEIGHT** 2⅞–6oz (80–170g)

**THIS BAT IS FOUND** widely but patchily throughout Africa, and also the Arabian Peninsula. The upperside is mid-gray-brown, the underside paler, and most show a yellowish collar. The long, broad snout gives a dog-like (rather than fox-like) profile. It uses tongue-clicking echolocation to navigate within the large, dark, roosting caves. Colonies can hold many thousand individuals, though some are far smaller, and mutual grooming and other social interaction occurs, especially as dusk approaches. They leave the roost at sunset to forage, eating ripe fruits (such as this mango). While eating, the bat shields the fruit with a wing to try to prevent theft by another bat. The species is a key pollinator and seed disperser, but colonies are often persecuted when mistakenly seen as a pest of fruit crops.

***Rousettus amplexicaudatus***

# Geoffroy's Rousette

**IUCN STATUS** Least Concern
**LENGTH** 3⅛–3⅜in (7.8–8.7cm)
**WEIGHT** unknown

**THIS SPECIES HAS** a broad distribution across southeast Asia, occurring from Yunnan in China through to Cambodia, Thailand, Myanmar, Vietnam, Laos, peninsular Malaysia, Indonesia (including Java and Bali), New Guinea, the Philippines, and other nearby island groups. It is a fairly small, gray fruit bat, paler and yellow-tinted on the underside. The wings are dark gray. Its snout is long and slender, the eyes large and dark, the ears medium-sized and wide-spaced. It forages in primary and secondary forest and roosts in large caves nearby, using tongue-clicking echolocation to navigate within them. Roosts can be vast: one cave system on Samal, Philippines (seen here), supports up to two million individuals, in cramped conditions that leave some at risk from rat predation. The bats roost at this unnatural density because they have been displaced from other cave roosts. This famous roost is protected as the Monfort Bat Sanctuary. It attracts many visitors who watch the bats emerge just after sunset. This species feeds on fruit, nectar, and pollen. Females have two pups a year, the first in mid-spring and the second in early fall.

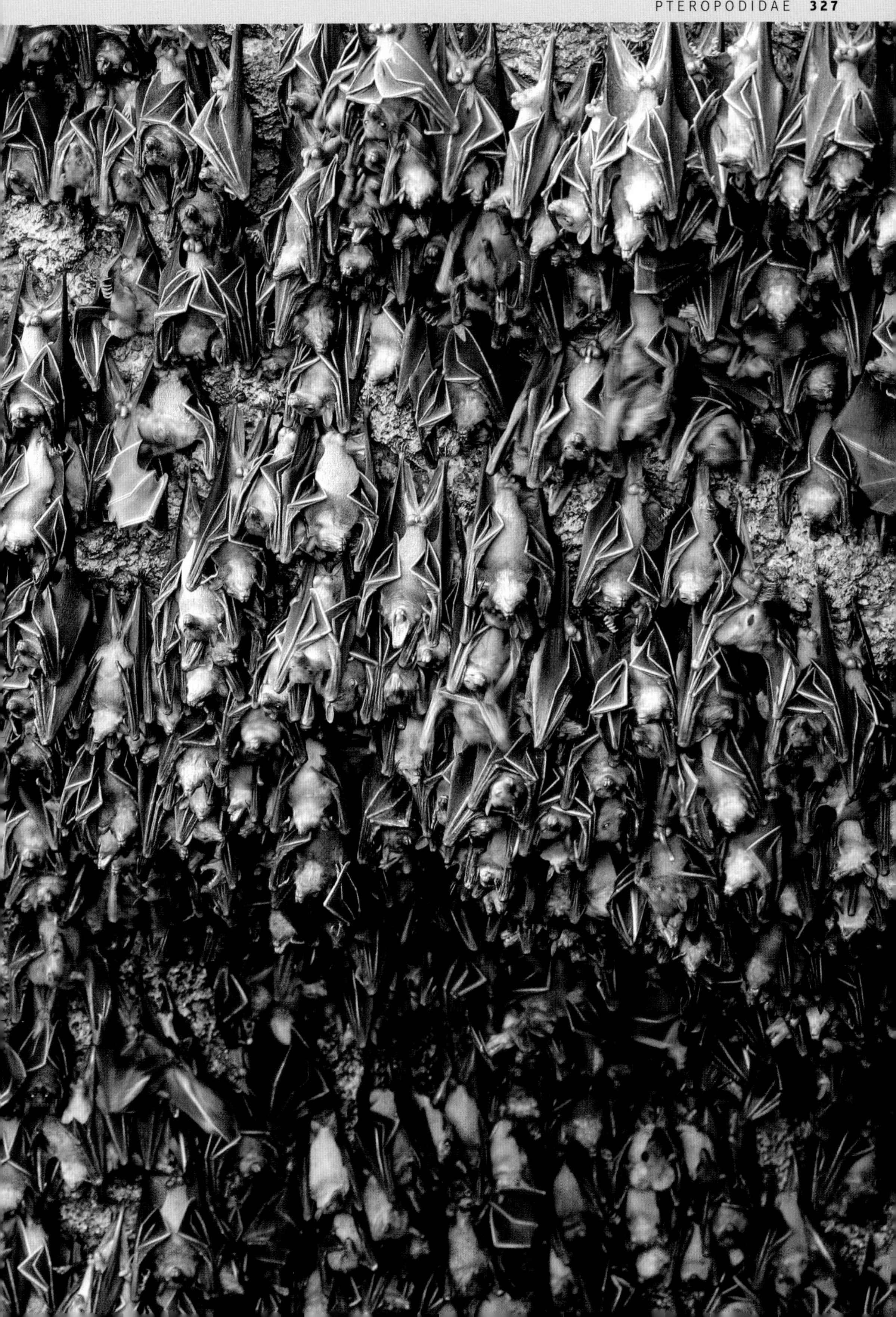

***Megaloglossus azagnyi***

# Western Woermann's Fruit Bat

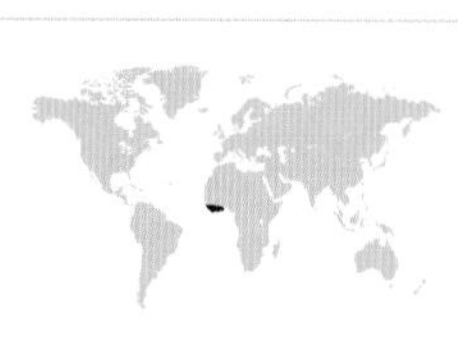

**IUCN STATUS** Vulnerable
**LENGTH** unknown
**WEIGHT** unknown

THIS SPECIES OCCURS in west Africa. It has recently been split from *M. woermanni*, which occurs further east, but may overlap in distribution. This is a small, dainty flying fox, with yellowish-brown fur and dark wings. It occurs in lowland, closed-canopy moist forest, and will visit nearby plantations and forest-grassland mosaic habitats to forage. It also occurs around human habitations and will roost in buildings (though more typically among foliage). It feeds exclusively on nectar and pollen, using its exceptionally long, brush-tipped tongue on plants such as this *Pentadesma butyracea*. It uses a small home range, though females range more widely than males. Lactating females have been observed in most months of the year. Usually one pup is born, but at least one record of twins exists.

***Nanonycteris veldkampii***

# Veldkamp's Bat

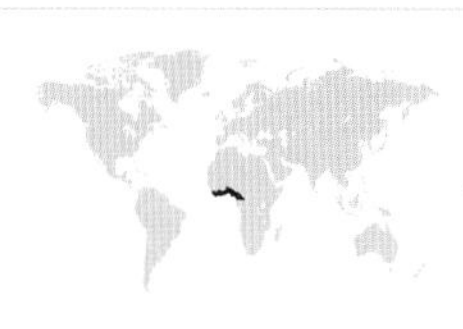

**IUCN STATUS** Least Concern
**LENGTH** 2¼–3in (5.7–7.5cm)
**WEIGHT** ⅝–1⅛oz (19–33g)

THIS SMALL FRUIT BAT occurs in west Africa. It has pale, creamy fur and light orange-brown wings; juveniles are grayer. It occurs in tropical and subtropical forests of all kinds and also in moist savanna, moving around through the year to keep up with availability of food in different habitats. This species roosts in trees, alone or in loose assemblages. Active at night, it feeds mainly on nectar, approaching its chosen flower feet-first, then settling and wrapping its wings around the flower for stability. After feeding, it rolls backward off the flower to take flight. It is an important pollinator, for example, for *Parkia* plants, and also eats fruit, such as this ripe fig. This bat may breed more than once a year, but births seem to peak at the start of the rainy season (late spring).

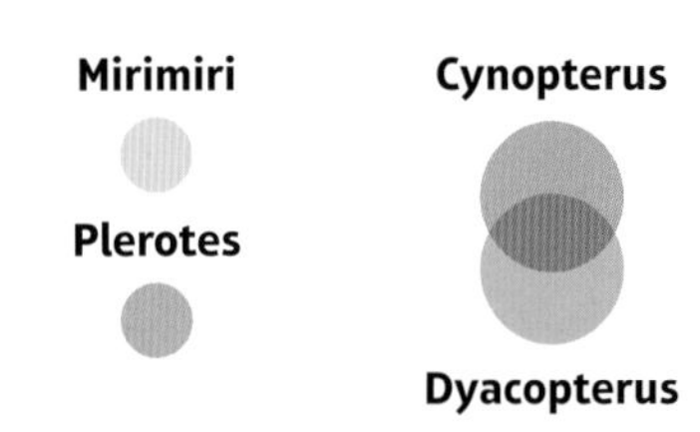


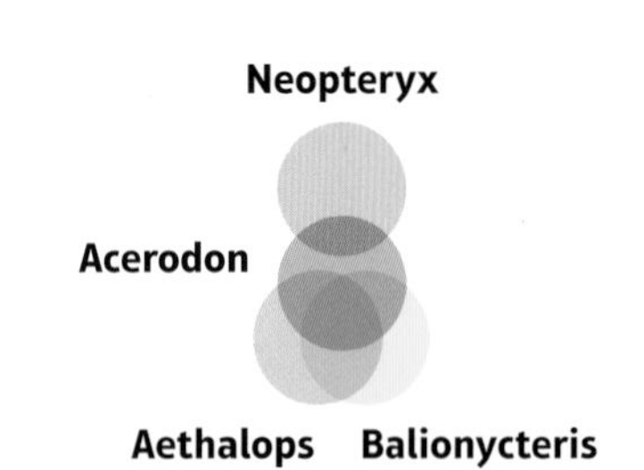


**Mirimiri, Neopteryx & Plerotes**

The final three monotypic genera include the endangered *Mirimiri acrodonta*, recently moved to its new genus from *Pteralopex*. It is known only from one mountain, Koroturanga, on the island of Taveuni, Fiji, where just a handful of specimens have been taken. *Neopteryx frosti* of Sulawesi, Indonesia is also endangered and has had no records from any of its known sites since 1991, though further surveys are needed. *Plerotes anchietae* (Central Africa) is little known.

**Cynopterus & Dyacopterus**

The genera *Cynopterus* and *Dyacopterus*, both of which are found in southeast Asia, are notable in that male lactation has been observed in some of their species. This may be due to a hormonal imbalance triggered by environmental contaminants. *Cynopterus* are the "short-nosed fruit bats": very small pteropodids, with relatively short snouts and large ears. The species this book recognizes are *C. horsfieldii* (see page 333), *C. sphinx* (page 334), *C. brachyotis*, *C. luzoniensis* (Indonesia and the Philippines), *C. minutus* (Malaysia and Indonesia), *C. nusatenggara* (Indonesia), and *C. titthaecheilus* (Indonesia). *Dyacopterus* are also small bats, and very little known. This book recognizes three species: *D. brooksi* (Sumatra in Indonesia), *D. rickarti* (Philippines), and *D. spadiceus* (widespread in southeast Asia).

**Aethalops**

The Asian genus *Aethalops* contains two species, though some authorities lump them together as one. They are very small pteropodids with rather dainty faces and long, dense fur, dark on the upperside and paler on the underside. These bats occur in upland forest regions, feed on fruit and flowers, and are able to hover while feeding. Besides *A. alecto*, which is present in peninsular Malaysia, Sumatra, Java, Bali, and Lombok, this book recognizes *A. aequalis* of Borneo as a separate species.

**Balionycteris**

The small bats of the genus *Balionycteryis* are known as the "spotted-winged fruit bats" because their patagia are marked with creamy spots; there are also pale facial markings. These are forest species that search for smaller fruits in the forest understory, and are also reported

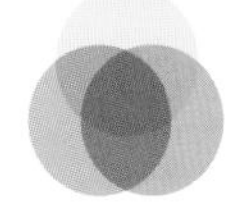


to take some insects. Many authorities only recognize *B. maculata* (see page 332) and consider the smaller *B. seimundi* of Sumatra and Durian island to be a subspecies. However, this book treats them as separate species based on their consistent morphological differences.

## Acerodon

The genus *Acerodon* is closely related to *Pteropus*. These bats look like the "true flying foxes" of *Pteropus*, and one of them, *A. jubatus*, rivals the very biggest *Pteropus* in size. The genus is confined to southeast Asia and contains five species. The attractive, golden-headed *A. jubatus* occurs on the Philippines and is quite widespread, but its population (no more than 20,000 individuals) is thought to be around 1 or 2 percent of what it was 200 years ago. The Philippines have suffered massive deforestation, and many fruit bats are hunted, despite this being illegal. The distribution of the species is not properly known yet, but at known roosts that are protected and have good habitat nearby it is holding its own. The other endangered species, *A. humilis*, is known only from two small islands in the Talaud group, Indonesia. It was missing entirely between 1909, when specimens were taken, and 2002, when it was rediscovered. Protecting the remaining forest on the islands is key to saving it. The three other species this book recognizes are *A. celebensis* (Indonesia), *A. leucotis* (Philippines), and *A. mackloti* (Indonesia).

## Thoopterus

The genus *Thoopterus* contains one, two, or three species, depending on authority. All are very similar and found in and near Sulawesi, Indonesia. These are little-known bats, with dusky dark gray fur and dark wings. This book recognizes two species—*T. nigrescens* and *T. suhaniahae.*

## Megaerops

Bats of the genus *Megaerops* are also Asian in distribution, and occur mainly in primary forest. These are small and rather short-nosed pteropodids, with projecting nostrils and rather small ears. The lips are quite prominent. Besides *M. ecaudatus*, described on page 332, this book recognizes *M. kusnotoi* (Java), *M. niphanae* (south and east Asia), and *M. wetmorei* (Malaysia, Indonesia, and the Philippines).

## Ptenochirus

The so-called "musky fruit bats" make up the genus *Ptenochirus*. There are two species by Simmons and Cirranello's taxonomy: *P. jagori* and *P. minor*. Both are cave-roosting bats which occur only in the Philippines but *P. jagori* is much more widespread. These are dark bats with shortish, sturdy snouts. They are both able to survive in secondary forest as well as primary, so have not been as badly impacted by deforestation as some forest Philippine species. *P. jagori* is even able to live in urban areas.

***Balionycteris maculata***

## Spotted-winged Fruit Bat

**THIS BAT OCCURS** on the Malaysian peninsula and Borneo, and on the Riau Archipelago. It is small and dark gray-brown, paler on the belly, with pale spots on its wing membranes. It has a broad snout with clearly separated nostrils, brown eyes, and small ears. It occurs in lowland forest—by preference primary forest, but also mangrove swamps and sometimes more disturbed habitats such as rubber plantations. It finds fruit by scent as well as sight. Roosts are normally located in the bases of hollowed-out root masses in epiphytic ferns. These can be very noisy, with a range of different calls. Nursery colonies consist of a male, nine females, and their young. The species breeds at any time of year, though most females only have one pup each year. It is threatened by forest clearance and forest fire.

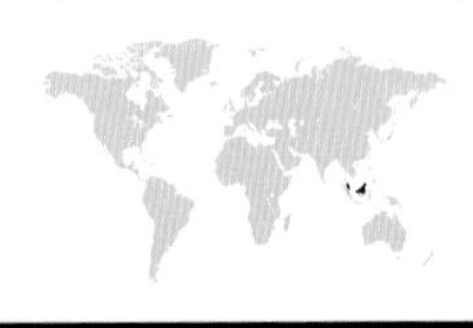

**IUCN STATUS** Least Concern
**LENGTH** 2–2⅝in (5–6.6cm)
**WEIGHT** ⅜–½oz (9.5–14.5g)

***Megaerops ecaudatus***

## Temminck's Tailless Fruit Bat

**THIS SMALL FRUIT BAT** occurs in Malaysia, Sumatra, Borneo, and Thailand, and possibly also north India (molecular DNA studies of the *Megaerops* bats found in India are needed to confirm which species they are). It has a short, fluffy, light brown coat, small dark ears, and a fairly short, robust snout; a prominently split nose tip has a whorl of skin around each nostril. The eyes are dark brown, large, and bulging. This species can be found in mature rainforest, both primary and secondary, mainly occurring at elevations of between 3,300 and 10,000ft (1,000 and 3,000m). The bats roost in foliage, moving their roosting sites around during the year, as different fruits become available. This bat is elusive and uncommon. Its ecology has, to date, barely been studied.

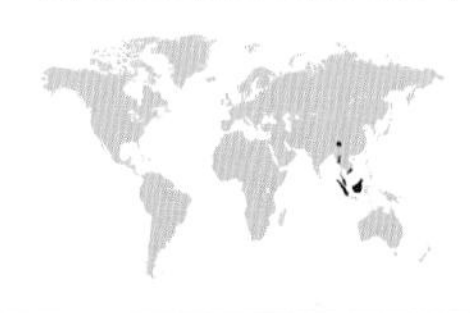

**IUCN STATUS** Least Concern
**LENGTH** unknown
**WEIGHT** unknown

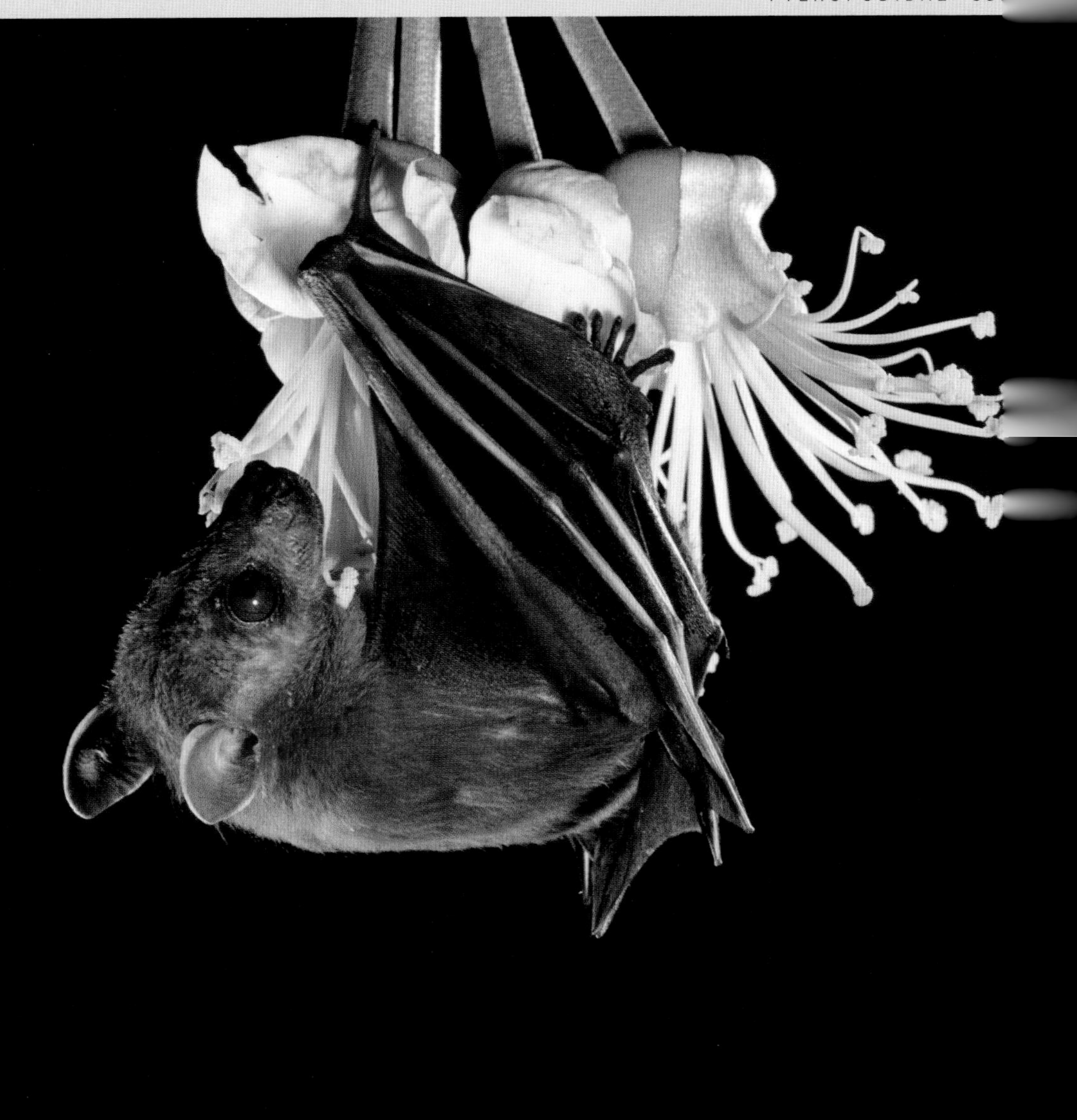

***Cynopterus horsfieldii***

# Horsfield's Fruit Bat

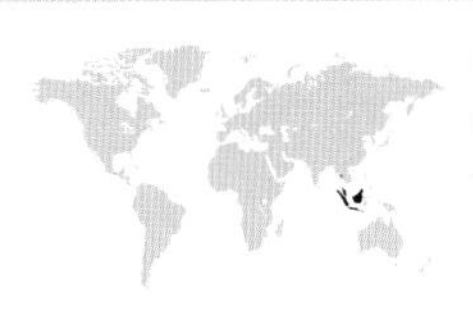

**IUCN STATUS** Least Concern
**LENGTH** 3⅛–3¾in (8–9.6cm)
**WEIGHT** 2⅛oz (60g)

**THIS BAT OCCURS** in Thailand, the Malay Peninsula, Borneo, Sumatra, Java, and Bali. It is a medium-sized fruit bat with light-brown fur, paler on the belly, and a variable gingery collar. The wing membranes are dark, contrasting with the whitish digits. It has brown eyes and medium-sized, white-edged ears. In Thailand this bat seems to occur only in primary forest, but in Indonesia and Malaysia it occurs in other habitats, including secondary forest, parks, and orchards. It roosts in caves, but also constructs leaf tents, and tends to live in small harem groups consisting of one male and about three females ("spare" males roost alone). The species mainly eats fruit. It also consumes some nectar and pollen, and has been observed intoxicated from drinking fermented palm juice.

*Cynopterus sphinx*

# Greater Shortnosed Fruit Bat

**IUCN STATUS** Least Concern

**LENGTH** 3½–4⅜in (8.9–10.9cm)

**WEIGHT** 2⅝oz (75g)

**THIS BAT IS WIDESPREAD** and common in southeast Asia, from Pakistan across India to southeast China, and south through the Malay peninsula to parts of Borneo and Indonesia. It has warm brown or gingery fur, grayer on the underside, and reddish eyes. It occurs mainly in tropical forest, but also visits farmland, mangrove forest, and fields. Same-sex groups of eight or nine roost together, except during the mating season when groups are mixed and larger; adult males may share palm tent-roosts with several females. It feeds on various wild and cultivated fruits and is an important pollinator and seed disperser for a variety of plants, including wild banana flowers. Among adults, females appear to greatly outnumber males. Across the range, females give birth to a single pup twice a year.

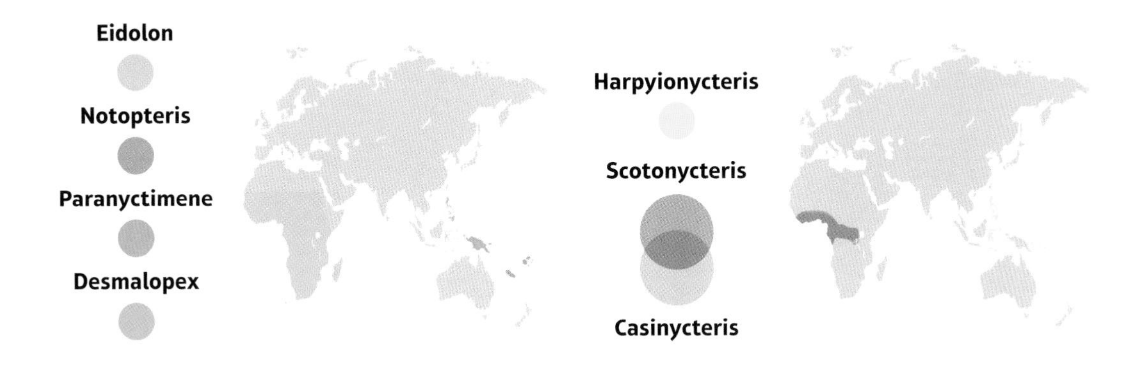


## Eidolon

The two *Eidolon* species, *E. dupreanum* and *E. helvum* (see page 338), are found in Africa and Madagascar, respectively.

## Scotonycteris & Casinycteris

The genus *Scotonycteris* occurs in west Africa, represented by *S. bergmansi*, *S.occidentalis*, and *S. zenkeri*. These are attractive, small, light-colored bats, with delicate faces and pale markings at the base of the ears, and around and between the eyes. Another west African genus, *Casinycteris*, is very similar—in fact *C. ophiodon* was placed in *Scotonycteris* until recent DNA studies indicated it belonged in the sister genus. The other two species in the genus are *C. argynnis* and *C. campomaanensis*.

## Harpyionycteris

*Harpyionycteris* species are known as "harpy fruit bats." *H. whiteheadi* is found in the Philippines, while *H. celebensis* occurs on Sulawesi. They are large, dark brown, tailless pteropodids, which inhabit primary forest and roost in trees. Neither species is well known, but *H. celebensis* in particular seems to be very rare.

## Notopteris

Yet another genus from southeast Asia is *Notopteris*, represented by *N. macdonaldi* in Fiji, and *N. neocaledonica* in New Caledonia. These are small, slender-snouted bats with long, free tails. They feed mainly, if not exclusively, on nectar. Both are rare and declining because of habitat loss and disturbance at their cave roosts.

## Paranyctimene

The genus *Paranyctimene,* like *Nyctimene*, are known as "tube-nosed bats" because of their prominent, tubular nostrils. With their broad muzzles, prominent mouths, and small ears, they have a curious, rather ape-like appearance. There are two species, *P. raptor* and *P. tenax*, both of which occur in New Guinea. The species are very similar, and occur in the same areas. It was only in 2001 that *P. raptor* was described as a distinct species. Both their distribution and ecology are almost unstudied.

## Desmalopex

The two flying foxes of the genus *Desmalopex* have only recently been separated from *Pteropus.* They are *D. leucopterus* and *D. microleucopterus.* Both species are found only in the Philippines, in dry forest, and are known as "white-winged flying foxes" because of the white spotting on their wing membranes. They are rather elusive bats, roosting in small groups, and flying high over the forest as they move from roosts to feeding sites. Consequently, their behavior and ecology are little studied.

Pteralopex

Styloctenium

Eonycteris

## Pteralopex

The monkey-faced bats that make up the genus *Pteralopex* are among the most unusual-looking of all bats, mainly because of their very small, low-set ears. This, coupled with their broad snouts and large, orange eyes, gives them a distinctly primate-like look, though perhaps closer to a lemur than a monkey. They are large bats and all occur only in the Solomon Islands in Melanesia. This book recognises five species, these are *P. flanneryi*, *P. pulchra*, *P. anceps*, *P. atrata*, and *P. taki*. All five face threats from habitat loss—all appear to require primary forest—and overhunting. *P. flanneryi* and *P. pulchra* have not been observed at all for some years, and may already be extinct.

## Styloctenium

*Styloctenium* comprises the "stripe-faced bats," named because of the prominent white stripes on their faces, above and between the eyes, and on the cheek sides. They also have white shoulder markings. They are medium-sized with mid red-brown or gray-brown fur, and rather long, slender snouts. *S. mindorensis* occurs only on the island of Mindoro, in the Philippines, and is little known, while *S. wallacei* occurs on Sulawesi and the Togian islands. There may have been another *Styloctenium* species in Australia. Rock paintings from the Kimberley region in Western Australia, dating from 20,000 years ago, show a bat with the distinctive white facial pattern.

## Eonycteris

The dawn bats, genus *Eonycteris*, occur in south and southeast Asia and comprise three species: the rare and range-restricted *E. major* of Borneo, *E. robusta* of the Philippines, and the very widespread *E. spelaea* (see opposite). These are smallish and elegant pteropodids, with long, slender noses for nectar-feeding. They have short, dense, pale gray-brown fur, large ears, and very large, dark eyes. They forage in forest and other wooded habitats, and roost in caves.

***Eonycteris spelaea***

# Dawn Bat

**IUCN STATUS** Least Concern
**LENGTH** 3⅜–5in (8.5–12.5cm)
**WEIGHT** 1⅛–2⅞oz (32–82g)

THIS SPECIES OCCURS across much of south and southeast Asia, including many Philippine islands. It is a dark brown bat, paler on its underside. The male has a "beard" of long, fine hairs—these are specialized, scent-dispersing hairs. It occurs in forests and on cultivated land. Primarily nectivorous, and an important pollinator of crops, it also feeds on fruit. This species shares its cave roosts with other bats. Its colonies are often very large—one roost on Palawan holds more than 50,000 individuals. Females usually give birth to singleton young, twice a year. In flight its wing-clapping sounds may provide basic echolocation for navigation, although it finds food by scent. Though viewed as common, this species has declined dramatically due to limestone extraction and bushmeat harvesting.

***Eidolon helvum***

# African Straw-colored Fruit Bat

**IUCN STATUS** Near Threatened
**LENGTH** 5⅝–8½in (14.3–21.5cm)
**WEIGHT** 8⅛–12⅜oz (230–350g)

**THIS BAT IS** widespread in sub-Saharan Africa, absent only from the Horn of Africa, the Namibian desert region, and the southernmost part of South Africa. It also occurs in southwest Arabia. A long-winged fruit bat, it has a straw-yellow upperside, and drabber-grayish underside. The face and membranes are dark. Most abundant in tropical forest, it also occurs in more arid areas and around towns. It roosts in trees, in colonies of 100,000 or more. The largest-known nursery colony assembles annually in Zambia's Kasanka National Park and includes between five and ten million individuals. It roams more than 18 miles (30km) from the roost when foraging at night, flying high and fast to feeding grounds. This species feeds mainly on fruit, but also on nectar and other parts of plants; it usually hangs from the plant as it feeds. It is a key seed disperser for economically important plants such as the African teak tree *Milicia excelsa*. Females give birth to singleton pups in spring. This bat is heavily hunted for bushmeat; this and deforestation have caused declines in west and central Africa.

## CRASEONYCTERIDAE

This bat family holds only one species, *Craseonycteris thonglongyai*—a remarkable species that is described opposite. The relationships of Craseonycteridae have been explored through molecular DNA studies, which determined the family belongs within the superfamily Rhinolophoidea. The families to which it is most closely allied are Hipposideridae and Rhinopomatidae.

*C. thonglongyai* was discovered only in 1973, in Thailand, by the ornithologist and mammalogist Kitti Thonglongya, also noted for discovering the white-eyed river martin (a species of swallow). The scientific and English names of the species both honor their discoverer. Unfortunately, this bat was targeted by souvenir hunters shortly after its roosts were discovered. It also suffers general disturbance, especially from regular burning of areas of forest during its breeding season. It suffers an ongoing decline and has a very small geographic range. Its population in Myanmar is estimated to be around 1,500, with 5,100 in Thailand.

In 2007, *C. thonglongyai* was designated a focal species by the EDGE (Evolutionarily Distinct and Globally Endangered) project. Biologists have been working on how the bats select and use their roosts, and how they react to varying levels of human disturbance. Work is also underway to encourage tourists to view the bats as they forage, rather than entering the caves.

*Craseonycteris thonglongyai*

# Kitti's Hog-nosed Bat

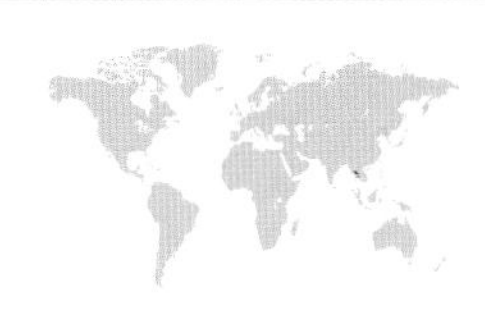

**IUCN STATUS** Vulnerable
**LENGTH** 1⅛–1¼in (2.9–3.3cm)
**WEIGHT** ⅛oz (1.7–2g)

**THIS BAT IS KNOWN** from two separate small regions in southeast Asia: one in southeast Myanmar and another, smaller area in Thailand, northwest of Bangkok. Studies are underway to assess the genetic distinctiveness of the two populations. This could be the world's smallest bat and perhaps its smallest mammal (the Etruscan Shrew and *Kerivoula minuta* are longer-bodied, but both average slightly lighter). It occurs in forest and roosts within limestone caves near rivers. An efficient hunter, it feeds on small, flying insects in the canopy, and can meet its energy requirements with two hunting forays per night—less than an hour of flight time. When not hunting, it becomes torpid to conserve energy. Just eight roosting sites are known in Myanmar and thirty-five in Thailand.

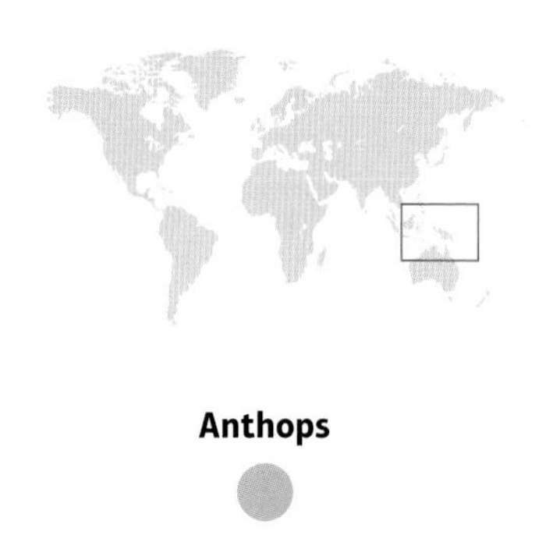

## HIPPOSIDERIDAE

The species that make up the family Hipposideridae are known as the "roundleaf bats." This is because of their nose-leaves, which tend to have a rounded bottom edge, often divided into two lobes. However, the nose-leaf's shape varies across the genera; in some groups it is very large and elaborate. These small to medium-sized bats also typically have funnel-shaped ears with tapering, pointed tips. The eyes are small and the wings usually quite broad, powering an energy-efficient, fluttering flight. The family is distributed across Africa, south Asia, Australia, the Philippine Islands, and the Solomon Islands. In total the family contains five genera, of which four are small. The fifth (*Hipposideros*) holds seventy-seven of the eighty-seven species.

The hipposiderids vary considerably in color, from brown or gray to rufous, with distinct white patches in some cases. Most have a paler underside than upperside. They are often gray in new molt, gradually bleaching to reddish or even bright orange. Their ears vary considerably in size. The hipposiderid ear does not have a tragus, but it does have a prominent antitragus—a fold that forms a projection opposite the ear opening. The nose-leaf has a pointed projection near the top center, or sometimes more than one; there may also be additional "leaflet" folds on the sides.

Hipposiderids are not known to form any kind of pair bond or stable partnership, though one species, *H. commersoni*, is known to form harems. Females give birth to one pup per year, usually in spring after having mated in the fall; implantation is delayed, and is likely triggered by the increase in daylight hours through late winter. It takes the pup between seven and twenty weeks to reach independence, depending on species. These bats vary considerably in roosting habits. Some are highly social and form colonies of many thousands, while others roost alone or in small groups. Roosts are usually in caves, sometimes buildings, or other shelters. These bats usually hang by their feet from a horizontal surface, wrapping their wings around their bodies as they sleep.

Hipposiderids produce their echolocation calls through their nostrils rather than the open mouth, and the calls are typically of very high frequency. The nose-leaf's shape helps them to direct the calls. They usually hunt by flying close to the ground around vegetation, and mainly capture flying insects. They have strong jaws and an up-and-down, side-to-side chewing action, which helps them to deal with hard-bodied prey.

### Anthops

The only species in the genus *Anthops* is *A. ornatus*. It is known in English as the Flower-faced Bat because of its elaborate

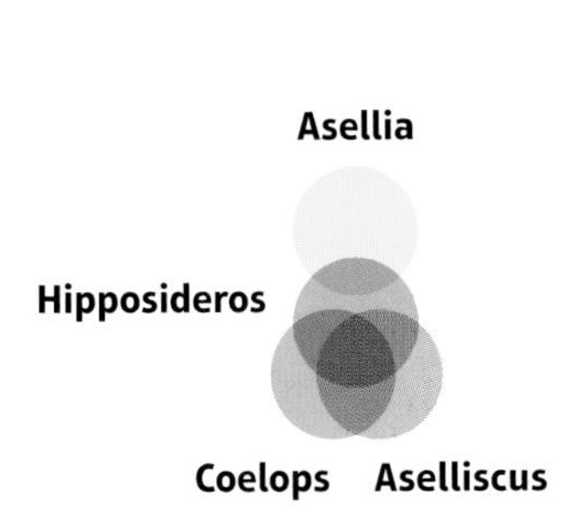


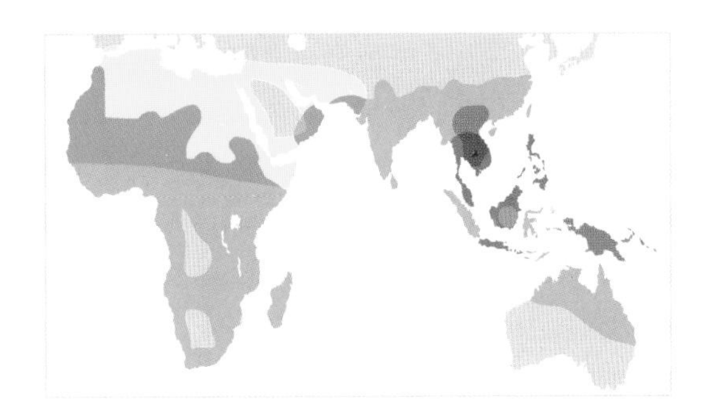

nose-leaf with its many side folds. This bat occurs in forest habitats on the North Solomons (Papua New Guinea) and the Solomon Islands. It is apparently rare and is little known.

## Asellia

The "trident bats" of the genus *Asellia* have a row of three vertical, blade-shaped points at the top of their nose-leaf; the central one is particularly pointed. They occur in variable colors. These bats are found in Africa and west Asia. They are arid-habitat specialists, forming large roosts in all kinds of sheltered areas and emerging well after sunset to hawk insects with a fast, agile, butterfly-like flight. This book recognizes four species, these include *A. patrizii* and *A. tridens*. *A. patrizii* occurs in Eritrea, Ethiopia, and Saudi Arabia, while *A. tridens* is a common species in desert areas of north Africa and the Middle East. The other two species are little known. *A. arabica*, an apparently rare semi-desert and savanna species, is found in Yemen and Oman. *A. italosomalica* occurs in Somalia and on Socotra Island (Yemen).

## Aselliscus

The bats of the genus *Aselliscus* are also known as "trident bats," although their "trident" is less apparent. Here the top part of the nose-leaf appears as a single, broad lobe, with two splits either side of the center; these form two broad side projections and a much narrower central one. The newly described *A. dongbacana*, from north Vietnam, is poorly known. The other two species are *A. stoliczkanus* (found extensively in mainland southeast Asia) and *A. tricuspidatus* (present on many Indonesian islands and also Papua New Guinea, Solomon Islands, and Vanuatu).

## Coelops

*Coelops* is a small genus of hipposiderids that are notable for lacking tails (hence their English name "tailless bats"). *C. frithii* occurs widely in southeast Asia, including Java, Bali, and Taiwan (China). *C. robinsoni* is found on peninsular Malaysia and on Borneo; it is affected by deforestation.

## Hipposideros

The fifth genus, *Hipposideros*, is very speciose and occurs throughout the family's total distribution. Its members have rounded nose-leaves, but lack the additional ornamentation of the other genera. The species covered in detail are *H. abae* (see page 346), *H. armiger* (page 348), *H. ater* (page 348), *H. atrox* (page 349), *H. bicolor* (page 347), *H. caffer* (page 349), *H. diadema* (page 352), *H. doriae* (page 352), *H. dyacorum* (page 353), *H. galeritus* (page 353), *H. pomona* (page 350), *H. ruber* (page 351), and *H. vittatus* (page 354).

Five species of *Hipposideros* are under threat. They are *H. commersoni* (Madagascar), *H. jonesi* (west Africa; threatened by disturbance at its cave roosts), *H. lekaguli* (Philippines, Thailand, and Malaysia; affected by disturbance at its roosts and also hunted), *H. pelingensis* (Sulawesi and other Indonesian islands; threatened by limestone extraction in the caves where it roosts), and *H. turpis* (Yaeyama Islands in Japan, also Thailand and Vietnam; threatened by disturbance at its roosts and loss of suitable feeding habitat).

A further twelve species are also considered to be in danger. They are *H. curtus* (Cameroon and Equatorial Guinea; rare and threatened by habitat loss and disturbance), *H. demissus* (Makira Island, Solomon Islands; has a very restricted distribution), *H. durgadasi* (India, with a very small range; threatened by habitat loss), *H. edwardshilli* (Papua New Guinea; known from very few sites), *H. halophyllus* (Malaysia and Thailand; restricted to very few roost sites because of exacting ecological needs), *H. inornatus* (north Australia; declining and losing roosting habitat), *H. khaokhouayensis* (Laos; newly discovered species with apparently very small natural range), *H. marisae* (Côte d'Ivoire, Guinea, Liberia; very rare with few known sites), *H. orbiculus* (Indonesia, Malaysia; affected by loss of its rainforest habitat), *H. ridleyi* (Borneo, Malaysia, and Singapore; dependent on primary forest), *H. scutinares* (Laos and Vietnam; few roosts are known and those are subject to disturbance and hunting), and *H. sorenseni* (Java, with only one known roost).

**Hipposideros commersoni**

*H. coxi* is known only from Borneo and with a very restricted distribution. A further two restricted species are *H. hypophyllus*, known from only a single roost site in Karnataka, India, and *H. lamottei*, which occurs on the slopes of Mount Nimba, on the border between Guinea, Liberia, and Côte d'Ivoire; the latter species is also threatened by mining activity.

The little-known species of *Hipposideros* are as follows: *H. boeadi* (recently discovered in Sulawesi), *H. breviceps* (known only from specimens taken in 1941 on North Pagai, Mentawai Islands, Sumatra), *H. camerunensis* (west, central, and east Africa; known from just three records), *H. coronatus* (Philippines, very few specimens exist), *H. crumeniferus* (poorly known species from Timor, Indonesia), *H. dinops* (North Solomons and Solomon Islands; rare and barely known), *H. inexpectatus* (Sulawesi, with very few specimens known), *H. macrobullatus* (recorded sparsely from Indonesia), *H. nequam* (known only from the type specimen, taken in Malaysia). Species yet to be assessed by the IUCN are *H. alongensis* (newly discovered in Vietnam), *H. einnaythu* (newly discovered in Myanmar), *H. griffini* (newly discovered in Vietnam), *H. khasiana* (northeast India), *H. nicobarulae* (Nicobar Islands), and *H. pendleburyi* (Thailand; a recent split from *H. turpis*).

The remaining species of *Hipposideros*, classified as Least Concern by the IUCN, include: *H. beatus* (west Africa), *H. calcaratus*

(New Guinea, Solomon Islands), *H. cervinus* (Malaysia, Indonesia, Philippines, New Guinea, Australia), *H. cineraceus* (south and southeast Asia), *H. corynophyllus* (New Guinea), *H. cyclops* (west, central, and east Africa), *H. fuliginosus* (Sierra Leone, Liberia, Democratic Republic of Congo), *H. fulvus* (Afghanistan, Pakistan, India, Sri Lanka, Vietnam), *H. gigas* (west, central, and east Africa), *H. grandis* (Myanmar, China, Thailand, Vietnam), *H. lankadiva* (south Asia), *H. larvatus* (southeast Asia), *H. lylei* (Myanmar, Vietnam, Thailand, Malaysia), *H. madurae* (Madura Islands, Java), *H. maggietaylorae* (New Guinea), *H. megalotis* (Saudi Arabia, east Africa), *H. muscinus* (Papua New Guinea), *H. obscurus* (Philippines), *H. papua* (Biak and other nearby islands, Indonesia), *H. pratti* (Myanmar, China, Thailand, Vietnam, the Malay peninsula), *H. pygmaeus* (Philippines), *H. rotalis* (Laos), *H. semoni* (eastern New Guinea, north Australia), *H. speoris* (India, Sri Lanka), *H. stenotis* (north Australia), *H. sumbae* (Indonesia), *H. tephrus* (north and west Africa, Yemen), *H. thomensis* (Sao Tomé), *H. wollastoni* (New Guinea).

***Hipposideros larvatus***

***Hipposideros abae***

# Aba Roundleaf Bat

**IUCN STATUS** Least Concern
**LENGTH** unknown
**WEIGHT** unknown

**THIS BAT OCCURS** in west and central Africa, in a narrow band from Guinea running east through the Central African Republic and South Sudan to Uganda. It has gingery fur, whitish on its underside, and very small eyes. Its range incorporates the Guinea savanna zone, where it lives in disturbed, secondary-growth rainforest as well as savanna. It roosts in caves or under rocky outcrop; roosting groups are usually fewer than a hundred individuals. Its ecology is not well known. An insect-eater, it finds prey through echolocation, probably specializing in short-range hunting and usually flying fairly near the ground. As well as active hunting, it will perch at a night roost and scan for moving prey before flying out to make a capture. Although uncommon, it can survive in areas of damaged rainforest.

***Hipposideros bicolor***

# Bicolored Leaf-nosed Bat

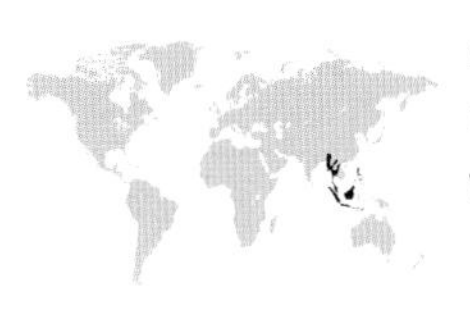

**IUCN STATUS** Least Concern
**LENGTH** unknown
**WEIGHT** unknown

**THIS FAIRLY SMALL** hipposiderid is found in parts of Thailand, and widely in the Malay Peninsula and Indonesia; it occurs across the whole of Borneo, and has been recorded from the Philippines (Luzon and Mindoro). Its taxonomy is complex, with several island forms showing distinctive genetic and behavioral traits. It is mainly found in undisturbed, primary forest, and is rare or absent in disturbed forest. It roosts in caves in at least part of its range. An agile insect-hunter, it takes beetles, moths, and other flying insects. Common in some of its range, this bat does occur in several protected areas, but its apparent reliance on primary forest makes it vulnerable in areas where deforestation is rapid. Its status in the Philippines is unclear and needs further study.

***Hipposideros armiger***

## Great Himalayan Leaf-nosed Bat

**THIS BAT OCCURS** in south and southeast Asia, from the Himalayan region across to the Malay Peninsula. It is large and dark brown, with large, broad-based ears and a mostly bare face. It has a round, complex, frilly-edged nose-leaf, bearing four leaflets at the sides (other *Hipposideros* species have three). It prefers upland habitats, foraging mainly at low levels in forest and bamboo thickets. Roosts can hold more than 1,000 individuals, and are in warm, humid caves or buildings, sometimes with other bat species. This species emerges early in the evening to hunt, looping and circling close to trees to hawk and glean insect prey; it also hunts around artificial lights. Females give birth once a year, usually to twins. The small, genetically distinct Hainan population is in particular need of conservation.

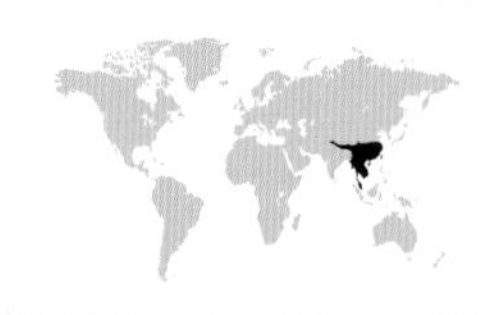

**IUCN STATUS** Least Concern
**LENGTH** unknown
**WEIGHT** 1⅝–2½oz (45–71.2g)

***Hipposideros ater***

## Dusky Leaf-nosed Bat

**THIS SMALL BAT OCCURS** patchily in India, across to Myanmar, and south through Thailand, Malaysia, Indonesia, and New Guinea; it is also present in north Australia. It has variable brown, rufous-ginger, or grayish fur with darker membranes. The ears are broad-based and upright with pointed tips, the eyes are small, and the pinkish face bears a simple, rather square-cut, black-edged nose-leaf. The species occurs in many habitat types, including wet and dry forest of various kinds, scrub, and mangroves. It roosts in caves, tunnels, wells, and tree holes. Most individuals leave the roost at least an hour after sunset; they hunt quite near the ground for diverse flying insects. Their flight is slow and highly maneuverable. Pups are born in the fall or early winter, and are independent by January.

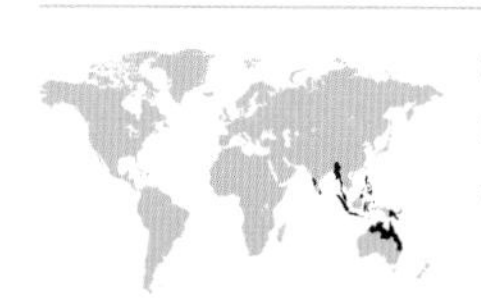

**IUCN STATUS** Least Concern
**LENGTH** 1¼–2in (3.3–5cm)
**WEIGHT** ⅛–⅜oz (4–10g)

*Hipposideros atrox*

## Lesser Bicolored Leaf-nosed Bat

**THIS BAT IS KNOWN** from Thailand and may also occur in neighboring countries. It is yet very little known, as it was only recognized as a distinct species from the Bicolored Leaf-nosed Bat (*H. bicolor*) after research in 2010. It is a typical *Hipposideros* species—an agile insect-catcher with woolly, light brown fur, large, wide ears with pointed tips, and a small, blunt face with an elaborate, flat nose-leaf. Its separation from the Bicolored Leaf-nosed Bat is justified by their consistently different call pitches: Bicolored calls at 131 kHz and Lesser Bicolored at 142 kHz. The task now is to map the distribution and habitat usage of the two species accurately, and to study any differences in their ecology. At the time of writing, this species has not been assessed by the IUCN.

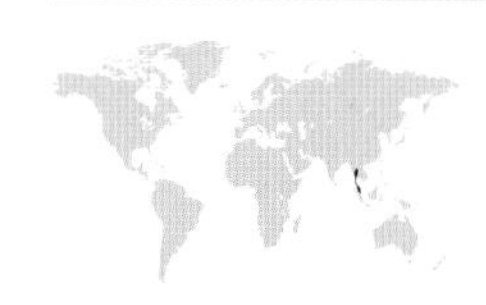

**IUCN STATUS** Not assessed
**LENGTH** unknown
**WEIGHT** unknown

*Hipposideros caffer*

## Sundevall's Roundleaf Bat

**THIS SPECIES IS** present extensively but patchily in sub-Saharan Africa, particularly west and southeast Africa; it is also found in Morocco, west Saudi Arabia, and Yemen. It is a rather pale hipposiderid that varies from gray to rufous-orange. The ears are relatively small and inward-angled, the eyes very small, and the nose-leaf simple, round, and quite large. The species occurs in the lowlands and up to 8,200ft (2,500m). It forages within dense vegetation, and roosts in limestone caves as well as mineshafts and other cavities. Roosts may hold hundreds or thousands of individuals (one large cave in Gabon holds 500,000), and may also be shared with other bat species. This bat has low wing-loading and uses a slow flight and refined echolocation to find moths, caddisflies, and other prey within the understory.

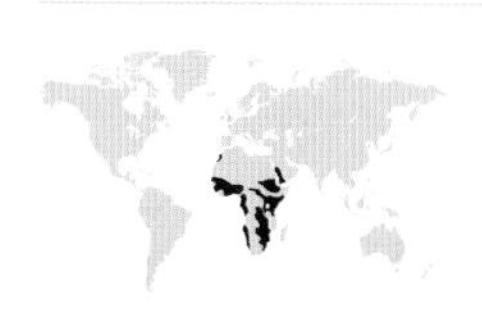

**IUCN STATUS** Least Concern
**LENGTH** 2⅝–3¾in (6.6–9.4cm)
**WEIGHT** ⅛–⅜oz (5–11g)

***Hipposideros pomona***

# Andersen's Leaf-nosed Bat

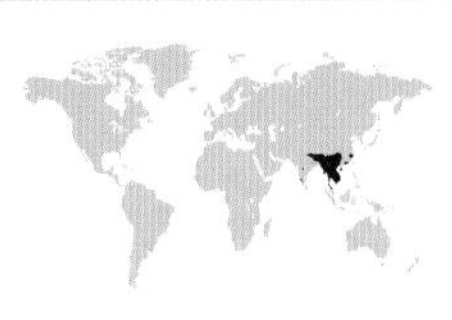

**IUCN STATUS** Least Concern
**LENGTH** unknown
**WEIGHT** unknown

**THIS BAT, ONLY RECOGNIZED** as a separate species from *H. bicolor* in the late twentieth century, is found patchily in south and southeast Asia. It appears to be willing to use disturbed forest and even suburban and urban habitats. It roosts in caves, but also uses buildings. Roosting groups usually hold fewer than ten individuals. It is insectivorous, hawking for flying prey. Females have a single pup, which remains in the maternity roost while the mothers hunt—studies have shown that pups can distinguish their mother's communication calls from those of other females, helping mother and pup reunite more quickly. This bat is common in most of its range, though is harvested for food and traditional medicine in Myanmar; tourist development of cave roosts is problematic in some other areas.

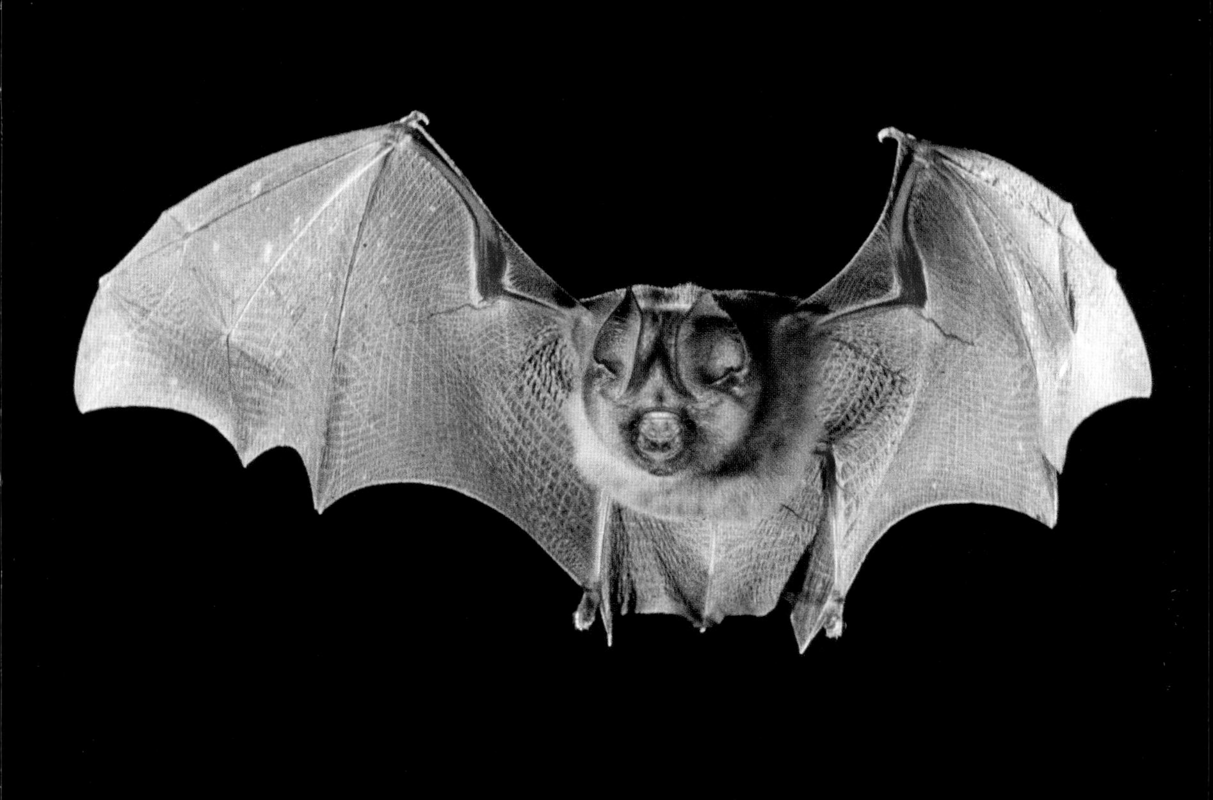

***Hipposideros ruber***

# Noack's Roundleaf Bat

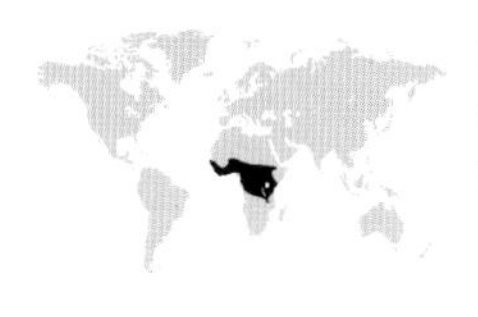

**IUCN STATUS** Least Concern
**LENGTH** unknown
**WEIGHT** unknown

THIS BAT IS WIDESPREAD in west and central Africa. Its fur ranges from gray to bright red with grayish bare parts, broad wings, a squashed face with a round-edged nose-leaf, and broad, triangular ears Its habitat is wet, lowland, tropical forest and riverside forests in drier habitats. It roosts in caves, crevices, and old mineshafts, sometimes in abandoned buildings, and, more rarely, within large tree hollows. Where space allows, colonies can contain half a million individuals. A specialized hunter, it uses echolocation calls at a frequency that picks up moths' wing-flutter motion, whether the prey is perched or flying. It almost exclusively targets moths with a wing length of ⅜–1in (1–2.5cm). Not particularly well studied, it is currently one of Africa's most abundant bat species.

***Hipposideros diadema***

## Diadem Leaf-nosed Bat

**THIS SPECIES OCCURS** extensively through southeast Asia, from Myanmar, Vietnam, and Cambodia south through Thailand, and across most of Indonesia to New Guinea and the Solomon Islands. It is also present in northeast Queensland, Australia. It has variable fur coloration, from pale straw or rufous to darker brown, often with white spots around the shoulders and a darker stripe on the center of the forehead. It has tiny eyes and a large round nose-leaf, with small leaflets at the sides. The ears are average-sized and pointed. This bat roosts in caves, in colonies sometimes hundreds or thousands strong. It forages in primary and disturbed forest, hunting low and often over water. Females produce one pup a year. Disturbance of its roosts is a problem in parts of southeast Asia.

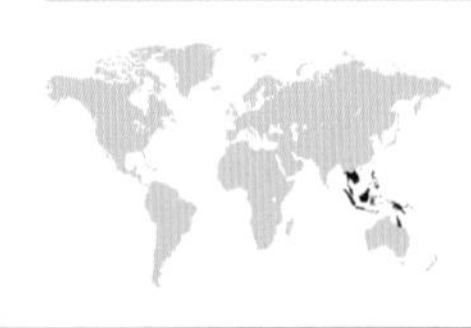

**IUCN STATUS** Least Concern
**LENGTH** 2⅜–3⅞in (6–10cm)
**WEIGHT** 1¼–1¾oz (34–50g)

***Hipposideros doriae***

## Bornean Leaf-nosed Bat

**THIS BAT OCCURS** in peninsular Malaysia, Sumatra, and Borneo. It has light grayish-brown or yellowish-brown fur, with browner membranes and ears. The face is flat, short and blunt, with a rounded nose-leaf. It has a very steep forehead; the eyes are relatively large for a hipposiderid and are set close to the snout tip. The ears are medium-sized. It is only known to occur in primary forest. This species is not well studied, though likely to be ecologically similar to others in the genus. It is rare throughout its extensive range, and its dependence on primary forest makes ongoing habitat loss—due to logging, conversion to agriculture, and uncontrolled forest fires—a threat. Given the projected rate of habitat loss in its range, this bat may soon require reclassifying as Vulnerable.

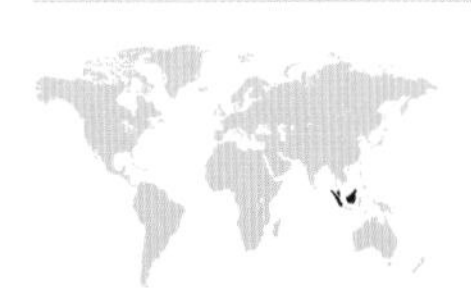

**IUCN STATUS** Least Concern
**LENGTH** unknown
**WEIGHT** ¼oz (6g)

***Hipposideros dyacorum***

## Dayak Leaf-nosed Bat

**THIS BAT OCCURS** on the Malay peninsula, where it is quite widespread; it is also present throughout the island of Borneo. It is a typical hipposiderid, with gray-brown to bright orange fur (a little paler on the underside), and broad wings that allow for agile flight that expends little energy. The ears are quite large, upright, dark-edged, wide, and funnel-shaped with pointed tips. It has a pink face and limbs, a steep, domed head, and a simple, round nose-leaf. The species occurs in lowland rainforest. It roosts in various sheltered spots including caves, but also spaces between boulders and in tree holes; hundreds may roost together in larger caves, hanging by their feet from the ceiling. A poorly known species, it may be declining as forest loss continues throughout its range.

**IUCN STATUS** Least Concern
**LENGTH** unknown
**WEIGHT** unknown

***Hipposideros galeritus***

## Cantor's Leaf-nosed Bat

**THIS SPECIES OCCURS** in Bangladesh, India, and Sri Lanka, west and south Thailand (including Terutau Island), Cambodia, Laos, south Vietnam, and peninsular Malaysia; it is also found in Sumatra, Bangka, Java, and Borneo. A brown-furred *Hipposideros*, it has a round nose-leaf, sometimes edged whitish, and narrow, pointed ear-tips. It is found in wet forest, sometimes in rubber plantations, and occurs from sea level to 3,600ft (1,100m). Its cave roosts can be thousands strong and it has been observed sharing roosting space with up to seven other bat species. Smaller roosts may be found in buildings and rock crevices. This bat hunts flying insect prey with a low-level, steady, and agile flight. It may be affected by deforestation in some regions, but is also found in a number of protected areas.

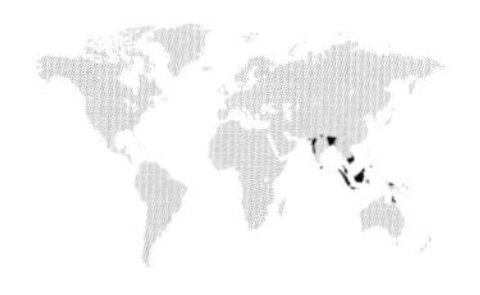

**IUCN STATUS** Least Concern
**LENGTH** 2¾–3½in (7–9cm)
**WEIGHT** ¼oz (6–7g)

***Hipposideros vittatus***

# Striped Leaf-nosed Bat

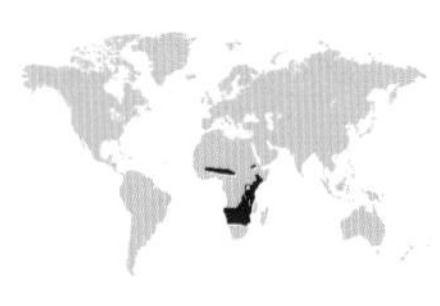

**IUCN STATUS** Near Threatened
**LENGTH** unknown
**WEIGHT** unknown

**THIS SPECIES HAS BEEN** recorded over a wide area of east and south Africa, also sparsely in central and west Africa. It is migratory, moving considerable distances between the dry and rainy seasons. It has light brown fur (darker and grayer in juveniles), with pale flank stripes and dark membranes. Its ears have elongated tips, relatively narrow and pointed for a *Hipposideros* species. The nose-leaf is broad and rounded with three accessory leaflets at the sides. It is associated with wooded savanna over most of its range, but uses forest habitats in west Africa. Several extremely large cave roosts (holding tens of thousands of individuals) are known, but it also roosts in smaller groups in tree hollows and buildings. An insectivore, it can feed on large, hard-bodied prey such as grasshoppers. Although this bat appears to be widespread, its tendency to concentrate in a few extremely large roosts makes it vulnerable—especially as it is widely and unsustainably harvested at its roosts (its body fat is used by local people to make candles). Roost caves also suffer disturbance from mining activity and tourism.

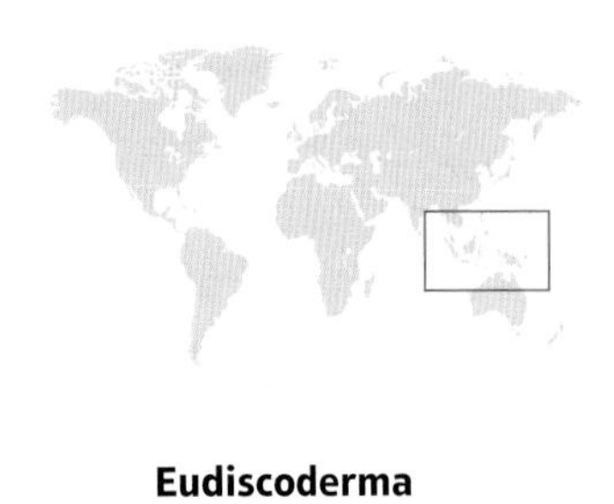

Eudiscoderma

## MEGADERMATIDAE

The bats that make up the small family Megadermatidae are also known as the "false vampire bats." They are most closely related to the leaf-nosed bat families Rhinolophidae, Hipposideriade, and their allies, together forming the superfamily Rhinolophoidea. These are large, striking, and distinctive species (*Macroderma gigas* is the largest of all microbats), and are confined to the tropics and subtropics of the Old World, from Africa through south and southeast Asia to Australia. There are six species within five genera. A number of fossil species have also been described, dating back to thirty-seven million years ago. Although the species within this family are quite diverse, there are various consistent features (particularly in skull morphology) that allow for reliable identification. The teeth in particular are distinctive, most notably in that the upper incisors are absent, while the canines are large and have an extra point. These bats also have a long, upward-pointing nose-leaf, and large, long, and parallel-edged ears with long, slim tragi. The ears are connected by their inner bases with a tall band of skin across the forehead. The eyes are large. The wings are large, long, and broad, as is the uropatagium, but the tail is short, or in some cases absent. Despite their name, they are not blood-drinkers, but some species prey on vertebrates, which they capture on the ground by "pouncing" from a perch. Most species have mid-brown or paler fur, and some have quite strongly colored bare parts. Two of the six megadermatids are covered with individual accounts. See page 358 for *Cardioderma cor* and page 360 for *Lavia frons*.

### Eudiscoderma

*Eudiscoderma thongareeae* was only described in 2015, following observations and collection of specimens in Bala Forest, south Thailand. It is a large bat with very large ears and a complex, large nose-leaf, the raised central part of which has the shape of a downward-pointing arrow. It has robust jaws with powerful teeth, adapted for dealing with hard-bodied prey such as large beetles. It is known only from one site with a total area of just

1½ square miles (4km$^2$), and apparently dependent on evergreen rainforest. Its habitat is, as yet, unprotected. However, the bat itself may soon be added to Thailand's list of protected wildlife.

### Macroderma

*Macroderma gigas* is a large, pale bat that occurs in Australia, in Northern Territory, Queensland, and Western Australia. It has suffered a dramatic range contraction over the last few decades.

### Megaderma

The megadermatid *Megaderma lyra* is widespread in south and southeast Asia. *M. spasma* is described on page 359.

***Macroderma gigas***

*Cardioderma cor*

# Heart-nosed Bat

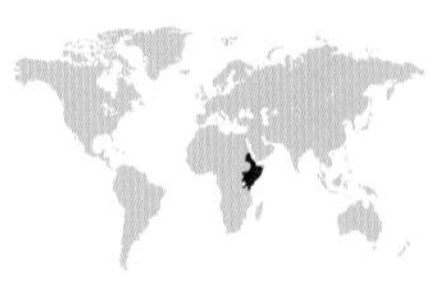

**IUCN STATUS** Least Concern
**LENGTH** 2¾–3in (7–7.7cm)
**WEIGHT** ¾–1¼oz (21–35g)

**THIS BAT OCCURS** in east Africa, from the northeast of Sudan north through Somalia and south to central Tanzania. It has light gray fur with darker membranes. The snout is short, with a heart-shaped nose-leaf. It has large eyes and ears. It mainly occurs in scrubby forest and dry savanna, and its day roosts may be in caves, tree hollows (particularly baobabs), or buildings. Roosts are typically small, but may hold up to eighty bats. This species specializes in large beetles, and also takes small vertebrates from the ground, honing in on prey-generated sounds. It gives distinctive, low-pitched calls while foraging, perhaps as territorial "song" to discourage other bats from foraging nearby, as these increase when prey is scarce. Some of its cave roosts suffer disturbance due to use in religious ceremonies.

***Megaderma spasma***

# Lesser False Vampire Bat

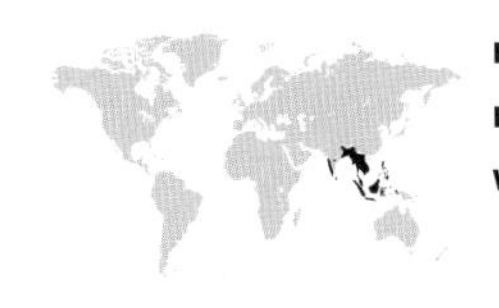

**IUCN STATUS** Least Concern
**LENGTH** 2½–3¾in (6.5–9.5cm)
**WEIGHT** unknown

**THIS BAT IS WIDESPREAD** in southeast Asia, as well as Sri Lanka and east India. In appearance it resembles *M. lyra*, with fluffy gray fur (whitish on the belly). The large ears connect at their bases. The skin folds where the ears' inner edges join have been shown to help focus sound entering the ear. It has relatively large eyes and a short snout with a large nose-leaf, formed by two leaflets with a central ridge.

This bat prefers dense and humid forests; it roosts in caves, tree holes, and buildings. It gleans prey from the ground or other surfaces rather than hawking in flight, feeding on both small vertebrates and large arthropods. Prey is found by passive listening; it has also been known to enter buildings to catch geckos on walls. This species may be declining due to deforestation.

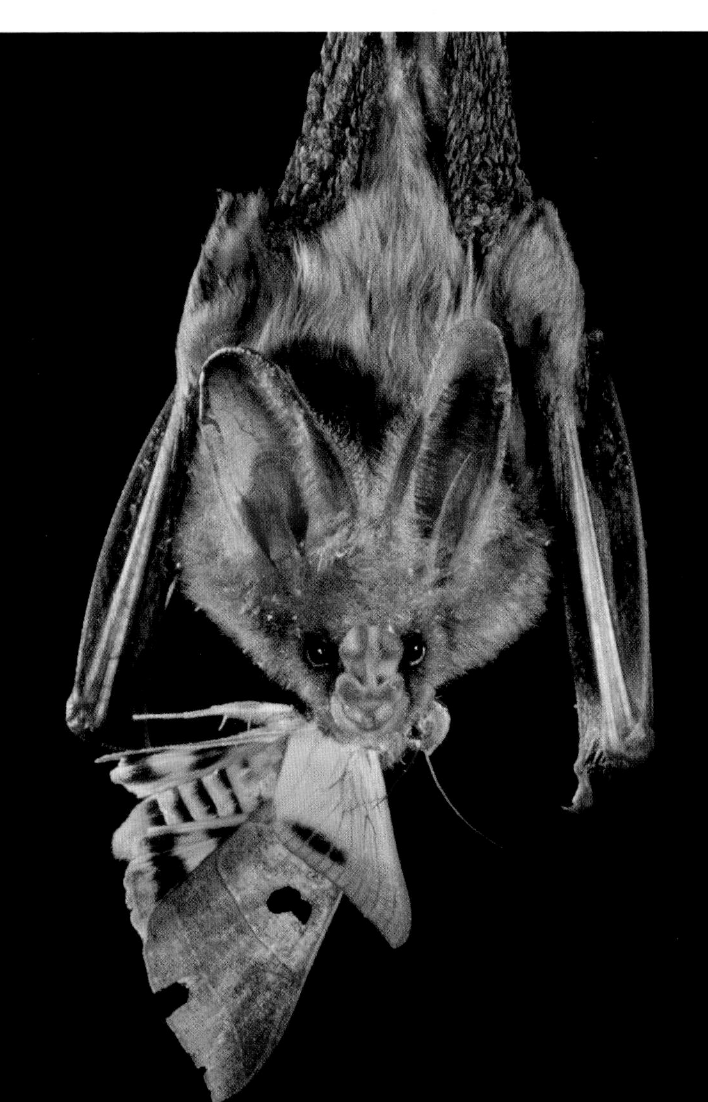

***Lavia frons***

# Yellow-winged Bat

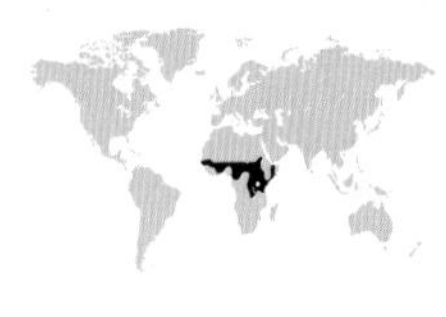

**IUCN STATUS** Least Concern
**LENGTH** 2¼–3⅛in (5.8–8cm)
**WEIGHT** 1–1¼oz (28–36g)

THIS SPECIES IS WIDESPREAD in sub-Saharan Africa, from Senegal across to Somalia, then south to Congo, Tanzania, and just into Zambia. It has long, pearly-gray fur, contrasting with bright yellow membranes, ears, and facial skin. The ears are long, round-ended, and upright; it has moderate-sized eyes and a long, upright nose-leaf. It is mainly found in scrubby savanna, close to rivers, where it roosts in bushes. It hunts by flying out to intercept large insects, detected by sound; it then carries them back to a perch to eat. This bat lives in monogamous pairs. These roost together, about 3ft (1m) apart, perform circling courtship flights, and forage together. The male leaves the roost first to seek out hunting areas, later guiding the female to a productive spot. They also engage in mutual grooming. A single pup is born in spring, which the mother carries with her on feeding flights. The parents tolerate the pup's presence in their territory for about a month after it has begun to hunt for itself. This bat has a stable population, occurring in several protected areas.

**Rhinolophus**

## RHINOLOPHIDAE

The family Rhinolophidae is made up of a single, very large genus, *Rhinolophus*. These species are known as the "horseshoe bats," and there are 104 of them—making this one of the largest of all bat genera. The species within *Rhinolophus* occur extensively throughout the Old World in both temperate and tropical regions. The earliest-known bat fossil from Australia was a rhinolophid, and the only bat species to occur in the British Isles that are not vespertilionids are two species of rhinolophids. The extinct genus *Palaeonycteris* is also classed within this family; fossil evidence shows that members of this genus were present in Europe at least five million years ago. These bats are difficult to identify and can be very cryptic in their habits—hence the continuing high rate of discovery of new species, and the large number of species for which good survey data has yet to be gathered.

Rhinolophids make up part of the superfamily Rhinolophoidea, along with the family Hipposideridae and allied families. In terms of ecology they show some overlap with the insectivorous members of the family Phyllostomatidae, of the New World.

The rhinolophids as a group are distinctive for their complex bare, fleshy nose-leaves. These typically have a horseshoe-shaped bottom lobe, and one or more pointed projections on the upper lobe, as well as extra leaflets on the sides. The central, forward-pointing projection is known as the lancet. The ears are usually large, closely set, and funnel-shaped, with tapering, pointed tips; they do not have tragi. The eyes are very small. These features give the bats' faces a very characteristic—and, to our eyes, quite bizarre—appearance. The legs are relatively small and weak, meaning that these bats are not capable of locomoting on feet and thumbs. They invariably roost by hanging from a horizontal surface, rather than climbing into narrow crevices.

This family of bats is closely related to Hipposideridae, and some taxomonists actually consider Hipposideridae to be a subfamily (Hipposiderinae) within Rhinolophidae. However, there are several consistent anatomical differences between the two groups, which support their separation into two families. Hipposiderids have a rounded nose-leaf, rhinolophids a more pointed one with a central flattened structure (sella), which hipposiderids lack. Rhinolophids have three bones in their toes except the first; hipposiderids have just two in each toe. The dental formula is different (rhinolophids have six lower premolars, hipposiderids have four). There are also consistent differences in the structure of the pelvis and shoulders.

As with hipposiderids, the rhinolophids emit their complex echolocation calls through their nostrils. The nose-leaf's shape helps them to direct the calls. They can simultaneously give a call and listen to the returning echo from the call just before, enabling them to call more often than most other bats can. A range of harmonics in the calls expands their frequency range, allowing them to pick up on a wide range of smaller and larger targets. Their flight is slow and fluttering, and highly maneuverable; these bats are able to hover, which enables them to capture prey from surfaces (they can even take items caught in spiders' webs), as well as catch flying insects in the open air. They carry larger prey items to a roost to consume. These bats feed exclusively on insects and similar-sized arthropods.

Most rhinolophids are gregarious in their roosting habits. Those that live in temperate regions hibernate over winter, typically in caves and tunnels, and are exacting in their requirements for hibernacula in terms of size, ambient temperature and humidity, whether the floor is dry or under water, and the type of access point. Lack of suitable hibernation sites can therefore limit the population size in some areas, and disturbance at the roosts, intentional or otherwise, can be a major conservation issue. Non-hibernating species can use a wider variety of sites for their day roosts, including buildings and tree holes, but many species do still enter a period of energy-saving torpor every day as they sleep. Like hipposiderids, they wrap their wings around themselves when hanging in the roost, rather than holding the wings folded at their sides as most bats do. It is not unusual for them to share roosts with a variety of other bat species. Females have one pup a year; many species mate in the fall, and store the male's sperm through winter, conceiving and gestating the embryo the following spring.

Of the 104 *Rhinolophus* species this book recognizes, several are covered in detail, with full accounts as follows: *R. affinis* (see page 366), *R. blasii* (page 366), *R. clivosus* (page 367), *R. creaghi* (page 367), *R. euryale* (page 368), *R. ferrumequinum* (page 370), *R. fumigatus* (page 368), *R. hildebrandtii* (page 369), *R. hipposideros* (page 371), *R. landeri* (page 372), *R. macrotis* (page 372), *R. malayanus* (page 374), *R. megaphyllus* (page 373), *R. mehelyi* (page 376), *R. microglobosus* (page 373), *R. monoceros* (page 377), *R. pearsonii* (page 374), *R. shameli* (page 375), *R. simulator* (page 375), *R. sinicus* (page 378), *R. smithersi* (page 378), *R. swinnyi* (page 379), *R. thomasi* (page 380), *R. trifoliatus* (page 379), and *R. yunanensis* (page 381).

Many of the other rhinolophids have a large or very large range and face no severe threats. They are as follows: *R. acuminatus* (widespread in mainland southeast Asia), *R. alcyone* (west and central Africa),

*R. arcuatus* (Borneo, Indonesia, Philippines, and New Guinea), *R. beddomei* (India and Sri Lanka), *R. bocharicus* (widespread in central Asia), *R. borneensis* (mainland southeast Asia, Borneo, Java, Kalimantan), *R. capensis* (South Africa), *R. celebensis* (Indonesia, including Java, Bali, and Sulawesi), *R. coelophyllus* (southeast Asia), *R. damarensis* (south Africa), *R. darlingi* (sub-Saharan Africa, particularly the south of Africa), *R. denti* (west, central, and the south of Africa), *R. eloquens* (west, central, and east Africa), *R. euryotis* (Indonesia, Papua New Guinea), *R. inops* (Philippines), *R. lepidus* (central to south and southeast Asia), *R. luctus* (south and southeast Asia), *R. marshalli* (southeast Asia), *R. mossambicus* (Mozambique and Zimbabwe), *R. paradoxolophus* (China, Thailand, Laos, Vietnam), *R. philippinensis* (Indonesia, Malaysia, Papua New Guinea, Philippines, Timor-Leste, and Australia), *R. pusillus* (south, east, and southeast Asia), *R. rex* (China), *R. rouxii* (south Asia), *R. shortridgei* (China, India, Myanmar), *R. siamensis* (China, Laos, Thailand, Vietnam), *R. stheno* (southeast Asia), *R. subbadius* (south Asia), *R. tatar* (Sulawesi), and *R. virgo* (Philippines).

Many of the little-known rhinolophids are known only from one or a few old specimens, and they may well be extinct today. Others are known to be extant, but have not yet been studied or surveyed in any detail. These species are *R. adami* (known only from type series, taken in Congo in 1968), *R. convexus* (Laos and peninsular Malaysia; only known from a few specimens), *R. indorouxii* (India; appears to be very rare and forest-dependent), *R. keyensis* (Indonesia; a very poorly known species), *R. maendeleo* (known only from two localities in Tanzania), *R. mcintyrei* (Papua New Guinea; a barely known species), *R. mitratus* (India; known only from the type specimen taken in 1844), *R. nereis* (Siantan Island and North Natuna Islands, Indonesia; very poorly studied), *R. osgoodi* (China, known only from a couple of specimens), *R. sakejiensis* (Zambia; known only from the type specimens, taken from their roost within tree foliage in 2002), *R. silvestris* (Congo and Gabon; known only from three localities), and *R. subrufus* (Philippines; apparently fairly widespread but very little known).

The rhinolophids that face certain threats are as follows: *R. deckenii* (Kenya, Tanzania; a rare and poorly known, forest-dependant species), *R. formosae* (Taiwan, China; threatened by deforestation), *R. hillorum* (west Africa; forest-dependent and rare with a small range), *R. robinsoni* (Malaysia and Thailand; threatened by deforestation), *R. rufus* (Philippines; its cave roosts suffer heavy disturbance), *R. sedulus* (Borneo and peninsular Malaysia; a rare species affected by rapid habitat loss), and *R. xinanzhongguoensis* (China; suffering disturbance at its roosting caves due to tourism).

Several more are *R. canuti* (Java and Bali; threatened by disturbance at the roosts and by deforestation), *R. cohenae* (South Africa; newly discovered and apparently rare and range-restricted), *R. guineensis* (west Africa; rare and affected by habitat loss), *R. madurensis* (Madura Island and the Kangean Islands, Indonesia; threatened by limestone extraction at its cave roosts, and by deforestation), *R. ruwenzorii* (Democratic Republic of Congo, Rwanda, Uganda; threatened by loss of the montane forest where it occurs).

The rhinolophids considered to be endangered are *R. belligerator* (Sulawesi, with only one specimen known, from 1987), *R. cognatus* (Andaman Islands, declining and possibly threatened by roost disturbance due to illegal collection of edible swiftlet nests), *R. maclaudi* (east Guinea; poorly known, but apparently rare and probably affected by loss of forest habitat), *R. montanus* (Timor-Leste in the Pacific, rare and threatened by habitat loss), *R. proconsulis* (Indonesia and Malaysia; only known from three locations), and *R. ziama* (west Africa; known only from four specimens). Finally, *R. hilli* in Rwanda is barely known, but appears to have a tiny range in an area badly affected by deforestation.

In addition, a number of *Rhinolophus* bats are understudied as they have been discovered recently, or have recently been split from other species on the strength of new molecular DNA evidence. These species are as follows: *R. chiewkweeae* (Malaysia; a recent split from *R. yunanensis*), *R. cornutus* (Japan; a recent split from *R. pusillus*), *R. francisi* (Borneo, newly discovered), *R. hirsutus* (Philippines; a recent split from *R. macrotis*), *R. horaceki* (Libya; a recent split from *R. clivosus*), *R. huananus* (from China, newly described), *R. imaizumii* (Japan; a recent split from *R. pusillus*), *R. kahuzi* (from the Democratic Republic of Congo, newly described), *R. lanosus* (China; a recent split from *R. luctus*), *R. luctoides* (Malaysia; a recent split from *R. luctus*), *R. mabuensis* (from Mozambique, newly described), *R. monticolus* (from China, newly described), *R. morio* (Malaysian peninsula; a recent split from *R. luctus*), *R. perditus* (Iriomote, Ishigaki, Taketomi, and Kohama islands in Japan, a recent split from *R. pusillus*), *R. refulgens* (southeast Asia; a recent split from *R. lepidus*), *R. schnitzleri* (from China, newly described), *R. thailandiensis* (from Thailand, newly described), and *R. willardi* (from the Democratic Republic of Congo, also newly described).

*Rhinolophus affinis*

## Intermediate Horseshoe Bat

**THIS BAT RANGES** from north India through Nepal, Bangladesh, and to the far east of mainland China, south through Myanmar, Cambodia, Laos, Vietnam, and Thailand, then through much of Malaysia and Indonesia. It has gray-brown fur and similar-colored membranes. The face is dominated by a large, horseshoe-shaped nose-leaf. The small eyes sit close to the nose-leaf base, and the ears are smallish with pointed tips. It is found in forest habitats in both highland and lowland areas, also visiting orchards, plantations, farmland, and gardens. The species roosts in mixed-sex colonies in caves, sometimes several thousand strong. Females produce one pup a year; however, there is evidence of two peaks of breeding activity each year. This bat is quite common and adaptable, though some colonies may be threatened by limestone mining and human disturbance.

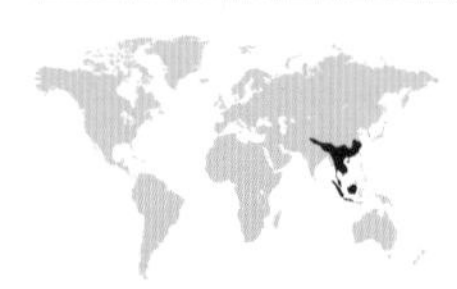

**IUCN STATUS** Least Concern
**LENGTH** 2¼–2½in (5.8–6.3cm)
**WEIGHT** unknown

*Rhinolophus blasii*

## Blasius's Horseshoe Bat

**THIS IS A VERY WIDESPREAD** species with several well-separated populations in Eurasia and Africa. It occurs in Greece and adjacent countries, across parts of the Middle East to north India. In Africa it is found in Ethiopia, Tanzania, Zambia, Zimbabwe, Mozambique, and eastern South Africa; it also occurs in northwest Africa. This bat is rather pale and sandy-gray in color; it has a large pink nose-leaf which, in profile, shows a double "spike" at the top. The ears are broad with sharply pointed tips. It occurs in scrub and woodland, sometimes desert edges, and roosts in caves. In some areas it hibernates over winter. The species is adaptable, and its population is increasing in some areas. Disturbance at cave roosts (often by farmers sheltering livestock) is sometimes an issue.

**IUCN STATUS** Least Concern
**LENGTH** 2½–3½in (6.2–9cm)
**WEIGHT** ¼–⅜oz (7–12g)

***Rhinolophus clivosus***

## Geoffroy's Horseshoe Bat

THIS RELATIVELY LARGE *Rhinolophus* bat occurs in the west of the Middle East, patchily in northeast and east Africa, and in south and southwest Africa—mainly close to the coast, especially in hotter climates. It has sleek, gray-brown fur, with blackish wings and ears. The nose-leaf is large but rather flattened, and not conspicuous at first glance. The bats leave their cave roosts soon after sunset to forage in semi-desert, savanna, grassland, and woodland, taking moths and beetles around and within dense vegetation. They often carry large prey to an open-air night roost in foliage, or an accessible building, leaving characteristic prey remains on the ground below. Their roosts can hold thousands of individuals. Though abundant, the species faces threats from disturbance and poisoning through heavy pesticide use in some areas.

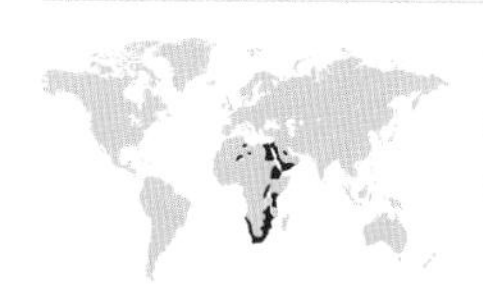

**IUCN STATUS** Least Concern
**LENGTH** 2⅞–4½in (7.4–11.3cm)
**WEIGHT** ⅜–⅞oz (12–26.1g)

***Rhinolophus creaghi***

## Creagh's Horseshoe Bat

THIS BAT OCCURS patchily on Borneo; it is also recorded from Palawan in the Philippines, and Nusa Penida, Semau, Roti, and Banggi Island in Indonesia. It has brown fur and dark bare parts. The nose-leaf is large, with two well-defined lower lobes and a large, pointed projection at the top. The ears have long, delicate points. The species occurs in primary forest, close to limestone caves suitable for roosting; some roosts hold thousands of individuals. It forages within the forest understory, where it hunts flying insects and gleans from the foliage. Although numbers can be very high in the right habitat, it is declining due to various threats, including limestone extraction and guano-mining at its caves; on Palawan it is hunted for food. Loss of forest is also a factor.

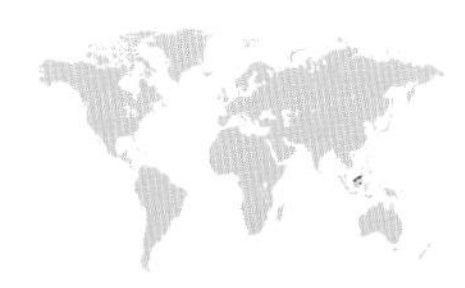

**IUCN STATUS** Least Concern
**LENGTH** unknown
**WEIGHT** ⅜–½oz (10.5–13.5g)

***Rhinolophus euryale***

## Mediterranean Horseshoe Bat

**THIS SPECIES OCCURS** in the Mediterranean region, in the south of Europe from Spain to Turkey (including many Mediterranean islands), northwest Africa and the west Middle East. It is also present further east, from the east of the Black Sea south through Iran. It has light sandy-brown fur and a light grayish-pink face; the wings are darker gray. Its nose-leaf is relatively small and neat. A cave-roosting species, it forages in scrubland and woodland edges, weaving through the vegetation with a slow-paced, agile, fluttering flight typical of the genus. Over winter this bat hibernates in deep parts of the cave. Summer roosts may be in buildings. It is declining within its European range and likely also elsewhere, due to loss of foraging habitat, overuse of organochlorine pesticides, and disturbance at its roosts.

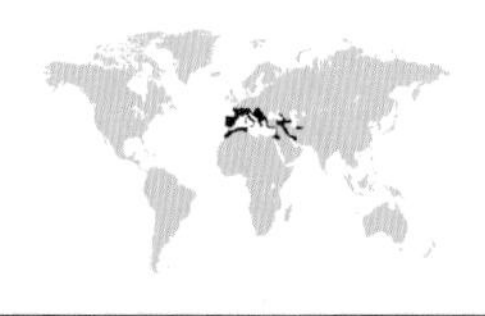

**IUCN STATUS** Near Threatened
**LENGTH** unknown
**WEIGHT** unknown

***Rhinolophus fumigatus***

## Rüppell's Horseshoe Bat

**THIS AFRICAN SPECIES** occurs from Senegal in west Africa, east through central Africa to Ethiopia, and south through east Africa to northern South Africa; it is also found in Zambia, Mozambique, Botswana, and Namibia. It is a light brown bat with rather concolorous bare parts. The ears are large, as is the complex, broad-based nose-leaf. It may be seen foraging in savanna (wet and dry) and around dry forest edges, but rarely any significant distance from the caves where it roosts. The roosts can be large (more than 500 individuals), especially in the north of its range. It preys mainly on beetles and moths, which it catches close to vegetation. Females produce a single pup per year, in the fall. This species is not particularly well studied. It occurs in many protected areas.

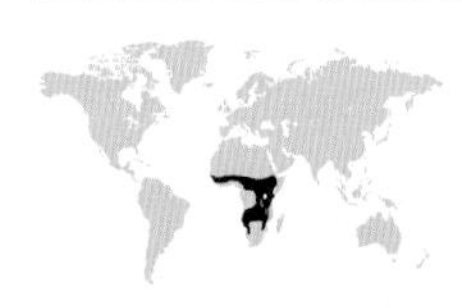

**IUCN STATUS** Least Concern
**LENGTH** 3½–4¼in (8.9–10.6cm)
**WEIGHT** ⅜–¾oz (11–20g)

***Rhinolophus hildebrandtii***

# Hildebrandt's Horseshoe Bat

**IUCN STATUS** Least Concern
**LENGTH** 4–5in (10.1–12.6cm)
**WEIGHT** ¾–1⅜oz (20–38.5g)

**THIS SPECIES OCCURS** in east Africa, from Ethiopia through to northeastern South Africa, and west almost as far as Angola. It is a typical *Rhinolophus* bat with grayish fur and membranes, and a large nose-leaf with a prominent, forward-projecting "horn" at the top. Individual variation in the nose-leaf size corresponds to a different frequency in the echolocation call. The species is most often found in wooded savanna, but also occurs in more open habitats. Roost sites include caves, mines, abandoned buildings, tree hollows, and disused animal burrows. Colonies usually consist of fewer than one hundred individuals. It hunts on the edge of—and within low-level—vegetation, catching insects, as well as invertebrates. Females produce a single pup per year, born at the start of the rainy season.

***Rhinolophus ferrumequinum***

# Greater Horseshoe Bat

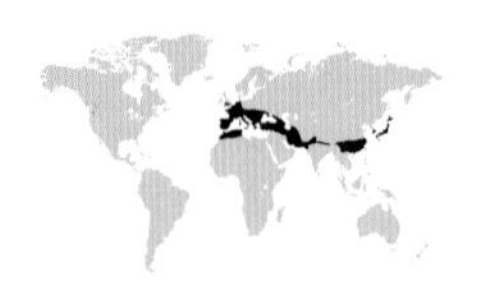

**IUCN STATUS** Least Concern
**LENGTH** 2¼–2¾in (5.7–7.1cm)
**WEIGHT** ⅝–1¼oz (17–34g)

THIS BAT HAS AN EXTENSIVE range, reaching across south and central Europe to northwest Africa, and through south Asia across to the east of China, South Korea, and Japan. It is also present in southwest England and Wales. It is a rather large *Rhinolophus* with pale gray or sandy-brown fur, tiny eyes, and a large nose-leaf; its large ears have tapering tips. This species forages in woodland clearings and edges, in grassland, and in scrub. Summer roosts are often in buildings, while in winter it hibernates in deep, cool caves. It is sensitive to small shifts in temperature and humidity—this can affect timing of births in spring, which in turn influences juvenile survival. In several European countries, this is a species of conservation concern, and its roosts are strictly protected.

***Rhinolophus hipposideros***

# Lesser Horseshoe Bat

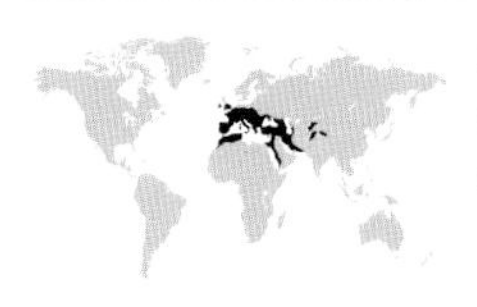

**IUCN STATUS** Least Concern
**LENGTH** 1⅜–1¾in (3.5–4.5cm)
**WEIGHT** ⅛–⅜oz (5–9g)

**THIS SPECIES IS MAINLY** found in Europe, reaching from Spain and the south of the British Isles across to Turkey. It is also found on the coast of the west Middle East, and in northwest Africa, and occurs patchily east from Iran, Uzbekistan, and Afghanistan along to north India and the extreme west of China. Much smaller than *R. ferrumequinum*, it is similarly grayish and plain, with tiny eyes and low-set ears. The nose-leaf is neat and rounded. It forages mainly in woodland and along woodland edges, hunting small and tiny insects. Summer roosts are usually small, and often in buildings or tree holes. This bat hibernates in small caves (seen here) and tunnels in groups of up to 500 individuals. It is protected over much of Europe, where it is of conservation concern.

*Rhinolophus landeri*

## Lander's Horseshoe Bat

**THIS BAT IS VERY WIDESPREAD** across sub-Saharan Africa, absent only from most of the Horn of Africa, part of the east coast, and the southwest of central Angola. It is a rather plain, gray-brown bat with a paler belly, though it can bleach to a bright, orange-red color. The bare parts are mid-gray and it has a round, dark-edged nose-leaf. It occurs in rainforest, savanna, and tall gallery woodland, and also in riverine forest and scrub. Roosts are caves, abandoned mines, and spaces between boulders. As with other *Rhinolophus*, it is adapted to forage in dense, cluttered environments, preying primarily on moths, but taking other insects too. Females bear one pup a year, the timing varying by latitude. It is not threatened and occurs in several protected areas.

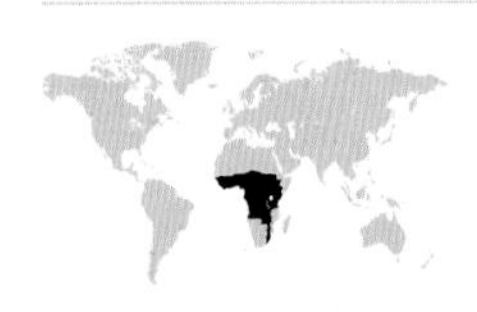

**IUCN STATUS** Least Concern
**LENGTH** 2¾–3⅛in (7–8cm)
**WEIGHT** ¼oz (6–10g)

*Rhinolophus macrotis*

## Big-eared Horseshoe Bat

**THIS BAT'S RANGE EXTENDS** from Bangladesh, north India, Nepal, and Pakistan, and reaches into China. It also occurs in southeast Asia, north Myanmar, and Thailand, as well as north Laos and Vietnam, peninsular Malaysia, Sumatra, and the Philippines. Its ears have pronounced, curving ridges inside, and large anti-tragi; the nose-leaf is also large. It has mid-brown fur with white tips and quite dark bare parts. The species is a forest specialist, occurring in primary but also disturbed forest areas, often at fairly high elevations; it roosts in caves and old mine workings. It flies fast for a rhinolophid and takes beetles, flies, and many other insects. An understudied species, it is threatened by deforestation in parts of its extensive range, and by deliberate persecution in some of its roosting caves.

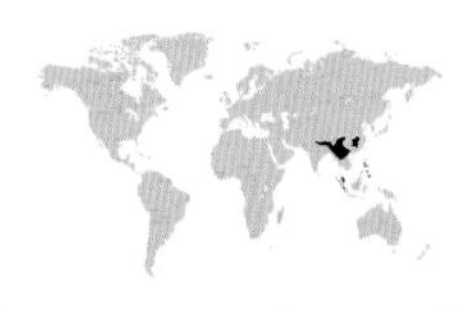

**IUCN STATUS** Least Concern
**LENGTH** 3–3⅜in (7.7–8.5cm)
**WEIGHT** ¼oz (6–8g)

***Rhinolophus megaphyllus***

## Eastern Horseshoe Bat

**THIS BAT IS FOUND** in east Papua New Guinea and the Bismarck Archipelago; it also occurs along the northeast and east coast of Australia, from north Queensland down to Melbourne. It has rather dark gray-brown fur, with similarly toned or darker bare parts. The face is typical of *Rhinolophus* species, with a prominent, horseshoe-shaped nose-leaf and wide-based, triangular ears with pointed tips. This bat forages in forest of all kinds, and also in more open environments including grassland and scrub. It roosts in humid caves, empty buildings, and rocky crevices; maternity colonies can hold more than 2,000 individuals, but are typically much smaller. They forage by fluttering around dense vegetation, using echolocation to pinpoint small flying insects. They may also fly out from a perch to intercept prey. This species is not threatened.

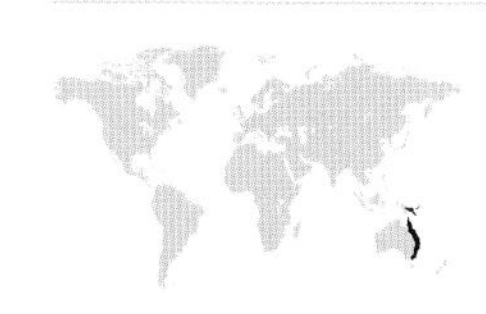

**IUCN STATUS** Least Concern
**LENGTH** 2⅝–3¼in (6.6–8.1cm)
**WEIGHT** ¼–⅜oz (7–10.5g)

***Rhinolophus microglobosus***

## Indo-Chinese Brown Horseshoe Bat

**THIS BAT OCCURS** through Myanmar, Cambodia, Laos, Thailand, and Vietnam. It is probably also present in south China. It has recently been split from *R. stheno*, on the basis of its consistently smaller size and higher-pitched vocalizations, and the fact that the two species occur together. It has pale brown, woolly-textured fur, with contrastingly darker membranes, face, and ears. The ear-tips are attenuated and fold slightly backward. The species is found in forested habitats of various kinds, and roosts in small or large colonies (up to 1,200 recorded) inside limestone caves. Feeding behavior and breeding biology resembles that of other *Rhinolophus* species, and pups are born in early spring. Quarrying activities threaten some of this bat's roosting caves, although it is of special conservation concern in Thailand.

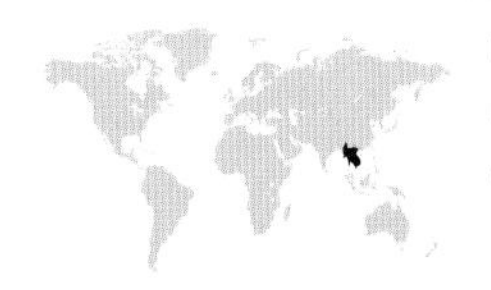

**IUCN STATUS** Least Concern
**LENGTH** unknown
**WEIGHT** unknown

***Rhinolophus malayanus***

## Malayan Horseshoe Bat

**THIS SPECIES IS IN** south Myanmar and west China, and ranges south through Laos, Cambodia, Vietnam, Thailand, and the Malay peninsula. It has a large, dish-shaped nose-leaf, its round bottom part in two distinct segments. Echolocation calls of populations in the north differ clearly and consistently from those in the south, likely suggesting they represent two distinct species, but no physical differences are apparent. Genetic investigation would clarify the picture. This is a versatile species that may be encountered foraging in forest, both primary and secondary, as well as farmland and other disturbed habitats. It roosts within limestone or other cave types. This bat is widespread and often common; it is not thought to be in decline at present.

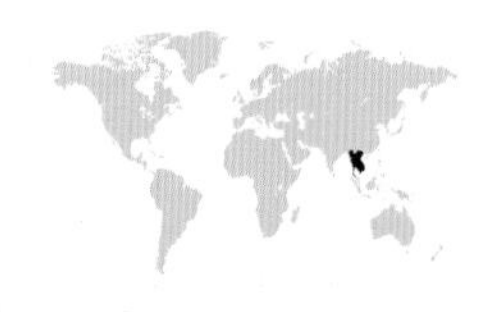

**IUCN STATUS** Least Concern
**LENGTH** unknown
**WEIGHT** unknown

***Rhinolophus pearsonii***

## Pearson's Horseshoe Bat

**THIS BAT OCCURS** in north India, Nepal, Bangladesh, and Bhutan, then extensively in south and southeast China, Myanmar, Laos, Vietnam, and Cambodia. It is a largish rhinolophid which may have dark brown or more tawny-rufous fur. It has a large, elaborate nose-leaf that conceals the mouth when viewed front-on. The upper part has several forward projections, it also has tiny eyes, and large ears with prominent, forward-tilted anti-tragi. The species tends to occur in more upland areas, foraging in forest, bamboo thickets, and farmland that adjoins forest. Roosts are typically inside limestone caves. This bat is not well studied, though ecologically it is likely similar to others in its genus. It is probably declining—at least in the west of its range—due to deforestation, and is not known to occur in any protected areas.

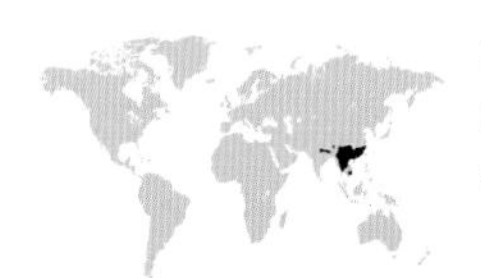

**IUCN STATUS** Least Concern
**LENGTH** unknown
**WEIGHT** unknown

***Rhinolophus shameli***

## Shamel's Horseshoe Bat

**THIS BAT'S RANGE** extends from north Myanmar southwest through Thailand, Laos, and Cambodia, just reaching west-central Vietnam. It also occurs on the Koh Chang islands off southeast Thailand. It tends to have rather orange-brown fur, with darker areas around the eyes and on top of the head. The bare parts are also darker. Its nose-leaf is long, though rather narrow; the eyes are very small, the ears dark and medium-sized. It forages in undisturbed lowland evergreen forest, both primary and secondary, and in some regions in secondary mixed deciduous forest and more disturbed areas. It roosts in limestone caves, often at quite high elevations of 3,300ft (1,000m) or more. The species is otherwise little known; studies are needed to clarify its ecology and population size. It is thought to face no significant threats at present.

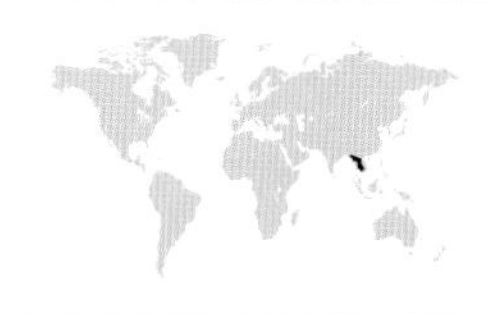

**IUCN STATUS** Least Concern
**LENGTH** unknown
**WEIGHT** unknown

***Rhinolophus simulator***

## Bushveld Horseshoe Bat

**THIS BAT OCCURS** in east Africa, from Ethiopia south through Uganda, Kenya, and Tanzania, and into Zambia, Zimbabwe, Mozambique, Botswana, and eastern South Africa. There are also outposts in west Africa (Nigeria and Cameroon). It is a rather light grayish or brown horseshoe bat, with relatively large ears and small nose-leaf. This bat forms cave roosts up to 300 strong; it may also roost alone in small crevices. It is found most often in wooded savanna, tending to forage over streams and rivers, or along drainage channels. Its diet is formed mainly of moths, but includes a fair proportion of beetles and assorted other insects. The birth period is in the fall or winter; females bear a single pup, which they feed for about seven weeks. The species is declining, due to habitat loss.

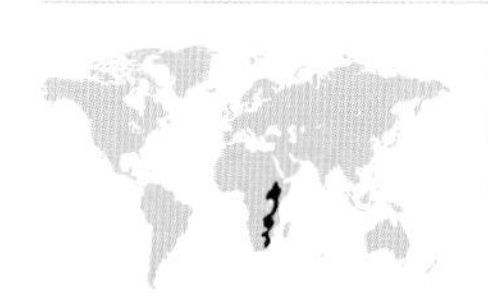

**IUCN STATUS** Least Concern
**LENGTH** 2⅜–3½in (6.1–8.8cm)
**WEIGHT** ¼–⅜oz (5.8–11g)

***Rhinolophus mehelyi***

# Mehely's Horseshoe Bat

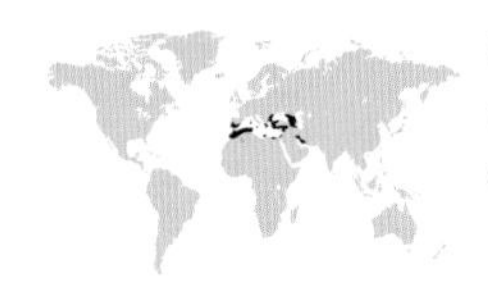

**IUCN STATUS** Vulnerable
**LENGTH** unknown
**WEIGHT** unknown

**THIS BAT OCCURS** patchily around the Mediterranean and Middle East, including Spain, Sardinia, Sicily, Greece, northwest Africa, and parts of northeast Africa; it is also found patchily further east, as far as Iran. It is a light-colored bat with darker fur on the crown of its head and around its eyes; the wing membranes are light gray. The species may be observed foraging for moths and other insects in dry woodland, scrub, and rockier, more arid areas, as long as some standing water is present. In summer it roosts in warmer caves, choosing deeper, cooler ones for winter hibernation. It is occasionally known to roost in buildings. The decline of this species is not fully understood. Likely causes include disturbance at roosting caves, and loss of suitable foraging habitat.

***Rhinolophus monoceros***

# Formosan Horseshoe Bat

**IUCN STATUS** Not assessed
**LENGTH** unknown
**WEIGHT** unknown

THIS BAT'S TAXONOMY is uncertain. It is sometimes regarded as a subspecies of *R. pusillus*, but this book treats it as a full species. It is endemic to the island of Taiwan (China), occurring in lowland regions. It has warm-toned brown fur, darker on the upperside than on the underside, and darker, grayish wing membranes. The ears are also gray with attenuated, folded-back tips and prominent anti-tragi.

The nose-leaf is large, especially the upper lobe, which has a high vertical projection. The eyes are very small and the lips are pink and prominent. The habitat requirements of this species are rather exacting: it occurs within dense interior areas of forest, but rarely in disturbed habitats. Roosts are usually in limestone caves, where the bats pack together in tight clusters.

***Rhinolophus sinicus***

## Chinese Horseshoe Bat

**A VERY WIDESPREAD** bat, this species occurs through north India, Nepal, and Bhutan into China; it reaches the east coast, where it is present on many small islands, and also north Myanmar and north Vietnam. A very small, darkish horseshoe bat, its chestnut-brown fur can become almost blackish on the crown and around the eyes, giving a masked appearance. The underside is a shade paler. This species has a relatively small nose-leaf, with a neatly folded upper lobe. It tends to occur in upland rainforest areas, roosting in variably sized colonies in caves, buildings, tree hollows, and other shelters. It is as yet rather poorly studied and its population trend is unknown. However, forests within its range are disappearing, and it is not recorded from any protected areas.

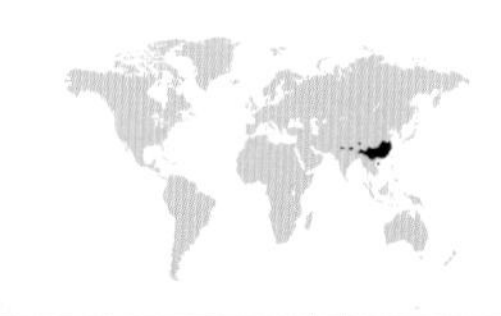

**IUCN STATUS** Least Concern
**LENGTH** 1⅜–1½in (3.5–3.7cm)
**WEIGHT** unknown

***Rhinolophus smithersi***

## Smither's Horseshoe Bat

**THIS SPECIES WAS** described to science in 2012, following its split from *R. hildebrandtii* on the basis of DNA, call, and body measurements. It has been recorded in two separate areas—Zimbabwe (just south of the Zambezi Escarpment) and South Africa in the Limpopo Valley, and around the Soutpansberg, Blouberg, and Waterberg Mountains. These two populations may themselves be distinct species. It is a pale sandy-gray bat with large, funnel-shaped, sharply pointed ears and a wide nose-leaf. It roosts in caves and old mineshafts and forages in savanna, close to water sources. This bat is classed as Near Threatened because, although its population is likely stable at present, it has a small population (possibly fewer than 1,000) and restricted range, so threats such as severe drought could have a serious impact.

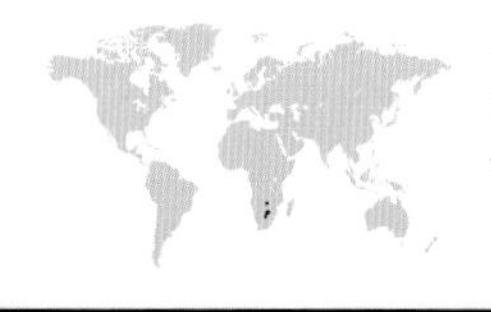

**IUCN STATUS** Near Threatened
**LENGTH** unknown
**WEIGHT** unknown

***Rhinolophus swinnyi***

## Swinny's Horseshoe Bat

**THIS SMALL AFRICAN** horseshoe bat is found from Tanzania and DR Congo south through Zambia and Zimbabwe into eastern South Africa to the coast; it also occurs in Swaziland. It is a dusky gray-brown species, with prominently large, wide-based ears and a medium-sized nose-leaf. This bat is very similar to *R. simulator*, though on average a little smaller. It occurs in rainforest, including in the high uplands, and in wooded moist and dry savanna. Roosts are in caves or mines and usually hold relatively few animals (often no more than ten); however, roost-site availability seems to dictate its distribution. The species uses a slow, fluttering flight to navigate through low vegetation, hunting moths and other insects. It occurs in some protected areas, but its population status is currently unknown.

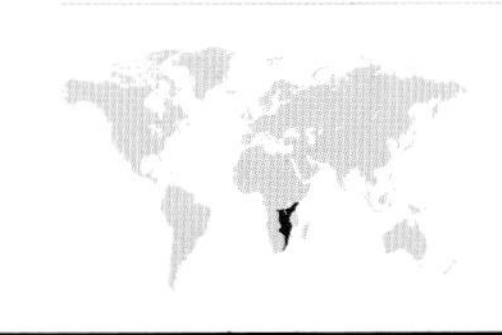

**IUCN STATUS** Least Concern
**LENGTH** 2⅛–3in (5.3–7.6cm)
**WEIGHT** ⅛–¼oz (5–8.3g)

***Rhinolophus trifoliatus***

## Trefoil Horseshoe Bat

**THIS DISTINCTIVE SPECIES** occurs from south Thailand through the Malay peninsula to Borneo and Sumatra. It is also found in parts of Java and other western Indonesian islands, with a few additional outposts in north India and southwest China. It has light gray-brown fur, against which its large, complex yellowish nose-leaf stands out. The limbs and ears are also yellow-tinted, while the wing membranes are light brown. This bat occurs in primary and secondary forest and mangrove swamp, roosting mainly in caves, but also sometimes singly among dead foliage. Like other *Rhinolophus* bats it hunts within dense, cluttered vegetation, relying on echolocation to navigate and find prey (all kinds of flying insects). Habitat loss is certain to be affecting it in some areas, but its population and ecology are poorly studied.

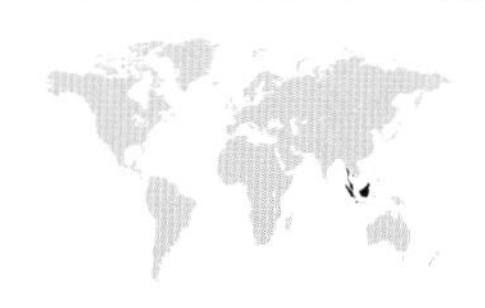

**IUCN STATUS** Least Concern
**LENGTH** 2–2½in (5.1–6.4cm)
**WEIGHT** ¼–¾oz (8–20g)

***Rhinolophus thomasi***

# Thomas's Horseshoe Bat

**IUCN STATUS** Least Concern
**LENGTH** 1⅞–2in (4.8–5cm)
**WEIGHT** unknown

**THIS BAT OCCURS** in south China; its range extends south through east Myanmar, Laos, Vietnam, north Cambodia, and into north and central Thailand. It is a brown bat with dark blackish bare parts. The ears are large and pointed, with slightly hook-shaped anti-tragi. Its nose-leaf is elaborately folded. It occurs from the lowlands to up to at least 3,000ft (1,000m), and roosts in limestone caves, often alongside other *Rhinolophus* species. It forages in nearby forest and other well-vegetated areas, including in degraded habitats. The species hunts within vegetation and probably has similar ecology to others in its genus, but it is almost unstudied. Its abundance appears to vary from common to rare in different regions, but it is not considered to be of conservation concern at present.

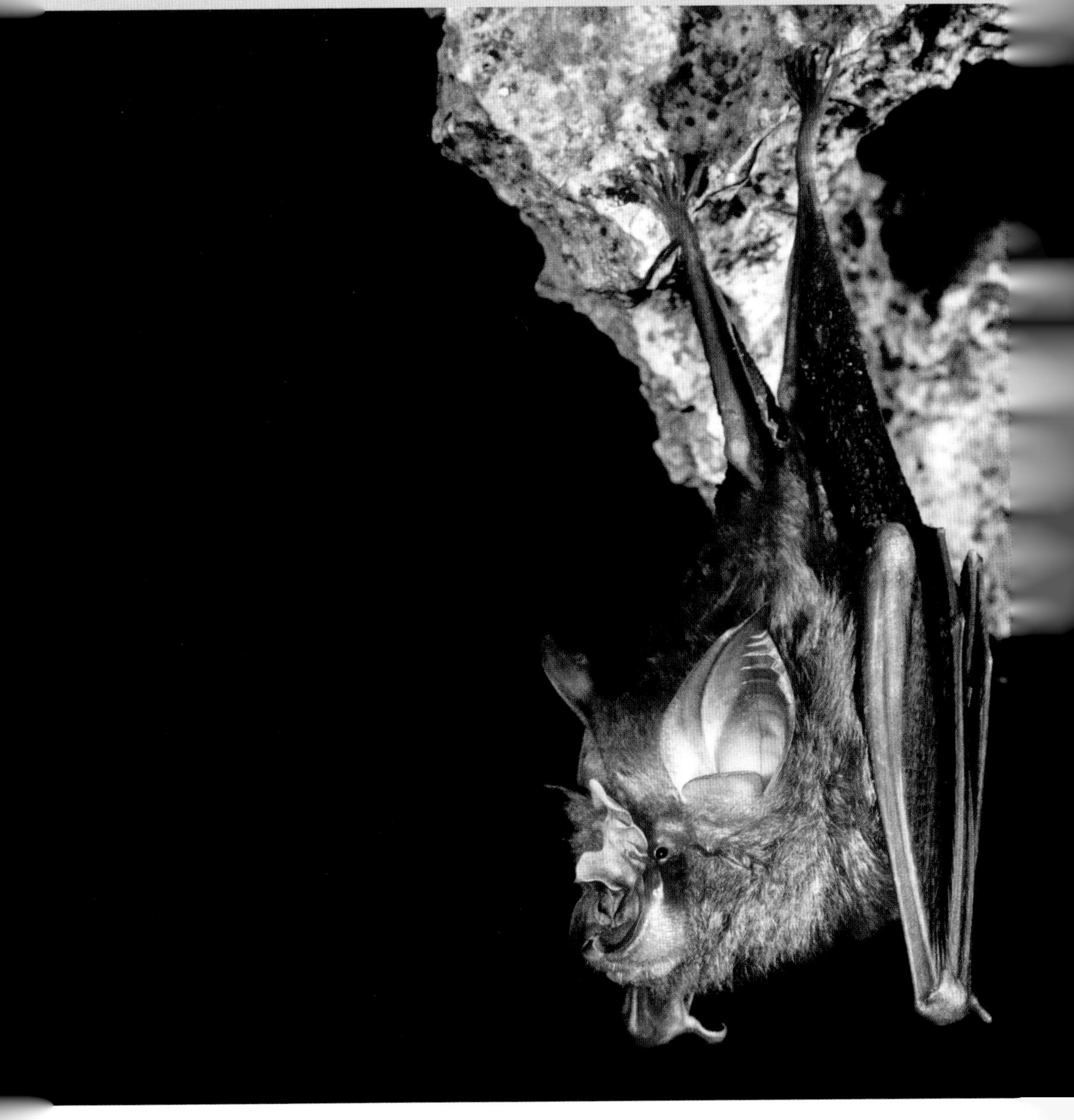

***Rhinolophus yunanensis***

# Dobson's Horseshoe Bat

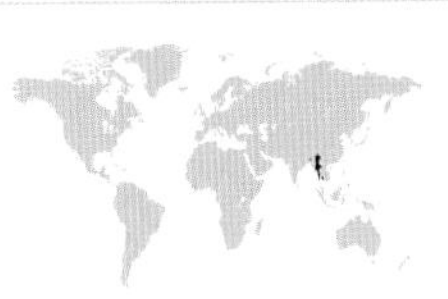

**IUCN STATUS** Least Concern
**LENGTH** 2⅜–2⅝in (6–6.8cm)
**WEIGHT** unknown

**THIS SPECIES IS PRESENT** in south China and through Myanmar and Thailand. It may occur also in west Laos, and there is a separate, small population in northeast India. It is rather similar to *R. pearsonii* in appearance, though a little larger. The fur is mid- or gray-brown with a woolly texture; the nose-leaf is broad. It has a central groove on the lower lip. This bat occurs in forest and within bamboo thickets, mainly at between 2,000 and 4,300ft (600 and 1,300m) elevation. Roosts are normally in limestone caves and hold a few hundred animals, but it is also known to use thatched roofs. Otherwise virtually unstudied, research is needed to determine the distribution limits and ecological requirements for this species. It is threatened by deforestation, especially the small Indian population.

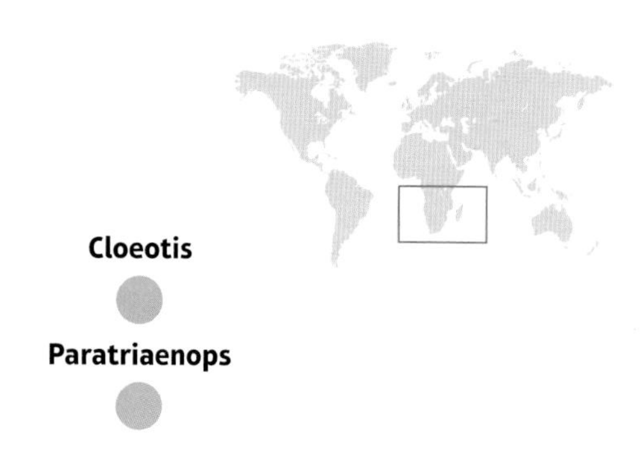

## RHINONYCTERIDAE

This is a family of Old World bats with prominent nose-leaves; it is related to Hipposideridae and to Rhinolophidae, and its genera were historically classified within Hipposideridae. However, molecular DNA studies carried out in 2015 show that the rhinonycterids, a "subtribe" within Hipposideridae, actually form a distinct, well-established grouping, which should be elevated to family level. The studies indicated that Rhinonycteridae diverged from its relatives somewhere around thirty-nine million years ago. Besides the DNA support, the family differs consistently from Hipposideridae in five particular features of nose-leaf structure.

***Cloeotis percivali***

These bats occur in Africa, Asia, and Australia. As a group they are sometimes known as "trident bats" because of the three-pronged upper part of the nose-leaf. This is structurally different to the trident-like structure seen in a few hipposiderids. Another name for the family is "Old World leaf-nosed bats." Several fossil genera have been assigned to this family, they are: *Archerops*, *Brachipposideros*, *Brevipalatus*, and *Xenorhinos*. Today, the family contains four small extant genera.

### Cloeotis

One of these genera is *Cloeotis*, which is monotypic. *C. percivali* is rather uncommon, but widespread in the south of Africa.

### Paratriaenops

The genus *Paratriaenops* holds three species. These possess very prominent nose-leaf tridents, the three "spears" being long, slender, and pointed. The ears are very broad at the base, and have pointed tips. Fur color varies from rather dull gray-brown to bright yellowish. *P. auritus* is described on page 384. The genus also includes *P. furculus*, which occurs in the lowlands on Madagascar, and also on Cosmoledo and Aldabra atolls in the Seychelles. It is abundant in parts of Madagascar. The final member of the group

Rhinonicteris

Triaenops

is *P. pauliani*, found only on Aldabra Atoll in the Seychelles. It is poorly known, with most observations being of individuals that had entered buildings to roost. The atoll has protection through its status as a UNESCO World Heritage site.

## Rhinonicteris

*Rhinonicteris* is another monotypic genus. Its sole species, *R. aurantia*, occurs in north and northwest Australia, where it is widespread. A disjunct population in the Pilbara and Gascoyne region of Western Australia ("Pilbara form") may warrant separation as a different species, as it shows several small differences in anatomy and vocalizations. This is a very striking bat with bright yellowish-orange fur. It forms very large roosts in very warm and humid caves—sometimes of more than 10,000 individuals, and often alongside other bat species. Management of its cave roosts in Australia has taught some valuable lessons in bat conservation. The erection of a mesh screen over the entrance to one well-known roost, designed to keep people out while allowing the bats access, proved very discouraging to the bats. Their numbers at this roost have still failed to recover, despite the mesh grille being removed.

## Triaenops

The final genus in the family is *Triaenops*, which holds five species. One is described in detail; see page 384 for the account on *T. menamena*. These bats have smaller nose-leaves, with shorter and wider tridents, than *Paratriaenops*; they also have smaller ears, with a distinct indentation on the outer edge near the tip. The fur color varies from gray-brown or mid-brown to bright orange in some cases.

The other species of *Triaenops* occur in Africa, Madagascar, and the Arabian Peninsula. *T. afer* occurs widely in east Africa. *T. persicus* is very widespread, occurring in east Africa, the Arabian Peninsula, and Iran. Another species is *T. rufus*, widespread in Madagascar. The final extant *Triaenops* is *T. parvus*, a newly discovered bat that occurs in Yemen and Oman, and is understudied.

Another *Triaenops* bat was formerly present on Madagascar. *T. goodmani* is known from three lower jaw bones, which were found in 1996 in a cave at Anjohibe, in the north of the island. The bones were estimated to be some 10,000 years old.

***Paratriaenops auritus***

## Grandidier's Trident Bat

**THIS BAT HAS A VERY** small range at the most northerly tip of Madagascar, and its distribution within that area is patchy. It has light yellowish-brown fur, a shade paler on the underside. Like other *Paratriaenops* bats, it is similar to the rhinolophids, with funnel-shaped ears and a round-bottomed nose-leaf. The upper lobe of the nose-leaf is divided into three straight, spear-shaped points (hence "trident bat"). It forages within dry forest, and roosts in narrow caves and abandoned mineshafts, the availability of which strongly dictates its distribution. The largest-known colony holds about 2,000 bats. To safeguard this species, it is important that known roosts are carefully monitored and protected against undue disturbance. This bat is present in three protected areas: Parc National d'Ankarana, Réserve Spéciale d'Analamerana, and Parc National de Loky-Manambato.

**IUCN STATUS** Vulnerable
**LENGTH** unknown
**WEIGHT** unknown

***Triaenops menamena***

## Rufous Trident Bat

**THIS BAT IS ENDEMIC** to Madagascar, occurring on the north, west, and south sides of the island, away from the central ridge. It has silky, rich, reddish-brown fur (though grayer individuals also occur), and darker bare parts. The ears are wide-based and smallish, with notches on their inner edges. The top lobe of its elaborate nose-leaf bears three points, the central one longer and straighter than the outer ones. Its habitat is mainly dry forest, but the species has been found in more humid areas, and also away from forest. Its roosts are inside caves and can hold several tens of thousands. The bulk of its relatively small population (about 121,000) is therefore concentrated at relatively few roosts, but it does occur in several protected areas.

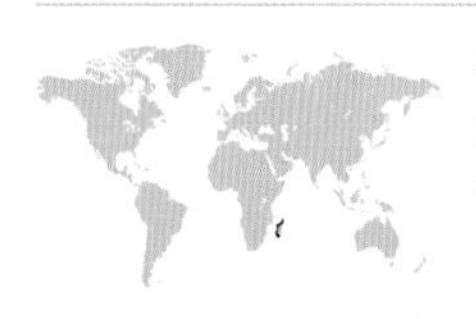

**IUCN STATUS** Least Concern
**LENGTH** 3⅜–4⅛in (8.6–10.4cm)
**WEIGHT** ¼–⅝oz (6.6–15.5g)

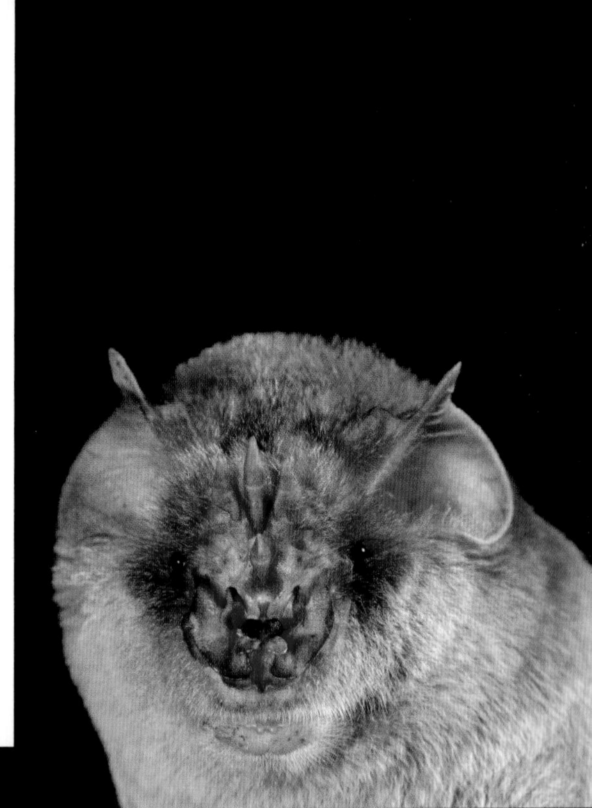

Rhinopoma

## RHINOPOMATIDAE

The small family Rhinopomatidae contains one genus (*Rhinopoma*), which comprises just six species. These distinctive small species are known as the "mouse-tailed bats," or "long-tailed bats" because of their very long free tail. This extends well beyond the relatively shallow uropatagium, and nearly equals the length of the animal's head and body put together.

Rhinopomatids are desert specialists, occurring in arid regions from north and west Africa across the Middle East to Pakistan, Afghanistan, and perhaps further east. Their bodies are adapted for this way of life, with traits including slit nostrils that can be sealed shut to keep out sand. The head is simple with a broad, robust muzzle; they also have rather large eyes and large ears with pronounced internal ridges. The ears connect at their inner edges by a ridge of skin across the forehead. The arms and legs are long and slender; the fur is short. They roost in caves and rocky crevices, and will store fat and enter torpor in the winter months to survive cold conditions. They are typically rather dull gray-brown on the upperside, a little paler on the belly. These bats have a fast flight that alternates between fluttering and gliding. They prey on flying insects, which they target using their relatively simple echolocation system.

The species *R. hardwickii* is described on page 386. *R. cystops* is present across north Africa and Arabia, in desert and semi-desert with sparse vegetation. Its roosts may be in caves, ruined buildings, underground tunnels, or among boulders. *R. hadramauticum* is the most threatened species in the genus. It is known from just one site: the town of Ash Shahar in Hadramaut province, southeast Yemen, where the only known colony (consisting of about 150 individuals) roosts in a single house. *R. microphyllum* has a very extensive range, from north Africa through Arabia and Afghanistan to north India, with a few old records from further east (Thailand and Sumatra). *R. muscatellum* occurs from Arabia across to central Asia. The final species is *R. macinnesi*—a poorly known bat from areas of hot desert in east Africa (most records are from Kenya).

***Rhinopoma hardwickii***

# Lesser Mouse-tailed Bat

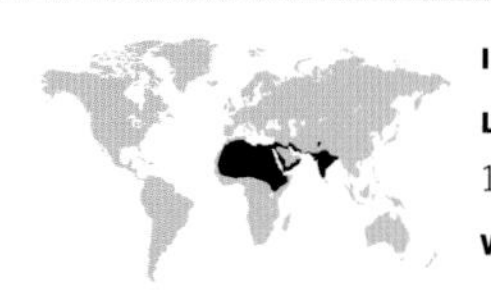

**IUCN STATUS** Least Concern
**LENGTH** 4¾–5⅝in (11.9–14.1cm)
**WEIGHT** ⅜oz (10–12g)

THIS SPECIES IS PRESENT across most of north Africa. Its range extends to Kenya in the east, and also reaches around the Middle East and across most of India to Nepal and Bangladesh. It has silver-gray fur and darker gray-brown membranes; the limbs are pink. It has long wings and a very long, free tail. It has an arched muzzle with a small, inconspicuous, triangular nose-leaf. This bat is found in desert areas with dry scrub, rocky outcrops, wadis, and oases, and is well-adapted to cope with low water supplies. It hibernates in winter in caves, and in cool crevices in summer to avoid heat stress. It flies fast, hawking insects in the open air. Heavy pesticide use is a serious threat.

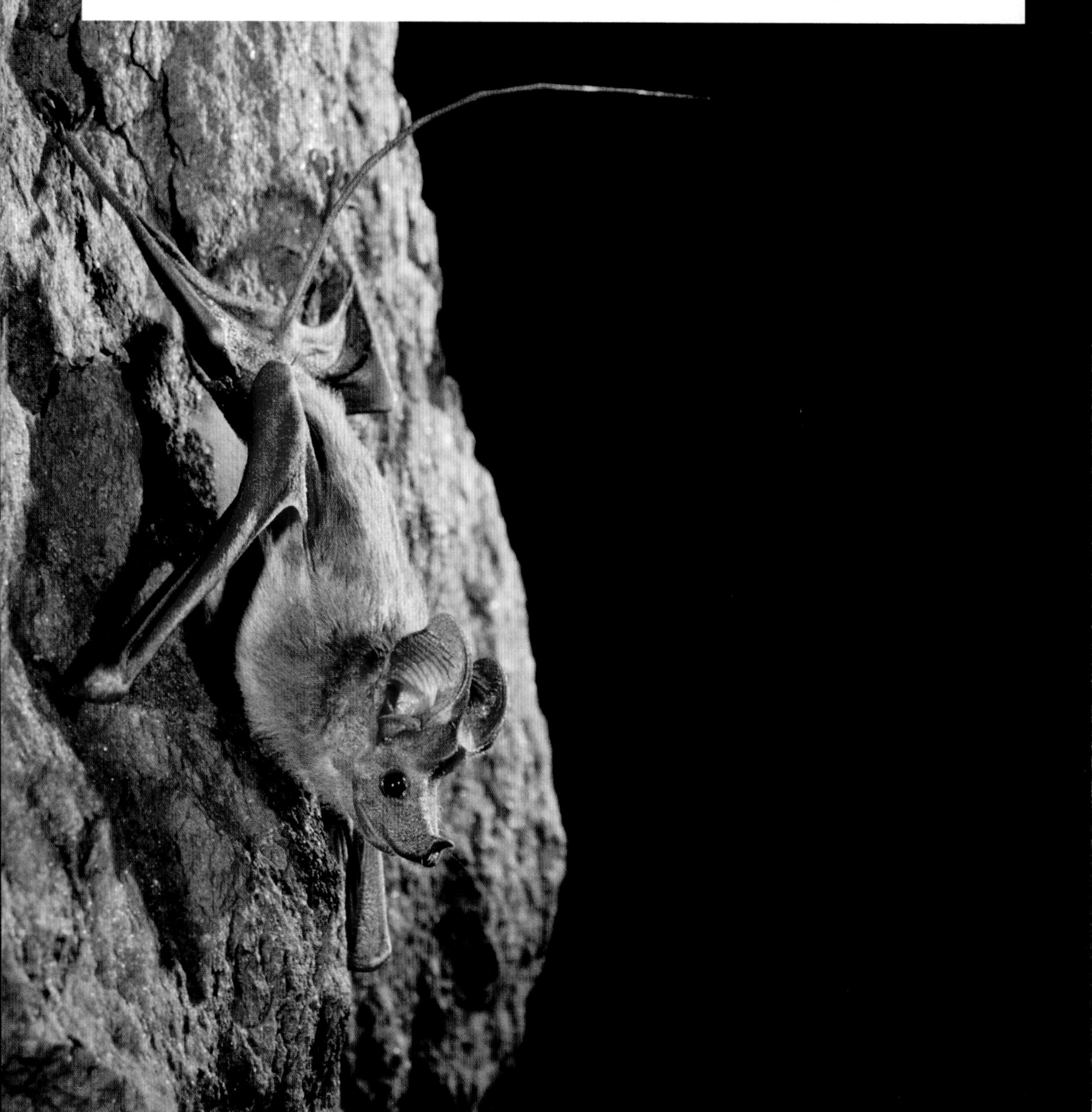

# INDEXES

## INDEX OF SCIENTIFIC NAMES

### A



### B



### C

**M**

**R**



**S**

**T**



**U**



**V**



**X**

## INDEX OF COMMON NAMES

## ACKNOWLEDGMENTS

I'd like to thank Susan Kelly and Kate Shanahan at Ivy Press for commissioning this book, and Tom Kitch, Stephanie Evans, and Elizabeth Clinton for seeing the project through to publication. I'd also like to thank Catherine Bradley, Caroline Eley, Michael Whitehead, James Lawrence, Heather Bowen, Richard Peters, and Ginny Zeal for their sterling work on the text and design.

I'm especially grateful to Merlin Tuttle for permitting the use of his spectacular photographs, and for bringing his exceptional knowledge and depth of experience to bear on the first drafts of the text—his guidance was absolutely invaluable.

This book would not be possible without the countless hours of hard graft in the field and the lab by biologists, researchers, and volunteers worldwide, who work in often extremely challenging conditions to shine a light into the mysterious midnight world of bats. I hope my words do justice to their efforts.

Finally, I'd like to thank my friends and family for supporting me through the long process of writing this book—for their forbearance in the face of yet another bat-related monologue or anecdote, and their willingness to offer distraction when it was most needed. I hope you all enjoy the end result.

The publisher would like to thank Nancy Simmons and Andrea Cirranello for the generous supply of their 2018 Chiroptera taxonomic list.

## BIOGRAPHIES

**Marianne Taylor** is a freelance writer, illustrator, photographer, and editor. Her interest in wildlife began at a very early age, when she became passionately interested in first butterflies, then birds, then everything else. Her love of the natural world has guided her life and work ever since. She worked for seven years in natural history publishing and began a new career as a writer in 2007. Since then she has written more than twenty books for adults and children on a range of natural history subjects, many illustrated with her own artwork or photographs.

**Dr Merlin Tuttle** is widely regarded as the father of modern bat conservation. He has studied and photographed bats for sixty years, is well known for his research on bat behavior and ecology and has led successful conservation efforts worldwide. His priceless collection of over 150,000 photos of bats forms the basis for this publication. He founded and built Bat Conservation International for thirty years. He retired, then founded Merlin Tuttle's Bat Conservation in 2014 due to growing demands for his unique experience. His website at www.MerlinTuttle.org is the sole source of his expanding photo collection and conservation leadership.